Error and the Growth of Experimental Knowledge

SCIENCE AND ITS CONCEPTUAL FOUNDATIONS

A series edited by David L. Hull

"I might recall how certain early ideas came into my head
as I sat on a gate overlooking an experimental blackcurrant plot . . . ."
—E. S. Pearson, "Statistical Concepts in Their Relation to Reality"

Deborah G. Mayo

Error and the Growth of Experimental Knowledge

THE UNIVERSITY OF CHICAGO PRESS

Chicago and London

Deborah G. Mayo is associate professor of
philosophy at the Virginia Polytechnic Institute
and State University.

The University of Chicago Press, Chicago 60637
The University of Chicago Press, Ltd., London
© 1996 by The University of Chicago
All rights reserved. Published 1996
Printed in the United States of America

05 04 03 02 01 00 99 98 97 96   1 2 3 4 5
ISBN: 0-226- 51197-9 (cloth)
0-226- 51198-7 (paper)

Library of Congress Cataloging-in-Publication Data

Mayo, Deborah G.
    Error and the growth of experimental knowledge / Deborah G. Mayo.
        p.   cm.—(Science and its conceptual foundations)
    Includes bibliographical references and index.
    1. Error analysis (Mathematics)   2. Bayesian statistical decision
theory.   3. Science—Philosophy.   I. Title.   II. Series.
QA275.M347   1996
001.4'34—dc20   95–26742
                                                               CIP

∞ The paper used in this publication meets the minimum requirements of the
American National Standard for Information Sciences—Permanence of Paper for
Printed Library Materials, ANSI Z39.48-1984.

To George W. Chatfield

for his magnificent support

Contents

Preface

DESPITE THE CHALLENGES TO AND CHANGES IN traditional philosophy of science, one of its primary tasks continues to be to explain, if not also to justify, scientific methodologies for learning about the world. To logical empiricist philosophers (Carnap, Reichenbach) the task was to show that science proceeds by objective rules for appraising hypotheses. To that end many attempted to set out formal rules termed inductive logics and confirmation theories. Alongside these stood Popper's method of appraisal based on falsification: evidence was to be used to falsify claims deductively rather than to build up inductive support. Both inductivist and falsificationist approaches were plagued with numerous, often identical, philosophical problems and paradoxes. Moreover, the entire view that science follows impartial algorithms or logics was challenged by Kuhn (1962) and others. What methodological rules there are often conflict and are sufficiently vague as to "justify" rival hypotheses. Actual scientific debates often last for several decades and appear to require, for their adjudication, a variety of other factors left out of philosophers' accounts. The challenge, if one is not to abandon the view that science is characterized by rational methods of hypothesis appraisal, is either to develop more adequate models of inductive inference, or else to find some new account of scientific rationality.

This was the problem situation in philosophy of science when, as a graduate student, I grew interested in these issues. With my background in logic and mathematics, and challenged by the problems formulated by Kyburg, Salmon, and others, I was led to pursue the first option—attempt to develop a more adequate account of inductive inference. .

I might never have sauntered into that first class on mathematical statistics had the Department of Statistics not been situated so closely to the Department of Philosophy at the University of Pennsylvania. There I discovered, expressed in the language of statistics, the very problems of induction and confirmation that were so much in the minds of the philosophers of science nearby. The more I learned, the more I suspected that understanding how these statistical methods worked would offer up solutions to the vexing problem of how we learn about the world in the face of error. But the similarity of the goals of philosophy of science and statistics, any more than the physical proximity of their departments on that campus, did not diminish the large gulf that existed between philosophical work on induction and the methods and models of standard statistical practice.

While logical-empiricist systems of inductive logic, despite a few holdouts, were largely being abandoned, most philosophers of statistics viewed the role of statistics as that of furnishing a set of formal rules or "logic" relating given evidence to hypotheses. The dominant example of such an approach on the contemporary philosophical scene is based on one or another Bayesian measure of support or confirmation. With the Bayesian approach, what we have learned about a hypothesis H from evidence e is measured by the conditional probability of H given e using Bayes's theorem. The cornerstone of the Bayesian approach is the use of prior probability assignments to hypotheses, generally interpreted as an agent's subjective degrees of belief. In contrast, the methods and models of classical and Neyman-Pearson statistics (e.g., statistical significance tests, confidence interval methods) that seemed so promising to me eschewed the use of prior probabilities where these could not be based on objective frequencies. Probability enters instead as a way of characterizing the experimental or testing process itself: to express how reliably it discriminates between alternative hypotheses and how well it facilitates learning from error. These probabilistic properties of experimental procedures are *error probabilities*.

Not only was there the controversy raging between Bayesians and error statisticians, but philosophers of statistics of all stripes were also full of criticisms of Neyman-Pearson error statistics and had erected a store of counterintuitive inferences apparently licensed by those methods. Before I could hope to utilize error statistical ideas in grappling with the problems of the rationality of science, clearly these criticisms would have to be confronted. Some proved recalcitrant to an easy dismissal, and this became the task of my doctoral dissertation. This "detour," fascinating in its own right, was a main focus for the next few years.

The result of grappling with these problems was a reformulation of standard Neyman-Pearson statistics that avoided the common misinterpretations and seemed to reflect the way these methods are used in practice. By the time that attempt grew into the experimental testing account of this book, the picture had diverged sufficiently from the Neyman-Pearson model to warrant some new name. Since it retains the centerpiece of standard Neyman-Pearson methods—the fundamental use of error probabilities—error-probability statistics, or just *error statistics*, seems about right.

My initial attempt to reformulate Neyman-Pearson statistics, however relevant to the controversy being played out within the confines of philosophy of statistics, was not obviously so for those who had largely abandoned that way of erecting an account of science, or so I was to learn, thanks to a question or challenge put forth by Larry Laudan in 1984.

In the new "theory change" movement that Laudan promoted, theory testing does not occur apart from appraising an entire *paradigm* (Kuhn), *research program* (Lakatos), or *tradition* (Laudan). In striking contrast to logical empiricist models and their contemporary (Bayesian) variants, the rational theory change models doubt that the technical machinery of inductive logic and statistical inference can shed much light on the problems of scientific rationality. It was perhaps Laudan's skepticism that drove me to pursue explicitly the task that had led me to philosophy of statistics in the first place—to utilize an adequate account of statistical inference in grappling with philosophical problems about evidence and inference. More than that, it was his persistent call to test our accounts against the historical record of science that led me to investigate a set of experimental episodes in science. I found, however, that little is learned from merely regarding historical episodes as instances or counterinstances of one or another philosophical thesis about science. That sort of historical approach fails to go deep enough to uncover the treasures often buried within historical cases.

What proved to be a gold mine for me was studying the nitty-gritty details of the data collection and analysis from experimental episodes. Here one can unearth a handful of standard or "canonical" strategies by which a host of noisy raw data may be turned into far more reliable modeled data. These investigations afforded me several shortcuts for how to relate statistical methods to full-bodied scientific inquiries. The rationale of statistical methods and models is found in their capacity to systematize strategies for learning from data, and thereby for furthering the growth of experimental knowledge.

In contrast to the global inductive approaches—a rule for any given data and hypothesis—so attractive to philosophers, I favor a model of experimental learning that is more of a piecemeal approach, whereby one question may be asked at a time in carrying out, modeling, and interpreting experiments—even to determine what "the data" are. The idea of viewing experimental inquiry in terms of a series of distinct models was influenced by experience with statistics as well as by early exposure to a seminal paper by Patrick Suppes (1969). By insisting on a global measure of evidential-relationship, philosophers have overlooked the value of piecemeal error-statistical tools for filling out a series of models that link data to experiments. But how are the pieces intertwined so that the result is genuinely *ampliative?* Unlocking this puzzle occupied me for some time. One way to describe how statistical methods function, I came to see, is that they enable us, quite literally, to learn from error. A main task of this book is to develop this view.

The view that we learn from error, while commonplace, has been little explored in philosophy of science. When philosophers of science do speak of learning from error—most notably in the work of Popper—they generally mean simply that when a hypothesis is put to the test of experiment and fails, we reject it and attempt to replace it with another. Little is said about what the different types of errors are, what specifically is learned when an error is recognized, how we locate precisely what is at fault, how our ability to detect and correct errors grows, and how this growth is related to the growth of scientific knowledge. In what follows, I shall explore the possibility that addressing these questions provides a fresh perspective for understanding how we learn about the world through experiment.

Readers who wish to read the concluding overview in advance may turn to chapter 13.

Recent trends in philosophy of science lead me to think that the time is ripe for renewing the debate between Bayesian and error statistical epistemologies of experiment. Two main trends I have in mind are (1) the effort to link philosophies of science to actual scientific practice and scrutinize methodologies of science empirically or naturalistically, and (2) the growing interest in experiment by philosophers, historians, and sociologists of science.

Although methods and models from error statistics continue to dominate among experimental practitioners who use statistics, the Bayesian Way has increasingly been regarded as the model of choice among philosophers looking to statistical methodology. Given the current climate in philosophy of science, readers unfamiliar with philosophy of statistics may be surprised to find philosophers (still) declaring invalid a widely used set of experimental methods, rather than trying to explain why scientists evidently (still) find them so useful. This has much less to do with any sweeping criticisms of the standard approach than with the fact that the Bayesian view strikes a resonant chord with the logical-empiricist gene inherited from early work in confirmation and induction. In any event, it is time to remedy this situation. A genuinely adequate philosophy of experiment will only emerge if it is not at odds with statistical practice in science.

More than any other philosophical field of which I am aware, the probability and statistics debates tend to have the vehemence usually restricted to political or religious debates. I do not hope to bring hard-core Bayesians around to my view, but I do hope to convince the large pool of tempered, disgruntled, and fallen Bayesians that a viable non-Bayesian alternative exists. Most important, I aim to promote a general change of focus in the debates so that statistical accounts are scrutinized according to how well they serve specific ends. I focus on three chief tasks to which statistical accounts can and have been put in philosophy of science: (1) modeling scientific inference, (2) solving problems about evidence and inference, and (3) performing a critique of methodological rules.

The new emphasis on experiment is of special relevance in making progress on this debate. Experiments, as Ian Hacking taught us, live lives of their own, apart from high level theorizing. Actual experimental inquiries, the new experimentalists show, focus on manifold local tasks: checking instruments, ruling out extraneous factors, getting accuracy estimates, distinguishing real effect from artifact, and estimating the effects of background factors. The error-statistical account offers a tool kit of methods that are most apt for performing the local tasks of designing, modeling, and learning from experimental data. At the same time, the already well worked out methods and models from error statistics supply something still absent: a systematic framework for making progress with the goals of the new experimentalist work.

This book is intended for a wide-ranging audience of readers interested in the philosophy and methodology of science: for practitioners and philosophers of experiment and science, and for those interested in interdisciplinary work in science and technology studies. My hope is to strengthen existing bridges and create some new bridges between these fields. I regard the book as nontechnical and open to readers without backgrounds in statistics or probability. This does not mean that it contains no formal statistical ideas, but rather that the same ideas will also be presented in semiformal and informal ways. Readers can therefore feel free to put off (temporarily or permanently) statistical discussions without worrying that they will miss the main arguments or the working of this approach. So, for example, the reader hungry for a full-blown illustration of the approach can jump to chapter 7 after wading through the first two sections of chapter 5.

I have attempted to design the book so that an idea not caught in one place will likely be caught in another. This attempt, I will confess, required me to diverge from a more usual linear approach wherein one defines the formal concepts needed, articulates an approach and then contrasts it with others, and so on. My critique of the Bayesian

Way—one key aim of the book—is woven through the book, which simultaneously addresses two interrelated aims: developing the alternative error statistical approach and tackling a number of philosophical problems about evidence and inference. In addition to this cyclical or "braided" approach, some may fault me for overlooking certain technical qualifications or for failing to mention so-and-so's recent result. Again, I admit my guilt, but this seemed a necessary trade-off to bring these ideas into the mainstream where they belong.

My intention is to take readers on a journey that lets them get a feel for the error probability way of thinking, links it to our day-to-day strategies for finding things out, and points to the direction in which a new philosophy of experiment might move. I want to identify for the reader the style of inference and argument involved—it is of a sort we perform every day in learning from errors. Once this is grasped, I believe, the appropriate way to use and interpret the formal statistical tools will then follow naturally. If anything like a full-blown philosophy of experiment is to be developed, it will depend as much on having an intuitive understanding of standard or "canonical" arguments from error as on being able to relate them to statistical models.

I began writing this book while I was a visitor at the Center for Philosophy of Science at the University of Pittsburgh in the fall of 1989. I am grateful for the stimulating environment and for conversations on aspects of this work with John Earman, Clark Glymour, Adolf Grünbaum, Nicholas Rescher, and Wesley Salmon.

My ideas were importantly shaped by Ronald Giere's defenses of Neyman-Pearson statistics in the 1970s, and he has been a wonderful resource over many years. Without his encouragement and help, especially early on, I might never have found the path or had the nerve to pursue this approach. I have benefited enormously from his scrupulous reading of earlier drafts of this manuscript and from his uncanny ability to distill, in a few elegant phrases, just what I am trying to say.

I am deeply grateful to Henry Kyburg, in whose (1984) NEH summer seminar on induction and probability many of my ideas on statistical inference crystallized. He has given me generous help with the revisions to early drafts and, more than anyone else, is to be credited with having provoked me to a bolder, clearer, and more direct exposition of my account; although he would have preferred that I perform even greater liposuction on the manuscript.

I owe the largest debt of gratitude to Wesley Salmon. My debt is both to his work, which has had a strong influence, as well as to his massive help throughout the course of this project. In countless conversations and commentaries, though I proffered only the roughest of drafts, he gave me the benefit of his unparalleled mastery of the problems with which I was grappling. I thank him for letting me try out my views on him, for his steadfast confidence and support, and for rescuing me time and again from being stalled by one or another obstacle.

I would like to acknowledge a number of individuals to whom I owe special thanks for substantive help with particular pieces of this book. I am indebted to Larry Laudan for lengthy discussions about my reworking of Kuhnian normal science in chapter 2, and I benefited greatly from Teddy Seidenfeld's recommendations and criticisms of the statistical ideas in chapters 3, 5, and 10. I thank Clark Glymour for help in clarifying his position in chapter 9. For insights on an early draft of chapter 11 on Peirce, I am grateful to Isaac Levi. An older debt recalled in developing the key concept of severe tests is to Alan Musgrave. My understanding of the views on novel predictions and my coming to relate them to severe tests (in chapters 6 and 8) grew directly out of numerous conversations with him while he was a visitor at Virginia Tech in 1986. I am grateful to communication with Karl Popper around that time, which freed me, in chapter 6, to clearly distinguish my severity concept from his.

Larry Laudan's influence, far more than the citations indicate, can be traced throughout most of the chapters; I benefited much from having him as a colleague from 1983 to 1987.

The epistemology of experiment developed in this book is broadly Peircean and I would like to acknowledge my debt to the scholarship of C. S. Peirce. Through the quandaries of virtually every chapter, his work served much like a brilliant colleague and a kindred spirit.

A portion of the rewriting took place while I was a visiting professor at the Center for Logic and Philosophy of Science at the University of Leuven, Belgium, in 1994, and I learned much from the different perspectives of colleagues there. I want to thank especially Herman Roelants, chair of the Department of Philosophy, for valuable comments on this work.

Many others gave helpful comments and criticisms on portions of this work: Richard Burian, George Barnard, George Chatfield, Norman Gilinsky, I. J. Good, Marjorie Grene, Ian Hacking, Gary Hardcastle, Valerie Hardcastle, Paul Hoyningen-Huene, William Hendricks, Joseph Pitt, J. D. Trout, and Ronald Workman.

At different times, I was greatly facilitated in the research and writing for this book by the support I received from an NEH fellowship for college teachers, an NEH summer stipend, and an NSF grant (Studies in Science, Technology, and Society Scholars Award). I am grateful to Virginia Tech and to Joseph Pitt, chair of the philosophy department, for helping to accommodate my leaves and for endorsing this project.

Portions of chapters 3, 8, 11, and 12 have appeared in previously published articles; I thank the publishers for permission to use some of this material: "The New Experimentalism, Topical Hypotheses, and Learning From Error," in *PSA 1994*, vol. 1, edited by D. Hull, M. Forbes, and R. Burian (East Lansing, Mich.: Philosophy of Science Association, 1994), 270–79; "Novel Evidence and Severe Tests," *Philosophy of Science* 58 (1991): 523–52; "Did Pearson Reject the Neyman-Pearson Philosophy of Statistics?" *Synthese* 90 (1992): 233–62; and "The Test of Experiment: C. S. Peirce and E. S. Pearson," in *Charles S. Peirce and the Philosophy of Science: Papers from the 1989 Harvard Conference*, edited by E. Moore (Tuscaloosa, Ala.: University of Alabama Press, 1983), 161–74.

I owe special thanks to Susan Abrams and the University of Chicago Press for supporting this project even when it existed only as a ten-page summary, and for considerable help throughout. I thank David Hull for his careful and instructive review of this manuscript, Madeleine Avirov for superb copyediting, and Stacia Kozlowski for a good deal of assistance with the manuscript preparation.

I obtained valuable feedback on this manuscript from the graduate students of my seminar Foundations of Statistical Inference in 1995, especially Mary Cato, Val Larson, Jean Miller, and Randy Ward. For extremely valuable editorial and library assistance over several years, I thank Mary Cato. For organizational help and superb childcare, I am indebted to Cristin Brew and Wendy Turner.

I am grateful to my father, Louis J. Mayo, for first sparking my interest in philosophy as a child; I regret that he passed away before completion of this book. I am thankful to my mother, Elizabeth Mayo, for understanding my devotion to this project, and to my son, Isaac, for not minding using the backs of discarded drafts as drawing paper for five years. My deepest debt is to my husband, George W. Chatfield, to whom this book is dedicated.

CHAPTER ONE

Learning from Error

The essays and lectures of which this book is composed are variations upon one very simple theme—the thesis that *we can learn
from our mistakes.*

Karl Popper, *Conjectures and Refutations*, p. vii

WE LEARN FROM OUR MISTAKES. Few would take issue with this dictum.
If it is more than merely a cliché, then it would seem of interest to
epistemologists to inquire how knowledge is obtained from mistakes
or from error. But epistemologists have not explored, in any serious
way, the basis behind this truism—the different kinds of mistakes that
seem to matter, or the role of error in learning about the world. Karl
Popper's epistemology of science takes learning from error as its linchpin, as the opening to his *Conjectures and Refutations* announces. In his
deductive model the main types of error from which scientists learn
are clashes between a hypothesis and some experimental outcome in
testing. Nevertheless, Popper says little about what positive information is acquired through error other than just that we learn an error
has somewhere been made. Since a great many current approaches
take Popper's problems as their starting place, and since I too make
learning from error fundamental, I begin by pursuing this criticism of
Popper.

1.1 POPPERIAN LEARNING THROUGH FALSIFICATION

For the logical empiricists, learning from experiment was a matter of
using observations to arrive at inductive support for hypotheses. Experimental observations were viewed as a relatively unproblematic
empirical basis; the task for philosophers was to build inductive logics
for assigning degrees of evidential support to hypotheses on the basis
of given statements of the evidence. Popper questioned the supposition
that experimental data were unproblematic and denied that learning
is a matter of building up inductive support through confirming in-

1

stances of hypotheses. For Popper, learning is a matter of deductive falsification. In a nutshell, hypothesis H is deductively falsified if H entails experimental outcome O, while in fact the outcome is $\sim O$. What is learned is that H is false.

Several familiar problems stand in the way of such learning. Outcome O, itself open to error, is "theory-laden" and derived only with the help of auxiliary hypotheses. The anomaly cannot be taken as teaching us that H is false because it might actually be due to some error in the observations or one of the auxiliary hypotheses needed to derive O. By means of the *modus tollens*, Popper remarks, "we falsify *the whole system* (the theory as well as the initial conditions) which was required for the deduction of [the prediction], i.e., of the falsified statement" (Popper 1959, 76). We cannot know, however, which of several auxiliary hypotheses is to blame, which needs altering. Often H entails, not a specific observation, but a claim about the probability of an outcome. With such a statistical hypothesis H, the nonoccurrence of an outcome does not contradict H, even if there are no problems with the auxiliaries or the observation.

As such, for a Popperian falsification to get off the ground, additional information is needed to determine (1) what counts as observational (and to decide which observations to accept in a particular experiment), (2) whether auxiliary hypotheses are acceptable and alternatives are ruled out, and (3) when to reject statistical hypotheses. Only with (1) and (2) does an anomalous observation O falsify hypothesis H, and only with (3) can statistical hypotheses be falsifiable. Because each determination is fallible, Popper and, later, Imre Lakatos regard their acceptance as decisions, driven more by conventions than by experimental evidence.

Lakatos sets out to improve Popper by making these (and other) decisions explicit, yielding "sophisticated methodological falsificationism." Nevertheless, Lakatos finds the decisions required by a Popperian falsificationist too risky, claiming that "the risks are daring to the point of recklessness" (Lakatos 1978, 28), particularly decision 2, to rule out any factors that would threaten auxiliary hypotheses. Says Lakatos: "When he tests a theory together with a *ceteris paribus* clause and finds that this conjunction has been refuted, he must *decide* whether to take the refutation also as a refutation of the specific theory. . . . Yet the decision to 'accept' a *ceteris paribus* clause is a very risky one because of the grave consequences it implies" (ibid., 110). Once the decision is made to reject alternative auxiliary factors, a mere anomaly becomes a genuine falsifier of the theory itself.

Lakatos regards such a decision as too arbitrary. Accepting what is often referred to as the Duhem-Quine thesis, that "no experimental result can ever kill a theory: any theory can be saved from counter-instances either by some auxiliary hypothesis or by a suitable reinterpretation of its terms" (Lakatos 1978, 32), Lakatos believes that an appeal procedure by which to avoid directing the *modus tollens* against a theory is always available.

Moreover, Lakatos, like Thomas Kuhn, finds Popper's picture of conjecture and refutation too removed from actual science, which often lives with anomalies and contains not just falsifications but also confirmations. Attempting to save something of Popper while accommodating Kuhn, Lakatos erects his "methodology of scientific research programmes." Lakatos suggests that there is a *hard core* against which *modus tollens* is not to be directed. In the face of inconsistency or anomaly, we may try to replace any auxiliaries outside this core (the "protective belt"), so long as the result is progressive, that is, predicts some novel phenomena. Determining the progressiveness of the theory change requires us to look not at an isolated theory, but at a series of theories—a *research program*. However, this holistic move does not really solve Popper's problem: as Lakatos recognizes, it enables any theory or research program to be saved—with sufficient genius it may be defended progressively, even if it is false (Lakatos 1978, 111). The cornerstone of the Popperian doctrine against saving theories from falsification is overturned. While Lakatos, like Popper, had hoped to avoid conventionalism, his solution results in making the growth of knowledge even more a matter of convention than did Popper's decisions. It is the unquestioned authority of the conventionally designated hard core, and not "the universe of facts," that decides where to direct the arrow of *modus tollens*. In Lakatos's view:

> The direction of science is determined primarily by human creative imagination and not by the universe of facts which surrounds us. Creative imagination is likely to find corroborating novel evidence even for the most "absurd" programme, if the search has sufficient drive. . . . A brilliant school of scholars (backed by a rich society to finance a few well-planned tests) might succeed in pushing any fantastic programme ahead, or, alternatively, if so inclined, in overthrowing any arbitrarily chosen pillar of "established knowledge." (Lakatos 1978, 99–100)

But let us pull back and recall the problems that set Lakatos off in the first place. Affirming experimental data? Ruling out alternative auxiliaries? Falsifying statistical claims? Why not see if there may not be perfectly good grounds for warranting the information that these tasks require without resorting to conventional decisions in the first place. This is where my project begins.

The key is to erect a genuine account of learning from error—one that is far more aggressive than the Popperian detection of logical inconsistencies. Although Popper's work is full of exhortations to put hypotheses through the wringer, to make them "suffer in our stead in the struggle for the survival of the fittest" (Popper 1962, 52), the tests Popper sets out are white-glove affairs of logical analysis. If anomalies are approached with white gloves, it is little wonder that they seem to tell us only that there is an error somewhere and that they are silent about its source. We have to become shrewd inquisitors of errors, interact with them, simulate them (with models and computers), amplify them: we have to learn to make them talk. A genuine account of learning from error shows where and how to justify Popper's "risky decisions." The result, let me be clear, is not a filling-in of the Popperian (or the Lakatosian) framework, but a wholly different picture of learning from error, and with it a different program for explaining the growth of scientific knowledge.

1.2 DAY-TO-DAY LEARNING FROM MISTAKES

The problem of learning from error in the sense of Popperian falsification, say Lakatos and others, is that learning from error itself is fraught with too much risk of error. But what grounds are there for thinking that such possible errors are actually problematic? How do scientists actually cope with them? It is not enough that mistakes are logically possible, since we are not limited to logic. Unless one is radically skeptical of anything short of certainty, specific grounds are needed for holding that errors actually occur in inquiries, that they go unnoticed, and that they create genuine obstacles to finding things out. No such grounds have been given. If we just ask ourselves about the specific types of mistakes we can and do make, and how we discover and avoid them—in short, how we learn from error—we would find that we have already taken several steps beyond the models of both Popper and Lakatos. Let me try to paint with broad brush strokes the kinds of answers that I think arise in asking this question, with a promise to fill in the details as we proceed.

1. After-trial checking (correcting myself). By "after-trial" I mean after the data or evidence to be used in some inference is at hand. A tentative conclusion may be considered, and we want to check if it is correct. Having made mistakes in reaching a type of inference in the past, we often learn techniques that can be applied the next time to check if we are committing the same error. For example, I have often discovered I was mistaken to think that A caused B when I found that B occurs even without *A*. In a subsequent inference about the effect of some factor *F*, I may deliberately consider what occurs without *F* in order to check this mistake. Other familiar after-trial checks are the techniques we develop for checking complex arithmetic operations or for balancing checkbooks.

In addition to techniques for catching ourselves in error are techniques for correcting errors. Especially important error-correcting techniques are those designed to go from less accurate to more accurate results, such as taking several measurements, say, of the length of wood or fabric, and averaging them.

2. Before-trial planning. Knowledge of past mistakes gives rise to efforts to avoid the errors ahead of time, before running an experiment or obtaining data. For example, teachers who suspect that knowing the author of a paper may influence their grading may go out of their way to ensure anonymity before starting to grade. This is an informal analogue to techniques of astute experimental design, such as the use of control groups, double blinding, and large sample size.

3. An error repertoire. The history of mistakes made in a type of inquiry gives rise to a list of mistakes that we would either work to avoid (before-trial planning) or check if committed (after-trial checking), for example, a list of the familiar mistakes when inferring a cause of a correlation: Is the correlation spurious? Is it due to an extraneous factor? Am I confusing cause and effect? More homely examples are familiar from past efforts at fixing a car or a computer, at cooking, and the like.

4. The effects of mistakes. Through the study of mistakes we learn about the kind and extent of the effect attributable to different errors or factors. This information is then utilized in subsequent inquiries or criticisms. One such use is to rule out certain errors as responsible for an effect. Perhaps putting in too much water causes the rice to be softer but not saltier.

Knowledge of the effects of mistakes is also exploited to "subtract out" their influences after the trial. If the effects of different factors can be sufficiently distinguished or subtracted out later, then the inferences are not threatened by a failure to control for them. Thus knowing the effects of mistakes is often the key to justifying inferences. In chapter 7 we will see how Jean Perrin debunked an allegation that his results on Brownian motion were due to temperature variations in his experiment. Such variations, he showed, only caused a kind of current easily distinguishable from Brownian motion.

5. *Simulating errors.* An important way to glean information about the effects of mistakes is by utilizing techniques (real or artificial) to display what it would be like if a given error were committed or a given factor were operative. Observing an antibiotic capsule in a glass of water over several days revealed, by the condition of the coating, how an ulceration likely occurred when its coating stuck in my throat. In the same vein, we find scientists appealing to familiar chance mechanisms (e.g., coin tossing) to simulate what would be expected if a result were due to experimental artifacts.

6. *Amplifying and listening to error patterns.* One way of learning from error is through techniques for magnifying their effects. I can detect a tiny systematic error in my odometer by driving far enough to a place of known distance. I can learn of a slight movement across my threshold with a sensitive motion detector. Likewise, a pattern may be gleaned from "noisy" data by introducing a known standard and studying the deviations from that standard. By studying the pattern of discrepancy and by magnifying the effects of distortions, the nature of residuals, and so forth, such deviations can be made to speak volumes.

7. *Robustness.* From the information discussed above, we also learn when violating certain recommendations or background assumptions does not pose any problem, does not vitiate specific inferences. Such outcomes or inferences are said to be robust against such mistakes. These are the kinds of considerations that may be appealed to in answering challenges to an inference. In some cases we can argue that the possibility of such violations actually strengthens the inference. (For example, if my assumptions err in specific ways, then this result is even *more* impressive.) An example might be inferring that a new teaching technique is more effective than a standard one on the basis of higher test scores among a group of students taught with the new technique (the treated group) compared with a group of students taught with the old (the control group). Whereas an assumption of the study might have been that the two groups had about equal ability, discovering that the treated group was actually less able than the control group before being taught with the new technique only strengthens the inference.

An important strategy that may be placed under this rubric is that of deliberately varying the assumptions and seeing whether the result or argument still holds. This often allows for the argument that the inference is sound, despite violations, that inaccuracies in underlying factors cannot be responsible for a result. For, were they responsible we would not have been able to consistently obtain the same results despite variations.

8. *Severely probing error.* Points 1 through 7 form the basis of learning to detect errors. We can put together so potent an arsenal for unearthing a given error that when we fail to find it we have excellent grounds for concluding that the error is absent. Having failed to detect a given infection with several extremely reliable blood tests, my physician infers that it is absent. The "error" inferred to be absent here is declaring that there is no infection when there is one.

The same kind of reasoning is at the heart of experimental testing. I shall call it *arguing from error.* After learning enough about certain types of mistakes, we may construct (often from other tests) a testing procedure with an overwhelmingly good chance of revealing the presence of a specific error, if it exists—but not otherwise. Such a testing procedure may be called a *severe (or reliable) test,* or a *severe error probe.* If a hypothesized error is not detected by a test that has an overwhelmingly high chance of detecting it, if instead the test yields a result that accords well with no error, then there are grounds for the claim that the error is absent. We can infer something positive, that the particular error is absent (or is no greater than a certain amount). Equivalently, we have grounds for rejecting the hypothesis, H', that the error is present, and affirming H, that it is absent. When we have such information, we say that H has passed a severe test. Alternatively, we can say that the test result is a *good indication* that H is correct.

Is it possible for such humdrum observations to provide a fresh perspective from which to address problems that still stand in the way of a satisfactory epistemology of science? I propose that they can, and that is the underlying thesis of this book.

To turn the humdrum observations into tools for experimental learning, they need to be amplified, generalized, and systematized. As I see it, this is the chief task of an adequate epistemology of experiment. I understand "experiment," I should be clear at the outset, far more broadly than those who take it to require literal control or manipulation. Any planned inquiry in which there is a deliberate and reliable argument from error may be said to be experimental.

How can these day-to-day techniques for learning from error take us beyond Popper's deductive falsification model?

1.3 ACCENTUATE THE POSITIVE, ELIMINATE THE NEGATIVE

I endorse many of Popper's slogans. Like Popper's, the present approach views the growth of knowledge as resulting from severe criticism—from deliberately trying to find errors and mistakes in hypotheses. It likewise endorses his idea that learning about a hypothesis is based on finding out whether it can withstand severe tests. Each of these slogans, however, is turned into a position where something positive is extracted from the severe criticism; for us, the focus is on *constructive criticism*, on learning from criticizing. It seems incumbent upon anyone mounting such an approach to dispel the ghosts of Popper's negativism right away, or at least sketch how they will be dispelled.

The most devastating criticism of Popper's approach is this: having rejected the notion that learning is a matter of building up probability in a hypothesis, Popper seems to lack any meaningful way of saying why passing severe tests counts in favor of a hypothesis.[1] Popper's account seems utterly incapable of saying anything positive. There are two variants to this criticism, which I shall take up in turn.

a. Why Should Passing Severe Tests Count in Favor of Hypotheses?

If the refuted hypothesis is rejected for one that passes the test it failed, that new hypothesis, Popper says, is preferable. But why should it be preferred? What is it that makes it better? The most Popper can say on its behalf is that it did better in passing the test the previous hypothesis failed and that "it will also have to be regarded as possibly true, since at the time *t* it has not been shown to be false" (Popper 1979, 14). Popper concedes that there are infinitely many other hypotheses that would also pass the tests that our current favorite has:

> By this method of elimination, we may hit upon a true theory. But in no case can the method *establish* its truth, even if it is true; for the number of *possibly* true theories remains infinite. (Popper 1979, 15)

Popper sees this as a way of stating Hume's problem of induction. Again,

> in my view, all that can possibly be *"positive"* in our scientific knowledge is positive *only* in so far as certain theories are, at a certain moment of time, preferred to others in the light of . . . attempted refutations. (P. 20)

For Popper, we should not *rely* on any hypothesis, at most we should prefer one. But why should we prefer best-tested hypotheses? It is altogether unsatisfactory for Popper to reply as he does that he simply does "not know of anything more 'rational' than a well-conducted critical discussion" (p. 22).

I, too, argue that hypotheses that pass genuinely severe tests gain merit thereby. How do I avoid Popper's problem? Popper's problem

---

1. It has been raised, for example, by Wesley Salmon (1966), Adolf Grünbaum (1978), and Alan Musgrave (1978).

here is that the grounds for the badge of "best-tested hypothesis of the moment" would also be grounds for giving the badge to an infinite number of (not yet even thought of) hypotheses, had they been the ones considered for testing. If a nonfalsified hypothesis H passes the tests failed by all the existing rivals, then H is best-tested, H gets the badge. Any other hypothesis that would also pass the existing tests would have to be said to do as well as H—by Popper's criteria for judging tests. But this is not the case for the test criteria I shall be setting out. These test criteria will be based on the idea of severity sketched above. A hypothesis H that passes the test failed by rival hypothesis H' (and by other alternative hypotheses considered) has passed a severe test for Popper—but not for me. Why not? Because for H to pass a severe test in my sense, it must have passed a test that is highly capable of probing the ways in which H can err. And the test that alternative hypothesis H' failed need not be probative in the least so far as the errors of H go. So long as two different hypotheses can err in different ways, different tests are needed to probe them severely. This point is the key to worries about underdetermination (to be discussed in chapter 6).

b. Corroboration Does Not Yield Reliability

There is a second variant of the objection that passing a severe test in Popper's sense fails to count in favor of a hypothesis. It is that saying a hypothesis is well tested for Popper says nothing about how successful it can be expected to be in the future.

According to Popper, the more severe the test a hypothesis has passed, the higher its corroboration. Popper regards the degree of corroboration of a hypothesis as "its degree of testability; the severity of tests it has undergone; and the way it has stood up to these tests" (Popper 1979, 18). Not only does Popper deny that we are entitled to consider well-corroborated claims as true, but we are not even to consider them as reliable. Reliability deals with future performance, and corroboration, according to Popper, is only a *report of past performance. Like preference, it is essentially comparative. . . . But it says nothing whatever about future performance, or about the 'reliability' of a theory*" (p. 18). (Nor would this point be affected, Popper adds, by the finding of a quantitative measure of corroboration.)

At least part of the reason for this criticism, as well as for Popper's admission, is the prevalence of the view that induction or ampliative inference requires some assignment of probability, credibility, or other evidential measure to hypotheses. This view is shared by the majority of Popper's critics, and it is one Popper plainly rejects. The present view of experimental learning, like Popper's, will not be in terms of assigning a degree of probability, credibility, or the like to any hypothesis or theory. But quite unlike Popper's view, this does not preclude our inferring a hypothesis reliably or obtaining reliable knowledge. The needed reliability assignment, I shall argue, is not only obtainable but is more in line with what is wanted in science.

Except in very special cases, the probability of hypotheses can only be construed as subjective degrees of belief, and I will argue that these yield an unsatisfactory account of scientific inference. As C. S. Peirce urged in anticipation of modern frequentists, what we really want to know is not the probability of hypotheses, but the probability with which certain outcomes would occur given that a specified experiment is performed. It was the genius of classical statisticians, R. A. Fisher, Jerzy Neyman, Egon Pearson, and others, to have developed approaches to experimental learning that did not depend on prior probabilities and where probability refers only to relative frequencies of types of outcomes or events. These relative frequency distributions, which may be called *experimental distributions,* model actual experimental processes.

Learning that hypothesis H is reliable, I propose, means learning that what H says about certain experimental results will often be close to the results actually produced—that H will or would often succeed in specified experimental applications. (The notions of "closeness" and "success" must and can be made rigorous.) This knowledge, I argue, results from procedures (e.g., severe tests) whose reliability is of precisely the same variety. My aim will be to show how passing a severe test teaches about experimental distributions or processes, and how this, in turn, grounds experimental knowledge.

The Emerging View of Experimental Knowledge

In summary, let me say a bit more about the view of experimental knowledge that emerges in my approach. I agree with Popper's critics that Popper fails to explain why corroboration counts in favor of a hypothesis—but not because such credit counts only in favor of a hypothesis if it adds to its credibility, support, probability, or the like. The problem stems from two related flaws in Popper's account: First, wearing the badge "best-tested so far" does not distinguish a hypothesis from infinitely many others. Second, there are no grounds for relying on hypotheses that are well corroborated in Popper's sense. I have also sketched how I will be getting around each flaw.

Popper says that passing a severe test (i.e., corroboration) counts in favor of a hypothesis simply because it may be true, while those that failed the tests are false. In the present view, passing a severe test counts because of the experimental knowledge revealed by passing. Indeed, my reason for promoting the concept of severity in the first place is that it is a test characteristic that is relevant as regards something that has passed the test. To figure out what an experiment reveals, one has to figure out what, if anything, has passed a severe test. The experimental inference that is licensed, in other words, is what has passed a severe test. What is learned thereby can be made out in terms of the presence or absence of an error. Even if the test cannot be regarded as having severely tested any claim, that fact alone is likely to be relevant.

Since the severe test that a hypothesis H passes is at the same time a test that fails some alternative hypothesis (i.e., H's denial), the knowledge gained from passing can also be expressed as learning from failing a hypothesis. (For example, passing H: the disease is present, is to fail H': the disease is absent.) So in failing as well as passing, the present account accentuates the positive.

The centerpiece of my account is the notion of severity involved. Unlike accounts that begin with evidence e and hypothesis H and then seek to define an evidential relationship between them, severity refers to a method or procedure of testing, and cannot be assessed without considering how the data were generated, modeled, and analyzed to obtain relevant evidence in the first place. I propose to capture this by saying that assessing severity always refers to a framework of *experimental inquiry*.

In my account of experimental testing, experimental inquiry is viewed in terms of a series of models, each with different questions, stretching from low-level theories of data and experiment to higher level hypotheses and theories of interest. (I will elaborate in detail in chapter 5.) Whether it is passing or failing, however, what is learned will always be in terms of a specific question in a given model of experimental inquiry. Later we will see how such bits of learning are pieced together.

By Popper's own admission (e.g., Popper 1979, 19), corroboration fails to be an indicator of how a hypothesis would perform in experiments other than the ones already observed. Yet I want to claim for my own account that through severely testing hypotheses we can learn about the (actual or hypothetical) future performance of experimental processes—that is, about outcomes that would occur with specified probability if certain experiments were carried out. This is *experimental knowledge*. In using this special phrase, I mean to identify knowledge of experimental effects (that which would be reliably produced by carrying out an appropriate experiment)—whether or not they are part of any scientific theory. The intent may be seen as providing a home for what may be very low-level knowledge of how to reliably bring about certain experimental results. To paraphrase Ian Hacking, it may be seen as a home in which experiment "lives a life of its own" apart from high-level theorizing. But it is to be a real home, not a life in the street; it has its own models, parameters, and theories, albeit experimental ones. And this is so whether the experimental effects are "in nature," whether they are deliberately constructed, or even whether they exist only "on paper" or on computers.

Popper's problems are insurmountable when hypothesis appraisal is considered as a matter of some formal or logical relationship between evidence or evidence statements and hypotheses; but the situation is not improved by appealing to larger units such as Lakatosian research programs. Appealing to an experimental framework and corresponding experimental strategies, I will argue, offers a fresh perspective and fresh tools for solving these problems.

The idea that focusing on experiment might offer new and largely untapped tools for grappling with problems regarding scientific inference is not new; it underlies a good deal of work in the philosophy of science of the last decade. As promising as this new experimentalist movement has been, it is not clear that the new attention to experiment has paid off in advancing solutions to these problems. Nor is it clear that those working in this movement have demarcated a program for developing a philosophy or epistemology of experiment. For sure, they have given us an important start: their experimental narratives are rich in illustrations of the role of experimentation and instrumentation in scientific inference. But something more general and more systematic seems to be needed to show how this grounds experimental knowledge and how this knowledge gets us around the problems of evidence and inference. Where we should look, I will argue, is to the already well worked out methods and models for designing and analyzing experiments that are offered in standard statistical practice.

Experimental knowledge, as I understand it, may be construed in a formal or informal mode. In its formal mode, experimental knowledge is knowledge of the probabilities of specified outcomes in some actual or hypothetical series of experiments. Its formal statement may be given by an *experimental distribution* (a list of outcomes and their associated probabilities), or by a standard "random" process such as a coin-tossing mechanism. Typically, the interest is only in some key characteristic of this distribution—a parameter—such as its arithmetic mean. In its informal mode, the one the practitioner is generally engaged in, experimental knowledge is knowledge of the presence or absence of errors. (For example, a coin-tossing model or its corresponding Binomial distribution might serve as a formal model of an informal claim about a spurious correlation.) I will stress the informal mode.

As we proceed, we will come to see the considerable scope of what can be learned from answers to questions about experimental processes and effects. How far experimental knowledge can take us in understanding theoretical entities and processes is not something that should be decided before exploring this approach much further, further even than I can go in this book. So, for example, I will not argue for or against different realist views. What I will argue is that experimental knowledge is sufficient and, indeed, that it is the key to answering the main philosophical challenges to the objectivity and rationality of science.

1.4 REVISITING THE THREE DECISIONS

The Popperian problems of the last section emanate from the concern with which we began: the three (risky) decisions required to get a Popperian test off the ground. By the various techniques of learning from error, I said, we can substantiate the information needed. This necessitates an approach to experimental learning radically different from Popper's. Nevertheless, it may be of interest to see how the three concerns translate into arguments for checking one or more experimental mistakes. The connections are these:

• The acceptance of observation or basic statements (decision 1) is addressed by arguments justifying the assumptions of the experimental data.
• The elimination of auxiliary factors (decision 2) is addressed by arguments that the experiment is sufficiently controlled.
• The falsification of statistical claims (decision 3) is accomplished by standard statistical tests.

The first two involve justifying assumptions about a specific testing context. In both cases the justifications will take the form either of showing that the assumptions are sufficiently well met for the experimental learning of interest or showing that violations of the assumptions do not prevent specific types of information from being obtained from the experiment. As important as it is to avoid error, the centerpiece of the approach I recommend is its emphasis on procedures that permit a justification of the second type—learning despite errors, or robustness. I will champion a third sort of justificatory argument: even if a mistake goes undetected, we will, with high probability, be able to find this out.

*"The Empirical Basis" Becomes the Assumptions of the
Experimental Data*

To arrive at the basic (or test) statements for Popper, two decisions are required. The first is to decide which theories to deem "observational," which for Popper means not that they are literally observational, but rather that they may be deemed unproblematic background knowledge for the sake of the test. Such information is often based on well-understood theories of instruments, for example, on theories of microscopes. The second is to decide which particular statements to accept—for example, that the instrument reads such and such.[2] Popper claims that although we can never accept a basic statement with certainty, we "must stop at some basic statement or other which we *decide to accept.*" Otherwise the test leads nowhere. "[W]e arrive in this way at a procedure according to which we stop only at a kind of statement that is especially easy to test. . . . at statements about whose acceptance or rejection the various investigators are likely to reach agreement" (Popper 1959, 104). Nevertheless, for Popper, we can no more rely on these than on other corroborated hypotheses. They, too, are merely conjectures, if at a lower level and easier to test. They are not literally basic statements, but more like "piles driven into a swamp."

But even these singular observation statements are not enough to get a Popperian falsification off the ground. We need, not singular observations, but observational knowledge; the data must warrant a hypothesis about a real or reproducible effect:

> We say that a theory is falsified only if we have accepted basic statements which contradict it. . . . This condition is necessary, but not sufficient; for we have seen that non-reproducible single occurrences are of no significance to science. Thus a few stray basic statements contradicting a theory will hardly induce us to reject it as falsified. We shall take it as falsified only if we discover a *reproducible effect* which refutes the theory. In other words, we only accept the falsification if a low-level empirical hypothesis which describes such an effect is proposed and corroborated. (Popper 1959, 86)

He calls this low-level hypothesis a *falsifying hypothesis* (p. 87).

Here Popper is recognizing what is often overlooked: the empirical data enter hypothesis appraisal in science as a hypothesis about the

---

2. Basic statements are "statements asserting that an observable event is occurring in a certain individual region of space and time" (Popper 1959, 103). As an example he gives "This clock reads 30 minutes past 3" (Popper 1962, 388).

data. Accounts of hypothesis appraisal that start with evidence *e* as given vastly oversimplify experimental learning. This recognition, however, only means trouble for Popper. In order for the acceptance of a falsifying hypothesis to be more than a conventional decision, there need to be grounds for inferring a reliable effect—the very thing Popper says we cannot have. We cannot rely on hypotheses about real or reproducible effects for Popper, because they are based on lower-level (singular) observation statements that may themselves be mistaken. "[A]nd should we try to *establish* anything with our tests, we should be involved in an infinite regress" (Popper 1962, 388).

Herein lies a presupposition commonly harbored by philosophers: namely, that empirical claims are only as reliable as the data from which they are inferred. The fact is that we can often arrive at rather accurate claims from far less accurate ones. Scattered measurements, for example, are not of much use, but with a little data massaging (e.g., averaging) we can obtain a value of a quantity of interest that is far more accurate than individual measurements. Our day-to-day learners from error know this fact but, to my knowledge, the only philosopher to attach deep significance to this self-correcting ability is C. S. Peirce.

The present approach rejects both the justificationist image of building on a firm foundation (e.g., protocol statements) and the Popperian image of building on piles driven into a swamp. Instead, the image is one of shrewd experimental strategies that permit detecting errors and squeezing reliable effects out of noisy data. What we rely on, I will urge, are not so much scientific theories but *methods* for producing experimental effects.

Ruling Out Auxiliaries as Arguing for Experimental Control

The need to rule out alternative auxiliary factors (decision 2) gives Popper the most trouble. In order for an effect (e.g., an anomaly or failed prediction) to be attributed to some flaw in a hypothesis *H*, it is required to affirm a ceteris paribus claim, that it is not due to some other possible factor. As Lakatos notes, one can test a ceteris paribus clause severely by assuming that there are other influencing factors, specifying them, and testing these assumptions. "If many of them are refuted, the *ceteris paribus* clause will be regarded as well-corroborated" (Lakatos 1978, 26). Lakatos found this too risky.

Ruling out auxiliaries is thought to be so problematic because it is assumed that there are always infinitely many causes for which we have not controlled. What is overlooked is the way in which experiments may be designed to deliberately isolate the effect of interest so that only a manageable number of causal factors (or types of factors) may produce the particular experimental outcome. Most important,

literal control is not needed: one need only find ways of arguing so as
to avoid the erroneous assignment of the cause of a given effect or
anomaly. Several ways were discussed in section 1.2. Another example
(under pretrial planning) would be to mimic the strategy of randomized treatment-control studies. The myriad of possible other causes—
even without knowing what they are—are allowed to influence the
treated and the control groups equally. In other cases, substantive alternative causes cannot be subtracted out in this manner. Then severe
tests against hypotheses that these causes are responsible for the given
experimental effect must be carried out separately.

In the present approach, ruling out alternative auxiliaries is tantamount to justifying the assumption either that the experiment is sufficiently well controlled or that the experiment allows arguing *as if* it
were sufficiently controlled for the purpose of the question of primary interest.

With effective before- and after-trial planning and checking, learning that an anomaly cannot be due to specific background factors may
finally show some primary hypothesis to be at fault. Even this rejection
has an affirmative side. Precisely because the background checks have
been required to be severe, such a rejection pinpoints a genuine effect
that needs explaining or that calls for a specific revision. One literally
learns from the error or anomaly. Note that no alternative hypothesis
to the one rejected is needed in this testing model.

Falsifying Statistical Claims by Statistical Hypothesis Testing

My approach takes as the norm the need to deal with the rejection
of statistical hypotheses—decision 3. Even where the primary theory
or hypothesis of interest is nonstatistical, a variety of approximations,
inaccuracies, and uncertainties results in the entry of statistical considerations in linking experimental data to hypotheses. A hierarchy of
models of experimental inquiry will be outlined in chapter 5.

In an interesting footnote, Lakatos remarks that statistical rejection
rules constitute "the philosophical basis of some of the most interesting
developments in modern statistics. The Neyman-Pearson approach
rests completely on methodological falsificationism" (Lakatos 1978, 25,
n. 6). Still, neither he nor Popper attempts to use the Neyman-Pearson
methods in his approach. By contrast, I shall make fundamental use of
this approach, albeit reinterpreted, as well as of cognate methods (e.g.,
Fisherian tests). My use of these methods, I believe, reflects their actual
uses in science and frees them from the confines of the particular philosophies of statistics often associated with them. Thus freed, these
methods make up what I call (standard) *error statistics*.

Although Popper makes no explicit attempt to appeal to error statistical methods, his discussion of decision 3 gets to the heart of a fundamental type of error statistical test. While extremely rare events may occur, Popper notes, "such occurrences would not be physical effects, because, on account of their immense improbability, *they are not reproducible at will. . . .* If, however, we find *reproducible* deviations from a macro effect . . . deduced from a probability estimate . . . then we must assume that the probability estimate is *falsified"* (Popper 1959, 203).

The basic idea is this: A hypothesis may entail only that deviations of a certain magnitude are rare, so that an observed deviation from what is predicted does not strictly speaking contradict the prediction. A statistical test allows learning that a deviation is not rare but is *reproducible at will*—that is, can be brought about very frequently. If we learn this, then we have found a real physical effect that is denied by and so, in this sense, contradicts the statistical hypothesis. Rather than viewing this as a conventional decision, it will be seen to rest on solid epistemological grounds. These grounds may be cashed out in two ways: (1) To construe such reproducible effects as unsystematic will very often be mistaken. So unreliable a method would be an obstacle to using data to distinguish real from spurious effects—it would be an obstacle to learning from error. (2) It is extremely improbable that one would be able to regularly reproduce an effect if in fact it was accidental. The hypothesis asserting that it is a "real effect" passes a severe test.

1.5 PIECEMEAL LEARNING FROM ERRORS

My account of the growth of experimental knowledge is a result of having explored the consequences of the thesis that we learn from our mistakes. It is not an attempt to reconstruct after-the-fact scientific inferences or theory changes, but to give an account of forward-looking methods for learning. These methods revolve around tests of low-level or local hypotheses. These hypotheses have their home in experimental models and theories. While experimental hypotheses may be identical to substantive scientific claims, they may also simply be claims about experimental patterns, real or constructed, actual or hypothetical. Local experimental inquiries enable complex scientific problems to be broken down into more manageable pieces—pieces that admit of severe tests. Even when large-scale theories are being investigated or tested, these piecemeal tests are central to the growth of experimental knowledge.[3]

3. Popper puts the burden on the hypothesis to have high information content and so be the most testable. The present approach puts the burden on the experimental test—it is the test that should be severe. The basis for tests with appropriately high severity is the desire to learn the most.

I propose that the piecemeal questions into which experimental inquiries are broken down may be seen to refer to standard types of errors. Strategies for investigating these errors often run to type. Roughly, four such standard or canonical types are

  a. mistaking experimental artifacts for real effects; mistaking chance effects for genuine correlations or regularities;
  b. mistakes about a quantity or value of a parameter;
  c. mistakes about a causal factor;
  d. mistakes about the assumptions of experimental data.

These types of mistakes are not exclusive (for example, checking *d* may involve checking the others), nor do they even seem to be on a par with each other. Nevertheless, they often seem to correspond to distinct and canonical types of experimental arguments and strategies.

I suggest that methodological rules should be seen as strategies for conducting reliable inquiries into these standard or "canonical" types of errors. Examples of methodological rules are the use of controlled experiments in testing causal hypotheses, the use of randomization, the preference for novel facts and the avoidance of "ad hoc" hypotheses, the strategy of varying the evidence as much as possible, and the use of double-blind techniques in experimenting on human subjects.

The overarching picture, then, is of a substantive inquiry being broken down into inquiries about one or more canonical errors in such a way that methodological strategies and tools can be applied in investigating those errors. One example (to be fleshed out later) is an inquiry into whether a treatment causes an increased risk of some sort. It might be broken down into two canonical inquiries: first, to establish a real as opposed to a spurious correlation between the treatment and the effect; second, to test quantitatively the extent of the effect, if there is one. One possible methodological strategy would be a treatment-control experiment and an analysis by way of statistical significance tests.

Normative Naturalism in Experimental Methodology

The present model for an epistemology of experiment is both normative and naturalistic. I have in mind this picture of experimental methodology: methodological rules for experimental learning are strategies that enable learning from common types of experimental mistakes. The rules systematize the day-to-day learning from mistakes delineated above. From the history of mistakes made in reaching a type of inference, a repertoire of errors arises; methodological rules are techniques for circumventing and uncovering them. Some refer to before-trial experimental planning, others to after-trial analysis of the data—or, more generally, learning from the data. The former includes rules about how specific errors are likely to be avoided or circumvented, the latter, rules about checking the extent to which given errors are committed or avoided in specific contexts.

Methodological rules do not rest on a priori intuitions, nor are they matters to be decided by conventions (e.g., about what counts as science or knowledge or rational). They are empirical claims or hypotheses about how to find out certain things by arguing from experiments. Accordingly, these hypotheses are open to an empirical appraisal: their truth depends upon what is actually the case in experimental inquiries. Hence, the account I propose is naturalistic. At the same time it is normative, in that the strategies are claims about how to actually proceed in given contexts to learn from experiments.

Since the rules are claims about strategies for avoiding mistakes and learning from errors, their appraisal turns on understanding how methods enable avoidance of specific errors. One has to examine the methods themselves, their roles, and their functions in experimental inquiry. A methodological rule is not empirically validated by determining whether its past use correlates with "successful" theories.[4] Rather, the value of a methodological rule is determined by understanding how its applications allow us to avoid particular experimental mistakes, to amplify differences between expected and actual experimental results, and to build up our tool kit of techniques of learning from error to determine how successful they have been in past applications.

No assignments of degrees of probability to hypotheses are required or desired in the present account of ampliative inference. Instead, passing a severe test yields positive experimental knowledge by corresponding to a strong argument from error. Accordingly, progress is not in terms of increasing or revising probability assignments but in terms of the growth of experimental knowledge, including advances in techniques for sustaining experimental arguments. Such features of my account stand in marked contrast to the popular Bayesian Way in the philosophy of science.

Next Step

The response to Popper's problems, which of course are not just Popper's, has generally been to "go bigger," to view theory testing in terms of larger units—whole paradigms, research programs, and a variety of other holisms. What I have just proposed instead is that the

4. This is essentially Larry Laudan's (1987, 1990b, 1996) normative naturalism.

lesson from Popper's problems is to go not bigger but smaller. More-over, I propose that this lesson is, in a sense, also Thomas Kuhn's, de-spite his being a major leader of the holistic movement. Let us there-fore begin our project by turning to some Kuhnian reflections on Popper.

CHAPTER TWO

Ducks, Rabbits, and Normal Science: Recasting the Kuhn's-Eye View of Popper

SHORTLY AFTER the publication of his enormously influential book *The Structure of Scientific Revolutions,* Thomas Kuhn offered "a disciplined comparison" of his and Popper's views of science in the paper "Logic of Discovery or Psychology of Research?" It begins with these lines:

> My object in these pages is to juxtapose the view of scientific development outlined in my book [*Structure*], with the better known views of our chairman, Sir Karl Popper. Ordinarily I should decline such an undertaking, for I am not so sanguine as Sir Karl about the utility of confrontations. . . . Even before my book was published two and a half years ago, I had begun to discover special and often puzzling characteristics of the relation between my views and his. That relation and the divergent reactions I have encountered to it suggest that a disciplined comparison of the two may produce peculiar enlightenment. (Kuhn 1970, 1)

"Peculiar enlightenment" is an apt description of what may be found in going back to Kuhn's early comparison with Popper and the responses it engendered. What makes my recasting of Kuhn peculiar is that while it justifies the very theses by which Kuhn effects the contrast with Popper, the picture that results is decidedly *un*-Kuhnian. As such I do not doubt that my recasting differs from the "peculiar enlightenment" Kuhn intended, but my task is not a faithful explication of what Kuhn saw himself as doing. Rather it is an attempt, at times deliberately un-Kuhnian, to see what philosophical mileage can be gotten from exploring the Kuhnian contrast with Popper. This exercise will serve as a springboard for the picture of experimental knowledge that I want to develop in this book.

Kuhn begins by listing the similarities between himself and Popper that place them "in the same minority" among philosophers of science of the day (Kuhn 1970, 2). Both accept theory-ladenness of observation, hold some version of realism (at least as a proper aim of science), and reject the view of progress "by accretion," emphasizing instead "the revolutionary process by which an older theory is rejected and replaced by an incompatible new one" (ibid., 2).

Despite these agreements Kuhn finds that he and Popper are separated by a "gestalt switch." Popper views the overthrowing and replacement of scientific theories as the main engine of scientific growth. Scientific knowledge, Popper declares, "grows by a more revolutionary method than accumulation—by a method which destroys, changes, and alters the whole thing" (Popper 1962, 129). Kuhn views such revolutionary changes as extraordinary events radically different from the "normal" scientific tasks of "puzzle solving"—extending, applying, and articulating theories. Moreover, the growth of scientific knowledge, for Kuhn, is to be found in nonrevolutionary or normal science. Although, in Kuhn's view, "normal science" constitutes the bulk of science, what has intrigued most philosophers of science is Kuhnian revolutionary science—with its big changes, gestalt switches, conversion experiences, incommensurabilities, and the challenges thereby posed to the rationality of theory change. Although there are several chapters on normal science in *Structure*, the excitement engendered by these revolutionary challenges has seemed to drown out the quieter insights that those chapters provide. Kuhn's description of normal science, when discussed at all, is generally dismissed as relegating day-to-day science to an unadventurous working out of "solvable puzzles" and "mopping-up" activities. In the view of Kuhn's critics, as Alan Musgrave (1980) puts it, normal science is either "a bad thing which fortunately does not exist, or a bad thing which unfortunately does exist" (pp. 41–42).

In this vein, Popper (1970) responds to Kuhn in "Normal Science and Its Dangers." Popper was aghast at Kuhnian normal science with its apparent call to "abandon critical discourse" and embrace unquestioning allegiance to a single accepted paradigm, encompassing theories as well as standards and values for their appraisal. Kuhnian normal science, were it actually to exist, Popper declares, would be pathetic or downright dangerous:

> In my view the "normal" scientist, as Kuhn describes him, is a person one ought to be sorry for. . . . The "normal" scientist, as described by Kuhn, has been badly taught. He has been taught in a dogmatic spirit: he is a victim of indoctrination. (Popper 1970, 52–53)

Most troubling for Popper is the alleged inability of normal scientists to break out of the prison of their paradigm or framework to subject it to (revolutionary) tests, and, correspondingly, Kuhn's likening theory change to a religious conversion rather than to a rational empirical appraisal.

Finding much to dislike about normal science, philosophers find little use for it when it comes to solving these challenges to the rationality of theory change. The big problems enter in revolutionary, not normal, science, and that is where those who would solve these problems focus their attention. I find Kuhnian normal science to be far more fruitful. It is here that one may discover the elements of Kuhn's story that are most "bang on," most true to scientific practice; and the enlightenment offered by the Kuhn-Popper comparison suggests a new way of developing those elements. Far from being the uncritical affair Popper fears, normal science, thus developed, turns out to offer an effective basis for severe testing. This, in turn, provides the key to getting around the big problems alleged to arise in revolutionary science, or large-scale theory change.

Let us pursue a bit further the contrasts Kuhn draws between his and Popper's philosophies of science. Except where noted, all references are to Kuhn 1970.

2.1 TURNING POPPER ON HIS HEAD

Kuhn asks, "How am I to persuade Sir Karl, who knows everything I know about scientific development and who has somewhere or other said it, that *what he calls a duck can be seen as a rabbit?* How am I to show him what it would be like to wear my spectacles?" (p. 3; emphasis added).

Kuhn's tactic is to take the linchpins of Popper's philosophy and show how, wearing Kuhnian glasses, they appear topsy turvy. While in Popper's view, what sets science apart from other practices is its willingness to continually subject its theories to severe and crucial tests, to the Kuhnian eye "it is normal science, in which Sir Karl's sort of testing does not occur, rather than extraordinary science which most nearly distinguishes science from other enterprises. If a demarcation criterion exists (we must not, I think, seek a sharp or decisive one), it may lie just in that part of science which Sir Karl ignores" (p. 6). But because normal science, for Kuhn, does not involve Popperian-style testing, Kuhn provocatively declares, "In a sense, to turn Sir Karl's view on its head, *it is precisely the abandonment of critical discourse that marks the transition to a science*" (p. 6; emphasis added).

If only we would view the highlights of the Popperian landscape through his spectacles, Kuhn proposes, we would see how Popper's view gets turned on its head. Specifically, we would see why, where Popper sees a fundamental theory failing a severe test, Kuhn sees a paradigm failing in its "puzzle solving ability" (crisis), and why, where

Popper sees a lack of testability, Kuhn sees a lack of puzzle solving. In so doing, Kuhn assures us, we would begin to see the sense in which "severity of test-criteria is simply one side of the coin whose other face is a puzzle-solving tradition," and with this, Kuhn proclaims, "Sir Karl's duck may at last become my rabbit" (p. 7).

I propose that we look at the high points of the Popperian landscape that the Kuhn's-eye view brings into focus. Why? First, I think that in identifying these points of contrast Kuhn makes a number of well-founded descriptive claims about scientific practice. Second, I think these claims have important normative epistemological underpinnings that have gone unnoticed. The highlights that interest me underlie the following portions of the above passages:

- "It is normal science, in which Sir Karl's sort of testing does not occur, rather than extraordinary science which most nearly distinguishes science from other enterprises." (P. 6)
- "It is precisely the abandonment of critical discourse that marks the transition to a science." (P. 6)
- "Severity of test-criteria is simply one side of the coin whose other face is a puzzle-solving tradition." (P. 7)

To extract the epistemological lessons I am after, however, it is not enough to turn our gaze toward the Kuhn's-eye view of Popper; spectacles capable of seeing the normative dimension are required.

Except for some tantalizing clues, Kuhn fails to bring out this normative dimension. This is not surprising given his view on what doing so would require. What more is there to say, Kuhn seems to ask, once one is done describing the behavior of practitioners trained in a certain way? Explaining the success or progress of science for Kuhn just *is* a matter of "examining the nature of the scientific group, discovering what it values, what it tolerates, and what it disdains. That position is intrinsically sociological" (p. 238).

Stopping with this kind of descriptive account has several shortcomings. First, it is not known which of the many features scientific communities share are actually responsible for scientific success. And even where we can correctly identify these features, there is an epistemological question that needs answering: why does following these practices yield reliable knowledge? Correlatively, why do enterprises not characterized by these practices turn out to be less successful sciences or not sciences at all?

Of Kuhn's key observations of normal scientific practice, I will ask: What is so valuable about normal science (as Kuhn describes it)? Why does the lack of a normal "puzzle-solving tradition" seem to go hand in hand with ineffective sciences or nonsciences? The bulk of my inquiry will be directed toward identifying the constraints and methods that enable what is learned in normal science to do what Kuhn rightly says it does: it extends, articulates, and revises theories, identifies genuine anomalies, functions creatively in developing alternative theories, and allows communication between paradigms (however partial).

2.2 POPPER IN LIGHT OF KUHN: J. O. WISDOM

J. O. Wisdom (1974), responding to Kuhn's "Logic or Psychology," offers an insightful way of assimilating Popper and Kuhn. He goes so far as to suggest that "Kuhn's theory and (when correctly interpreted) Popper's (when developed) are identical" (p. 839). In Wisdom's reading, the difference between Kuhn's picture of revolutionary science as something that occurs only after periods of normal science followed by crisis and Popper's emphasis on testing and falsifying theories is only apparent. (He is far less sanguine about accommodating Kuhn's "sociology of acceptance.")

How does Wisdom manage to convert Kuhn's rabbit into a Popperian duck? As we saw in chapter 1, Popper himself is well aware that for a theory genuinely to fail a test it is not enough that one of its consequences turns out wrong. Even for Popper, Wisdom rightly notes, before a failed prediction is taken as falsifying a theory, "a whole crop of loopholes have to be investigated: checking mathematics, accidental disturbances, instruments in working order, bungling, unsuspected misconceptions about the nature of the instruments, adequate articulation of the fundamental theory" (ibid., 838), and such investigations are the sorts of things done in normal science. Normal science is called upon, in effect, to pinpoint blame. First it is needed to determine if a genuine anomaly is at hand, to obtain a Popperian "falsifying hypothesis." Even then the anomaly may be due to other errors: faulty underlying assumptions or faulty initial conditions. With enough anomalies, however, a (Kuhnian) crisis arises, and as it becomes more and more unlikely that initial conditions are to blame for each and every anomaly, there are finally grounds for pinning the blame on the fundamental theory. Out of Kuhnian crises spring forth Popperian severe tests!

Wisdom's insightful analysis is in my view completely correct as far as it goes. But it goes only so far as to show how Kuhnian normal science can serve as a handmaiden to Popperian testing. By starting from the Popperian point of view of what the task is and utilizing Kuhnian ideas to fill out that task, he gets the order wrong and misses most of the uses of normal science. Normal science is needed to get extraordinary science off the ground all right, but starting with the latter as primary fails to do justice to Kuhn's main point of contrast with Popper.

Kuhn is quite clear about the main point of contrast (e.g., in his "Reflections"). He says, "my single genuine disagreement with Sir Karl about normal science" is in holding that when a full-bodied theory is at hand "the time for steady criticism and theory proliferation has passed" (Kuhn 1970, 246). Scientists can instead apply their talents to the puzzles of normal science. "[Sir Karl] and his group argue that the scientist should try at all times to be a critic and a proliferator of alternate theories. I urge the desirability of an alternate strategy which reserves such behaviour for special occasions" (p. 243).

Evaluating Kuhn's alternative research strategy requires us to re-analyze the aims of normal science. The result will neither be to turn Kuhn's rabbit into Popper's duck (as Wisdom has), nor to turn Popper's duck into Kuhn's rabbit (as in Lakatos's rational reconstruction of Kuhn), but to convert Kuhn's sociological description of normal science to a normative epistemology of testing.

2.3 NORMAL SCIENCE AS NORMAL TESTING

Let us begin by asking what is involved when practitioners turn their attention and apply their talents to the tasks of normal science. Kuhn identifies three classes of problems. "These three classes of problems— determination of significant fact, matching of facts with theory, and articulation of theory—exhaust, I think, the literature of normal science, both empirical and theoretical" (Kuhn 1962, 33). Kuhn elaborates on each:

1. *Determination of significant fact.* This concerns "that class of facts that the paradigm has shown to be particularly revealing of the nature of things" (Kuhn 1962, 25). Examples include stellar position and magnitude, the specific gravities of materials, wave lengths, electrical conductivities, boiling points, acidity of solutions. A good deal of work goes into the goal of increasing the accuracy and scope with which such facts are known.

2. *Matching of facts with theory.* This concerns arriving at data that "can be compared directly with predictions from the paradigm theory" (ibid., 26). Typically, theoretical and instrumental approximations limit the expected agreement, and it is of interest to improve upon it. Find-

ing new methods and instruments to demonstrate agreement constantly challenges experimentalists.

   3. *Articulating the paradigm theory.* This has three parts:

• Determining physical constants (e.g., gravitational constant, Avogadro's number, Joule's coefficient, the electronic charge).

• Determining quantitative laws. Kuhn's examples include determining Boyle's law relating gas pressure to volume, Coulomb's law of electrical attraction, and Joule's formula relating heat to electrical resistance and current.

• Conducting experiments to "choose among the alternative ways of applying the paradigm" (ibid., 29) to some closely related area. Each of these three classes of experiment has an analogous theoretical task.

In each class, Kuhn stresses, some current background theory or paradigm is needed to define the normal problem and provide means for determining if it is solved. Conjectured solutions to each problem may be viewed as hypotheses within a larger-scale background theory. Clearly, finding and evaluating hypotheses of these sorts is not a matter of uncreative hack science, nor does Kuhn suggest otherwise. What normal science does require, for Kuhn, are shared criteria for determining if problems are solved:

> No puzzle-solving enterprise can exist unless its practitioners share criteria which, for that group and for that time, determine when a particular puzzle has been solved. The same criteria necessarily determine failure to achieve a solution, and anyone who chooses may view that failure as the failure of a theory to pass a test. (Kuhn 1970, 7)

I do. But to emphasize that these are tests of the hypotheses of Kuhnian normal science, I will refer (in this chapter) to such testing as *normal testing.* Kuhn observes that "Tests of this sort are a standard component of what I have elsewhere labeled 'normal science' . . . an enterprise which accounts for the overwhelming majority of the work done in basic science" (ibid., 4). For reasons that will later become clear, I think that all experimental testing, if construed broadly, is based on normal testing. From experimental tests we acquire *experimental knowledge* (as characterized in chapter 1).

   To digress about my terminology: The appellation "normal testing" of course stems from Kuhn's normal science, with which it correlates. Later, when it is more developed and the deviations from Kuhn are clarified, I will call it *"standard testing."* But Kuhn sees normal science in contrast to nonnormal science, whereas I shall reject the idea that there are two kinds of science. If all science is normal (or even standard), is it not peculiar to use such an adjective altogether? No. In logic we talk about normal systems and corresponding standard models, and, while various nonnormal systems and nonstandard models exist, one may debate whether the latter are needed for some purpose, for example, the logic of arithmetic. Likewise, various nonstandard accounts of testing exist. My position is that for scientific reasoning they are either not needed or offer inadequate models.

Normal Hypothesis Testing: The Test of Experiment

Perhaps to achieve a stark contrast with Popperian tests, Kuhn calls normal problems "puzzles," where the loser when a conjectured solution fails is not the fundamental theory but the practitioner who was not brilliant enough. But since "blaming the practitioner," even for Kuhn, means only that the practitioner's conjectured solution fails to hold up to testing, it is less misleading to talk in terms of testing conjectured solutions to normal problems.[1] Kuhn supports this reading:

> There is one sort of "statement" or "hypothesis" that scientists do repeatedly subject to systematic test. I have in mind statements of an individual's best guesses about the proper way to connect his own research problem with the corpus of accepted scientific knowledge. He may, for example, conjecture that a given chemical unknown contains the salt of a rare earth, that the obesity of his experimental rats is due to a specified component in their diet, or that a newly discovered spectral pattern is to be understood as an effect of nuclear spin. (P. 4)

These are certainly the kinds of substantive hypotheses that philosophers of science want a theory of testing to address, something that tends to be lost in calling them "puzzles." The next steps, the test of experiment, have a hypothetico-deductive flavor:

> The next steps . . . are intended to try out or test the conjecture or hypothesis. If it passes enough or stringent enough tests, the scientist has made a discovery or has at least resolved the puzzle he had been set. If not, he must either abandon the puzzle entirely or attempt to solve it with the aid of some other hypothesis. (P. 4)

---

1. Hilary Putnam (1981) proposes that Kuhn's puzzles follow the form of an explanatory scheme that he calls Schema II in contrast to the pattern of a test based on a prediction (Schema I). In Schema II, both a theory and a fact are taken as given, and the problem is to find data to explain the fact on the basis of the theory. I agree with Putnam that Schema II captures the form of many normal problems, and concur that this normal task has been too little noticed by philosophers of science. Still, the task of checking or arguing for a conjectured explanation falls under what I would term a normal test.

Further, Kuhn finds it "unproblematic" to call successful conjectured solutions "true":

> Members of a given scientific community will generally agree which consequences of a shared theory sustain the test of experiment and are therefore true, which are false as theory is currently applied, and which are as yet untested. (P. 264)

So, as I read Kuhn, to infer that a hypothesis sustains the test of experiment it must have passed "enough or stringent enough tests," and, correspondingly, to regard a normal hypothesis as true is to accept it as correctly solving the associated normal problem.

Several characteristics of normal tests emerge from these passages. The tests are directed at a specific normal hypothesis, say, H. The test criteria determine whether H passes or fails (further divisions could be added). If H passes "enough or stringent enough tests," then H is taken to solve the puzzle, that is, to be correct or true. If H fails, it is concluded that H does not solve the puzzle (that H is incorrect or false). Blaming the background theory is tantamount to changing the puzzle and is disallowed. Indeed, in my reading, the main purpose of calling a normal problem a "puzzle" is to call attention to the fundamental restriction on what counts as an admissible solution: if a conjectured solution fails the test, only the conjecture and "not the corpus of current science is impugned" by the failure (p. 5).

Kuhn cites, as an example, how some eighteenth-century scientists, finding anomalies between the observed motions of the moon and Newton's laws, "suggested replacing the inverse square law with a law that deviated from it at small distances. To do that, however, would have been to change the paradigm, to define a new puzzle, and not to solve the old one" (Kuhn 1962, 39). This solution was not admissible. The normal scientist must face the music.

These are useful clues, but not quite enough. After all, a community could agree on any number of rules or criteria to pass or fail hypotheses, to call a test stringent, and so on. Frustratingly, Kuhn does not spell out what more is required for normal test criteria to pass muster, leading some to suppose that community consensus is enough for him. Yet a careful look at the demands Kuhn sets for normal tests suggests how to flesh out the earlier clues.

Underlying the stringency demand, I propose, is the implied requirement that before a hypothesis is taken to solve a problem it must have stood up to scrutiny: it must be very unlikely that the hypothesis really does not solve the problem. This *reliability (or severity) requirement*, discussed in chapter 1, will be developed as we proceed. For now, to avoid prematurely saddling Kuhn with my notion, the "reliability" of a test will be interchangeable with its "severity." That a test is reliable, for me, describes a characteristic of a *procedure* of testing, or of test criteria.[2] It does not say that the test never errs in declaring a problem solved or not, but that it does so infrequently. In other words, the normal scientist declares a problem solved only if the conjectured solution has withstood a scrutiny that it would very likely have failed, were it not correct.

It is only by some such reliability or severity requirement, I maintain, that Kuhn is right to locate the growth of knowledge in normal science. We can simultaneously unpack this requirement and motivate it on Kuhnian grounds by continuing our normative questioning.

Justifying the Pursuit of Normal Science

Kuhn observes that the bulk of scientific practice is directed at normal testing, rather than at Popperian testing, where Kuhn construes Popperian testing as criticizing fundamental theories. What interests me is whether there is an epistemological rationale for this focus on one sort of activity rather than another. The answer, I propose, is that one learns much more this way. Focusing on normal testing is a better research strategy.[3]

To bring the answer into focus we first must ask, Why, in the face of a rich enough theory to "support a puzzle-solving tradition," is it fruitful to concentrate on normal problems? (No answer precludes there also being something to be gained by seeking alternative theories.) When I look at Kuhn while wearing my spectacles, I discern this reply: if one has an interesting theory, one with predictions, suggestions for improvement, challenging puzzles, and so on, then taking up its challenges *will teach us a great deal,* and *a portion of what is learned will remain despite changes in theory.* With respect to the solved problems in normal research, Kuhn says that "at least part of that achievement always proves to be permanent" (Kuhn 1962, 25). To ignore its challenges is to forfeit this knowledge.

One can go further. Pursuing normal problems is a good strategy because, if there are anomalies that call for changes in theory, doing so will reveal them as well as help point to the adjustments indicated.

> In the developed sciences . . . it is technical puzzles that provide the usual occasion and often the concrete methods for revolution. . . . Be-

2. The reliability of a test procedure must be distinguished from the reliability of any particular hypothesis that passes the test, as that is generally understood.
3. Kuhn (1970, p. 243) identifies the key difference between himself and Popper as one of research strategy.

cause they can ordinarily take current theory for granted, exploiting rather than criticizing it, the practitioners of mature sciences are freed to explore nature to an esoteric depth and detail otherwise unimaginable. Because that exploration will ultimately isolate severe trouble spots, *they can be confident that the pursuit of normal science will inform them when and where they can most usefully become Popperian critics.* (Kuhn 1970, 247; emphasis added)

The rationale for pursuing normal problems is that (if done right) some positive payoff is assured. If normal science yields problem solutions, then new knowledge is brought forth. If normal testing determines that an anomalous result is real—that "it will not go away"—then there is knowledge of a real experimental effect (i.e., experimental knowledge). Further normal testing will indicate whether adjustments and revisions are called for. If, alternatively, the effect is a genuine anomaly for the underlying theory, normal science will let this be found out by means of gross or repeated failures (Kuhnian crisis). Even such crises, Kuhn notices, serve a creative function in developing alternative theories. Finally, normal science is the source of the most effective and severe tests of fundamental or basic theory:

> Though testing of basic commitments occurs only in extraordinary science, it is normal science that discloses both the points to test and the manner of testing. (P. 6)

and

> Because the [theory] test arose from a puzzle and thus carried settled criteria of solution, *it proves both more severe and harder to evade than the tests available within a tradition whose normal mode is critical discourse rather than puzzle solving.* (P. 7; emphasis added)

This last sentence gets to the heart of why, from the Kuhn's-eye point of view, "severity is the flip side of puzzle solving," and why one finds the most severe tests of theories, just what Popper seeks, in practices that have been engaged in the puzzle solving of normal science (normal testing). There is a clear epistemological ground for this, and a bit more spadework will uncover it. The last part of the quotation points the way. We have to ask what is wrong with "a tradition whose normal mode is critical discourse rather than puzzle solving"?

2.4 THE KUHN'S-EYE VIEW OF DEMARCATION

It is important to keep in mind that the critical discourse Kuhn is disparaging is the special kind of criticism that he imagines Popper to be championing: a relentless attack on fundamentals. It helps, in grasping

Kuhn here, if each time we read "critical discourse," we tack on the phrase "rather than puzzle solving." For Kuhn, finding a practice whose normal mode is critical discourse rather than puzzle solving is the surest tip-off that its scientific status is questionable. Hence Kuhn's provocative claim that a demarcation criterion may be found in the portion of science badly obscured by Popperian spectacles (normal science). Nevertheless, as Kuhn remarks (p. 7), Popper's demarcation line and his often coincide, despite the fact that they are identifying very different processes.

On the face of it, their two demarcation criteria are nearly opposite. For Popper, the hallmark of science is criticism and testability, whereas Kuhn, in deliberate contrast, declares that "it is precisely the abandonment of critical discourse that marks the transition to a science" (p. 6). To call what goes on in good normal testing an abandonment of critical discourse is highly misleading, because normal science itself is based on severe and critical normal tests. However, on my reading, what Kuhn takes good sciences to abandon is not normal testing—where all the fruitful learning really takes place—but rather "critical-discourse-rather-than-puzzle-solving." Good sciences do not and should not do what Kuhn takes Popper to be championing: relentlessly attacking fundamental theories, looking always for rival theories, and doing so to the exclusion of the positive learning of normal science. Although I do not endorse this provocative, idiosyncratic usage of "critical discourse," for the purposes of this chapter it helps us to construe Kuhn's demarcation criterion plausibly.

Astrology

To illustrate his contrast with Popper, Kuhn chooses astrology, out of a wish to avoid controversial areas like psychoanalysis (p. 7). His focus, he says, is on the centuries during which astrology was intellectually respectable. The example functions not only to make out his demarcation but also to show "that of the two criteria, testing and puzzle solving, the latter is at once the less equivocal and the more fundamental" (p. 7). Astrology was unscientific, says Kuhn, not because it failed to be falsifiable, nor even because of how practitioners of astrology explained failure. The problem is that astrologers had no puzzles, they could not or did not engage in normal science.

Engaging in normal science requires a series of puzzles and strict criteria that virtually all practitioners agree to use to tell whether the puzzles are solved. But a practice does not automatically become scientific by erecting such a series of puzzles and rules to pronounce them solved or not. Becoming a genuine science is not something that can occur by community decree, nor does Kuhn think it is.[4] Kuhn balks at those who would find in him recipes for becoming scientific, apparently unaware of how he invites this reading by failing to articulate the kinds of tests needed to carry out normal science legitimately and why only these tests qualify. Still, in several places Kuhn hints at the criteria normal testing requires (reliability or stringency). The most telling of all, I find, is his critique of astrology.

With astrology, Kuhn observes, not only are the predictions statistical, but there is a tremendous amount of "noise" from background uncertainties.

> Astrologers pointed out, for example, that . . . the forecast of an individual's future was an immensely complex task, demanding the utmost skill, and extremely sensitive to minor errors in relevant data. The configuration of the stars and eight planets was constantly changing; the astronomical tables used to compute the configuration at an individual's birth were notoriously imperfect; few men knew the instant of their birth with the requisite precision. No wonder, then, that forecasts often failed. (P. 8)

Kuhn's point seems to be this: astrology, during the centuries when it was reputable, did not fail to be scientific because it was not testable nor because practitioners did not take failures as grounds to overthrow astrology. Plenty of perfectly good sciences act similarly. The reason the practice of astrology was unscientific is that *practitioners did not or could not learn from failed predictions.*[5] And they could not learn from them because too many justifiable ways of explaining failure lay at hand. They could not use failures or anomalies constructively.

> The occurrence of failures could be explained, but particular failures did not give rise to research puzzles, for no man, however skilled, could make use of them in a constructive attempt to revise the astrological tradition. There were too many possible sources of difficulty, most of them beyond the astrologer's knowledge, control, or responsibility. Individual failures were correspondingly uninformative. (P. 9)

The above passage is the most revealing of all. For failed predictions to "give rise to research puzzles," a failure must give rise to a fairly well defined problem; specifically, the problem of how to explain it. It must be possible, in other words, to set up a reliable inquiry to

4. This is stressed by Hoyningen-Huene (1993) in his analysis of Kuhn's *Structure of Scientific Revolutions.* (See, for example, p. 193.)

5. My reading is not affected by the fact that Kuhn thinks it wrongheaded to call failed solutions "mistakes", and that he limits mistakes to errors in applying some rule, e.g., mistakes in addition.

determine its cause and/or the modifications called for. This is the kind of information normal tests can provide.

> Compare the situations of the astronomer and the astrologer. If an astronomer's prediction failed and his calculations checked, he could hope to set the situation right. Perhaps the data were at fault. . . . Or perhaps theory needed adjustment. . . . The astrologer, by contrast, had no such puzzles. (P. 9)

To "set the situation right" one needs to be able to discriminate between proposed explanations of the failure. Unless one can set up a stringent enough test of a hypothesized explanation, so that its passing can reliably be attributed to its being correct, that failed prediction will be unconstructive or uninformative.

By the same token, so long as there is no way to cut down these alternative explanations of failure, there is no ground for arguing that the failures *should* have been attributed to the falsity of astrology as a whole. In other words, if failed predictions do not give rise to research puzzles (reliable inquiries into their cause), then one cannot come to learn whether and, if so, how they can be explained within the global background theory. Thus, they cannot warrant discrediting the whole theory; they cannot warrant (Popperian) critical discourse.

It becomes clear, then, that mere critical discourse is not enough for genuine science. In fact, Kuhn must see the case of astrology as one in which the normal day-to-day practice is critical discourse (i.e., critical discourse rather than puzzle solving). Constructive criticism, one might say, requires at least being able to embark on an inquiry toward solving the Duhemian problems that will arise.

Unwarranted Critical Discourse

The practitioners of astrology, Kuhn notes, "like practitioners of philosophy and some social sciences . . . belonged to a variety of different schools, and the inter-school strife was sometimes bitter. But these debates ordinarily revolved about the *implausibility* of the particular theory employed by one or another school. Failures of individual predictions played very little role" (p. 9, n. 2). Practitioners were happy to criticize the basic commitments of competing astrological schools, Kuhn tells us; rival schools were constantly having their basic presuppositions challenged. What they lacked was that very special kind of criticism that allows genuine learning—the kind where a failed prediction can be pinned on a specific hypothesis. Their criticism was not constructive: a failure did not *genuinely indicate* a specific improvement, adjustment, or falsification.

Thus, I propose to construe the real force of Kuhn's disparaging practices "whose normal mode is critical discourse" as disparaging those practices that engage in criticism even where the criticism fails to be driven by the constrained tests that exemplify good normal science.[6] What is being disparaged, and rightly so, is unwarranted and unconstructive criticism. When the day-to-day practice is criticism that is not the result of the stringent constraints of normal testing, then that criticism is of the unwarranted or unconstructive variety. It is mere critical discourse. Nonsciences engage in mere critical discourse, not genuine criticism that allows learning from empirical tests.

Learning from tests requires being able to learn not only from failed predictions but also from successful ones. For practitioners of astrology, both failed and successful predictions were uninformative. They could not learn from successful predictions because they would not provide a warrant for crediting any astrological theory. Credit does not go back to any astrological theory because there were no grounds for attributing a successful prediction to some astrological cause, for example, to the stars and planets being in particular positions. Successful astrological predictions are likely, even if astrology is false: the tests are not severe. We will later see how to make this notion of severity concrete.

Astrology exemplifies an extreme situation in which severe tests are precluded. The situation might be described in modern statistical terms as having too much uncontrolled variability, or as lacking a way to distinguish the "signal" from the noise. The situation, Kuhn notes, is typical of practices that one might call "crafts," some of which eventually make the transition to sciences (e.g., medicine). (Current theories of the stock market might be said to be crafts.)

The transition from craft to science, Kuhn observes, correlates with supporting normal science or normal testing. However, Kuhn's demarcation slogan makes him appear to be saying that the transition comes about by "abandoning critical discourse." Kuhn fails to identify the kind of abandonment of criticism that is actually conducive to making a practice more scientific. It is unwarranted and unconstructive criticism that should be abandoned and replaced by the warranted criticism of normal testing.

Let us go back to a practice that, unlike astrology, is sufficiently developed to support normal testing (puzzle solving). If a hypothesized

6. This should be qualified to refer only to enterprises for predicting, controlling, or understanding the physical world, in short, to intended sciences. It would not be a disparagement, say, of art.

solution to a normal problem fails a test, it could, theoretically, be accounted for by alleging a fundamental flaw in the underlying theory—but such a criticism would very likely be unwarranted (at least if just from this one failure). Thus, to regularly proceed this way would very often be in error, thereby violating the reliability requirement of normal testing. On these grounds normal science calls for abandoning this type of criticism. For the same reason it admonishes the practice of dealing with a failed solution (failed hypothesis) by changing the problem it was supposed to solve. An enterprise that regularly allowed such a cavalier attitude toward failure would often be misled. Likewise, in the case where a hypothesis passes a test: if there is too much leeway allowed in explaining away failures, passing results teach little if anything.

Changing the problem, blaming one's testing tools or the background theory *where these are unwarranted*, is the kind of criticism that should be disallowed. Only then can the practice of hypothesis appraisal be sufficiently constrained to identify correctly genuine effects, gain experimental knowledge—or more generally, to accomplish the tasks of normal science *reliably*.[7] Thus recast, Kuhn's demarcation criterion intends to pick out those practices that afford experimental learning. I suggest we view such a demarcation criterion as indicating when *particular inquiries*, rather than whole practices, are scientific. It becomes, roughly:

> *Demarcating scientific inquiry:* What makes an empirical inquiry scientific is that it can and does allow learning from normal tests, that it accomplishes one or more tasks of normal testing *reliably*.[8]

This criterion becomes more specific when particular types of normal testing results are substituted. For example, an important type of normal test result is a failed prediction. The difference between a scientific and an unscientific treatment of a failed prediction is the extent to

7. Popper makes it clear that he allows the critical method to refer to the minor tests that Kuhn counts as puzzles. Popper (1974) addresses this point with an example. The heating engineer needing to figure out how to install a central heating system under unusual conditions may have to throw away the rule book of normal practice to find the solution. "When he works by trial and the elimination of error, and when he eliminates the error by a *critical* survey of tentative solutions, then he does not work in this routine manner; which for me makes him a scientist. But Kuhn . . . should either say that he was not a scientist, or an extraordinary one" (Popper 1974, 1147).

8. I am assuming that the empirical inquiry is aiming to find out about the world.

which it is used to learn (about its cause, about needed modifications, and so on).

So far our analysis has brought us to the following recasting of the Kuhnian observations with which we began: To understand the nature of the growth of scientific knowledge, one should look to tests of local hypotheses (normal experimental testing). An adequate account of normal testing should serve each of the functions Kuhn accords it, with the additional proviso that it do so reliably and with warrant. From our vantage point, what distinguishes Kuhn's demarcation from Popper's is that for Kuhn the aim is not mere criticism but constructive criticism.[9]

2.5 PARADIGMATIC MODELS OF ERROR

The Kuhnian picture of the role of background paradigms in normal science provides a useful framework for pursuing these ideas about experimental testing. While I would deny that a practitioner needs to work within a single paradigm, it will do no harm to see a given inquiry as within a single paradigm and it will make it easier to gain access to Kuhn's story. By the time the story is completed, delineating paradigms will not matter anyhow.

Kuhn's notion of paradigm is notoriously equivocal.[10] We may allow that a Kuhnian paradigm includes theories, specific hypotheses, and an ontology, as well as research aims and methods both for directing normal research and testing hypotheses. (In "Reflections on My Critics," p. 271, Kuhn says he would prefer to use the term "disciplinary matrix.") For Kuhn, sharing a paradigm is what accounts for "relative unanimity in problem-choice and in the evaluation of problem-solutions" (p. 271). However, we must carefully distinguish what Kuhn runs together here. What goes into choosing a problem is quite different from what goes into criticizing proposed solutions. We need, in short, to distinguish the paradigm's role in providing *research guides*—a source of problems and guides for solving problems or puzzles—from *experimental testing models*—tools for testing hypothesized solutions or for normal hypothesis testing. The second category includes tools for criticizing such tests. These two categories of tasks may overlap, but no matter. The point of the distinction is to see why

9. I hope it is clear that my use of the term "constructive criticism" differs radically from Lakatos in his attempt to revise Popper in light of Kuhn. For Lakatos, constructive criticism means replacing large-scale theories (progressively). For me, it means obtaining experimental knowledge by local arguments from error.

10. See, for example, M. Masterman 1970 and D. Shapere 1984, chaps. 3 and 4.

changing one's research program is not the same as changing one's experimental testing tools.

Paradigm as Research Guide

Let us begin with the role of a paradigm in providing a set of research guides. Its role is to supply the questions that need answering and suggest the kinds of answers afforded by given instruments, experiments, and tests. For Kuhn, a paradigm also creates a situation that fosters exploring subjects in esoteric depth and satisfies the psychological and practical requirements for sustaining such exploration.

This activity requires researchers to "accept" the program, but only in the sense that they choose to work on its problems and utilize the tools it offers for doing so. In Larry Laudan's terminology, they choose "to pursue it"; and, as he rightly urges, it is important to distinguish acceptance in this sense from taking an epistemological stance toward the theory.

Kuhn, in contrast, often suggests that the paradigm must have a grip on the minds of those scientists working within (i.e., accepting) it, allowing them to perceive the world through the paradigm. Is it true that working on a research program demands total immersion in the paradigm? Kuhn is right to insist that solving the problems of normal science requires creative brilliance (it is not hack science by any means), and perhaps total immersion is the most effective way to attain solutions. This could probably be investigated. Nevertheless, the results of such an investigation would be irrelevant to the questions about the epistemological warrant a theory might be required or entitled to have.

The factors that enter into choosing to take up a research program, pledging allegiance to it, living in its world—all the things Kuhn associates with accepting a global theory or paradigm—include psychological, sociological, pragmatic, and aesthetic values. Only by assuming these to be inextricably entwined with theory testing does Kuhn perceive the latter to turn on these values as well. Before considering global theory testing we need to focus on the second role of the global background theory or paradigm: its role in normal testing.

Experimental Testing Models: Error Paradigms

Significantly, Kuhn remarks that he was originally led to the concept of a "paradigm" in thinking of the concrete problem solutions or exemplars that practitioners share and that enable them to agree whether a problem is solved (p. 272). This is the role I propose to give to certain *experimental testing models* or *testing exemplars*. At one level,

I am saying precisely what Kuhn says here about the role of shared exemplars. Through them one gains the ability to see a number of experimental problems or contexts in the same way, permitting the application of similar techniques. One grasps standard ways to ask questions to arrive at experimentally determinable answers, and one learns what does and does not count as a satisfactory solution, an adequate fit, and a "good approximation." Nevertheless, I depart from Kuhn in several important ways. While this results in a view of normal science very different from Kuhn's official position, it is in keeping with my normative recasting of Kuhn. The main differences (to be discussed more fully in later chapters) are these:

1. Normal tests are not algorithms or routines. If one unearths what actually goes on in normal testing, one sees that even sharing experimental test examples does not secure the relatively unproblematic means of testing normal hypotheses that Kuhn imagines. Uncertainties about experimental assumptions, significance levels, the appropriateness of analogies, and so on have to be confronted. As a result, there is often much disagreement about the results of normal testing, and there is plenty of opportunity for biases, conscious and unconscious, to enter. Where consensus is reached, it is not because of anything like a shared algorithm. It is because, in good scientific practices, the very problems of interpreting tests, critiquing experiments and such, themselves "give rise to research puzzles" in the sense we have discussed. Nor do these mechanisms for reliable tests in the face of threats from biases become inoperative in large-scale theory appraisal.

2. Normal testing exemplars correspond to canonical experimental models. In my view, standard examples or normal testing exemplars are not a set of tools available only to those working within a given global theory or paradigm. Instead they consist of any models and methods relevant for testing solutions of normal problems, and these come from various background theories, from mathematics and statistics, and from theories of instruments and experiments.

It is no objection to this idea of a pool of shared models that in a particular paradigm the model comes dressed in the special clothing of that paradigm. What is shared is the corresponding bare-bones or *canonical version* of the model. For example, a particular instrument used to test a predicted quantity may be characterized as having a specific Normal distribution of errors, and the rule for declaring the prediction "successful" might be that the result be within two standard error units. The corresponding bare-bones or canonical model would be the mathematical family of Normal distributions and the corresponding statistical rules for declaring a difference "statistically significant." That canonical model, once articulated sufficiently, is available for use for a host of experimental inquiries—it is not paradigm specific.[11] It is like a standard instrument; indeed, a physical instrument is often at the heart of a canonical model.

3. *The use of exemplars in normal testing is open to objective scrutiny.* Paradigmatic experimental examples are exemplary because they exemplify cases where the kinds of errors known to be possible or problematic in the given type of investigation are handled well—that is, ruled out. Include also examples of infamous mistaken cases, especially those thought to have surmounted key problems. Kuhn's use of astrology is an excellent example of what I have in mind; it is a classic case of a nonscience. As we proceed, other examples will arise.

The use of paradigmatic exemplars is open to a paradigm-independent scrutiny, that is, a scrutiny that is not relative to the background theory within which their use takes place. Although they may be applied in a routine manner, their appropriateness typically assumed, the exemplars are used in *arguments* appraising hypotheses, and these arguments have or fail to have certain properties (e.g., reliability, severity). For example, the rule for determining whether agreement with a certain kind of parameter (with a certain distribution of error) is "good enough" may be the two-standard-deviation rule. Normal practitioners can and do criticize such rules as appropriate or not for a given purpose.[12]

Interestingly, Kuhn's attitude toward the exemplars of normal tests is analogous to Popper's treating the decisions required for testing as mere conventions. They simply report the standards the discipline decides to use to declare a problem solved or not.[13] By Kuhn's own lights, however, before normal practitioners may take a puzzle as solved the hypothesized solution must have passed stringent enough tests. The arsenal needed for normal testing, then, is a host of tools for detecting

11. It does not follow that scientists will recognize that the canonical model is appropriate to a given problem. As we will see (e.g., in chapter 7), important mistakes could have been avoided in Brownian motion tests if the applicability of certain statistical models had been noticed.

12. Kuhn's use of the examples of astrology and astronomy is a good case in point. Suppose he had used the latter to exemplify a practice *lacking* a puzzle-solving tradition. Such a use would be criticized as allowing erroneous characterizations of good scientific practices.

13. Laudan (1996) makes this point in "The Sins of the Fathers." Kuhn's relativism about standards, Laudan says, is the exact counterpart of Popper's methodological conventionalism or Carnap's principle of tolerance.

whether and how conjectured hypotheses (of a given type) can fail. They call for methods capable not only of determining whether a hypothesis correctly solves a problem, but also of doing so reliably. Appraising the use of exemplars in normal testing turns on how well they promote these aims—the very aims we extricated in our normative recasting of Kuhn.

In this connection one might note how Kuhn, while claiming that a gestalt switch separates him from Popper, is able to criticize Popper's research strategy on the grounds that one does not learn much through critical discourse rather than puzzle solving.

2.6 GETTING SMALLER—A CONSEQUENCE OF SEVERITY AND INFORMATIVENESS

The most valuable idea that comes out of the testing-within-a-paradigm concept at the same time gets to the heart of Kuhn's contrast with Popper. The idea is that testing a hypothesis, if the test is to be informative, is not to test everything all at once but to test piecemeal. The necessity of proceeding piecemeal follows from the desiderata of normal tests: that they be reliable (severe) and informative (constructive). This requires two things of tests: a hypothesis H is taken to solve a problem only if it passes sufficiently stringent or severe tests, and some hypothesis (or other) is likely to be taken to solve the problem if it actually does so (at least approximately).

A major flaw in Popper's account (recall chapter 1) arises because he supplies no grounds for thinking that a hypothesis H very probably would not have been corroborated if it were false. ("Not-H" included all other hypotheses that were not yet considered.) Satisfying the aims of good normal testing, in contrast, directs one to select for testing the hypothesis where "H is false" (not-H) does not refer to all other hypotheses in the domain in question. Rather it refers to a specific way in which a conjectured solution, H, could be wrong—could be erroneously taken as actually solving the problem. (For example, H might assert that an effect is systematic, not an artifact, and not-H that it is an artifact.) I will elaborate on this in chapter 6.

The twin desiderata of normal tests, which we can express as "be stringent, but learn something," compel a localization of inquiry in another sense, one that David Hull (1988) precisely puts his finger on. In the section "Summa Contra Kuhn," Hull explains that "scientists are willing to accept certain problems as solved and proceed to new problem areas" even on the basis of an apparently small number of tests "because they are confident that *error ramifies.* If the hypotheses that they are accepting in order to attack new problems are mistaken, the results of related, though partially independent, research are likely to signal that something is wrong" (p. 496; emphasis added).

Later I will have much more to say about the points of the last two paragraphs. Here my main concern is how my reading of Kuhn substantiates his contrast with Popper. The aims of normal testing give an entirely new impetus to a slew of microinquiries that from behind Popperian spectacles[14] might appear unexciting, without risk.

Far from desiring boldness in Popper's sense, Kuhn observes how normal practitioners often set out to achieve an outcome that is already anticipated or seek to redetermine a known result—bolstering the impression of normal science as lacking novelty. A very different impression arises if it is seen that a central aim of normal science is to improve on its own tools. It is not so much the new information about the scientific domain that is wanted but new ways of minimizing or getting around errors, and techniques for ensuring Hull's point, that important errors "ramify rapidly."

As Kuhn puts it, "Though its outcome can be anticipated . . . the way to achieve that outcome remains very much in doubt" (Kuhn 1962, 36). A good deal about method is likely to be learned by finding out how to achieve the expected outcome. Consider, for example, the continued interest in using eclipse results to estimate the deflection of light. As will be seen in chapter 8, a main problem was finding improved instrumental and analytical techniques. Seeking alternate ways to elicit a known solution is often an excellent route to discovering a new and clever mode of interacting with nature. One of the most effective ways to test and learn from such interaction is through quantitatively determined effects.

Quantitative Anomalies

If we continue to look at Kuhn with normative glasses, we see that he has put his finger on the rationale of normal tests. They provide experimental constraints that allow learning from tests—both to ground the three types of normal hypotheses needed to extend and flesh out theories (section 2.3), as well as to substantiate a crisis.

A crisis emerges when anomalies repeatedly fail to disappear. Crises, while heralding troubling times, as Kuhn stresses, also have an important positive role to play: "[T]hough a crisis or an 'abnormal situation' is only one of the routes to *discovery* in the natural sciences, it is prerequisite to *fundamental inventions of theory*" (Kuhn 1977, 208).

14. Perhaps I should say wearing what Kuhn takes to be Popper's spectacles.

While "prerequisite" is probably too strong, crises have an especially creative function because of the knowledge they embody. Examining Kuhn's views on crises, however exceptional and rare he feels them to be, reveals a lot about the kind of knowledge that normal science is capable of bringing forth.

True, when Kuhn turns his gaze to crisis and theory change, he sees the problem as calling for a sociopsychological solution. It calls for, he thinks, a study of what *scientists consider* an unevadable anomaly and "what scientists will and will not give up" to gain other advantages (Kuhn 1977, 212). The problem, put this way, Kuhn says, "has scarcely even been stated before." For philosophers of science, the problem would usually be posed as asking after the nature and warrant of discrediting and testing theories. Nevertheless, what emerges from Kuhn's point of view is relevant for the more usual problem (which is, of course, the problem that interests me), provided we keep on our normative-epistemological spectacles.

What emerges from Kuhn's descriptive inquiry is that quantitative measurement and knowledge of quantitative effects are of key importance for both crisis and testing. This is particularly clear in Kuhn 1977. What makes an anomaly so "persistently obtrusive" and unevadable as to precipitate a crisis, Kuhn stresses, is that it be quantitatively determined:

> No crisis is, however, so hard to suppress as one that derives from a quantitative anomaly that has resisted all the usual efforts at reconciliation. . . . Qualitative anomalies usually suggest *ad hoc* modifications of theory that will disguise them, and . . . there is little way of telling whether they are "good enough." An established quantitative anomaly, in contrast, usually suggests nothing except trouble, but at its best it provides a razor-sharp instrument for judging the adequacy of proposed solutions. (P. 209)

This can be supplied with a very plausible normative basis: the quantitative anomaly identifies a genuine experimental effect, in particular, a discrepancy of a specified extent, and rather precise statistical criteria can determine whether a proposed solution adequately accounts for it. For example, if a two-standard-deviation discrepancy has been identified, then a hypothesis, say H, that can account for an effect of at most 0.1 standard deviation fails to explain the identified discrepancy. Thus the quantitative aspect of the discrepancy makes it clear that hypothesis *H fails to solve the problem.*

Here the Kuhnian function of quantitative anomalies sounds very Popperian, and it is. But for Kuhn the quantitative knowledge arose from a normal problem, an effort to learn something—not from a test of a large-scale theory.

True to his perspective on the problem, Kuhn sees himself as reporting on the behavior and attitudes of scientists. Yet the respect scientists show for quantitative effects is not just a sociopsychological fact about them. It is grounded in the fact that to try to blunt the sharp criteria quantitative information affords would be to forfeit accuracy and reliability, to forfeit learning. As Kuhn reports:

> I know of no case in the development of science which exhibits a loss of quantitative accuracy as a consequence of the transition from an earlier to a later theory. . . . Whatever the price in redefinitions of science, its methods, and its goals, scientists have shown themselves consistently unwilling to compromise the numerical success of their theories. (Kuhn 1977, 212–13)

Fortunately, our spectacles allow us to get beyond merely noticing that scientists appear unwilling to give up "quantitative accuracy" and "numerical success," and enable us to discern when and why this obstinacy is *warranted*.

This discernment is a task that will engage us throughout this book. Two things should be noted right off: First, it is not numerical success in the sense of doing a good job of "fitting the facts" that warrants clinging to. Only the special cases where numerical success corresponds to genuine experimental effects deserve this respect. One, but not the only, way of demonstrating such experimental knowledge is Hacking's favorite practice: intervening in phenomena. Second, if we press the normative "why" question about what makes quantitative effects so special and quantitative knowledge so robust, we see that what is desirable is not quantitative accuracy in and of itself. What is desirable is the strength and *severity of the argument* that is afforded by a special kind of experimental knowledge. As such, it makes sense to call all cases that admit of a specifiably severe or reliable argument "quantitative," so long as this special meaning is understood. (This is how I construe C. S. Peirce's notion of a "quantitative induction," a topic to be taken up in chapter 12.)

Quantitative knowledge teaches not only about the existence of certain entities but also about the properties of the process causing the effect. Such knowledge has primacy, as Kuhn recognizes, "whatever the price in redefinitions of science, its methods, and its goals."

Pause

Our recasting of normal science, I believe, substantiates the three highlights of Kuhn's contrast with Popper with which we began: (1)

"It is normal science, in which Sir Karl's sort of testing does not occur, rather than extraordinary science which most nearly distinguishes science from other enterprises" (p. 6); (2) "It is precisely the abandonment of critical discourse that marks the transition to a science" (p. 6); and (3) "Severity of test-criteria is simply one side of the coin whose other face is a puzzle-solving [i.e., a normal science] tradition" (p. 7).

Briefly, our gloss on them went as follows: The fundamental features of scientific inquiries are to be found in the criteria of normal testing, and these criteria demand stringent normal tests, not (uninformative) attacks on fundamental theory. Because anomalies that are reliably produced in normal tests indicate real effects that will not go away, they provide the most severe tests of theories—when these are warranted. This explains Kuhn's promise that scientists "can be confident that the pursuit of normal science will inform them when and where they can most usefully become Popperian critics" (p. 247), that normal science will tell them when and where to find fault with the underlying theory. But Kuhn, we shall see, reneges on his promise. Once having brought normal scientists to the crisis point, Kuhn still will not let them be Popperian testers!

This gets to the heart of my critique of what Kuhn says about theory testing. Before beginning that critique, I might warn the reader against a certain misconstrual of the relation of that critique to the account of experimental learning that I am after. My aim is not to provide an account of large-scale theory appraisal. (I am not looking to fulfill Wisdom's idea of turning Kuhn's rabbit [normal science] into Popper's duck. Nor do I want to turn Popper's duck into Kuhn's rabbit—which is Lakatos's way.) Theory appraisal and theory testing, as distinguished from theory choice and theory change, do turn on experimental knowledge grounded in normal testing, but that is not the main reason for acquiring such knowledge. In my view, scientific progress and growth is about the accumulation of experimental knowledge.[15]

Although experimental knowledge is not all there is to science, it holds the key to solving important philosophical problems about science. For instance, it is important to show why large-scale theory appraisal is objective and rational—in the very senses that Kuhn rejects. I think, however, that we should reject Kuhn's depiction of the problem of theory testing, namely, as a comparative appraisal of rival large-scale paradigm theories. Nevertheless, since I want to consider how our results about normal science force a revised view of Kuhn's own

15. An important reason Kuhn despaired of progress is his rejection of the Popperian syntactic measures of how theory change represents progress.

story about large-scale theory change, I am willing to be swept up in
Kuhn's story a while longer.

2.7 THEORY CHOICE VERSUS THEORY APPRAISAL: GESTALT SWITCHES AND ALL THAT

Why, having brought normal scientists to the crisis point, to the point
of a warranted criticism of theory, will Kuhn still not let them be Popperian testers? (Granted I am being unclear about whether the theory
at stake is medium-sized, large-scale, or a full disciplinary matrix, but
that is because Kuhn is unclear.) According to Kuhn, the products of
normal science are never going to be decisive for falsifying or for adjudicating between global theories. Testing and changing global theories
or paradigms turns out not to be a matter of reasoned deliberation at
all. Colorful passages abound in Kuhn's *Structure*. One such passage
declares that

> the proponents of competing paradigms practice their trades in differ-
> ent worlds. . . . [T]he two groups of scientists see different things
> when they look from the same point in the same direction. . . . [B]e-
> fore they can hope to communicate fully, one group or the other must
> experience the conversion that we have been calling a paradigm shift.
> Just because it is a transition between incommensurables, the transi-
> tion between competing paradigms cannot be made a step at a time,
> forced by logic and neutral experience. Like the gestalt switch, it must
> occur all at once (though not necessarily in an instant) or not at all.
> (Kuhn 1962, 149)

> What were ducks in the scientist's world before the revolution are
> rabbits afterwards. (Ibid., 110)

This picture of revolutionary science has been convincingly criticized
by many authors (e.g., Laudan 1984b, 1990c; Scheffler 1982; Shapere
1984), and perhaps by now no further criticism is called for. But what
I wish to consider, if only briefly, is how our recasting of normal science
tells against Kuhn's view of global theory change.

Viewing global theory change as switching all elements of the paradigm, Kuhn supposes there to be no place to stand and scrutinize two
whole paradigms, as a genuine paradigm test would require. Lacking
an "empirically neutral system of language," Kuhn holds that "the proposed construction of alternate tests and theories must proceed from
within one or another paradigm-based tradition" (1962, 145). This
would be all right if it allowed that testing from within a paradigm
could rest on something like our interparadigmatic canonical models

of experiment. Failing to disentangle the experimental testing portion of the paradigm from immersion in its research program, Kuhn not surprisingly winds up viewing global theory change as arational— quite like the (experimentally) unwarranted critical discourse he attributes to nonsciences. It is as if the very process that allows practices to become scientific had shifted into reverse, until we are back to "mere" critical discourse:

> Critical discourse recurs only at moments of crisis when the bases of the field are again in jeopardy. Only when they must choose between competing theories do scientists behave like philosophers. That, I think, is why Sir Karl's brilliant description of the reasons for the choice between metaphysical systems so closely resembles my description of the reasons for choosing between scientific theories. (Pp. 6–7)

Indeed, the values Kuhn appeals to in theory change—simplicity, scope, fruitfulness, and the like—are precisely the criteria Popper claims we must resort to in appraising metaphysical systems.

The Circularity Thesis

Kuhn supposes that subscribers to competing global theories necessarily interpret and weigh these factors differently; hence, inevitably, one's own global theory gets defended. This *circularity thesis* is most clearly stated in *Structure*:

> Like the choice between competing political institutions, that between competing paradigms proves to be a choice between incompatible modes of community life. . . . When paradigms enter, as they must, into a debate about paradigm choice, their role is necessarily circular. Each group uses its own paradigm to argue in that paradigm's defense. (1962, 93)

> They will inevitably talk through each other when debating the relative merits of their respective paradigms. In the partially circular arguments that regularly result, each paradigm will be shown to satisfy more or less the criteria that it dictates for itself and to fall short of a few of those dictated by its opponent. (Ibid., 108–109)

While such circular defenses are possible, none of the requirements of paradigm theories, even as Kuhn conceives them, make their role in theory appraisal necessarily circular. On the contrary, there is much in what normal science requires that militates against such circularity—even in times of crisis. After all, "the criteria that [the paradigm] dictates for itself" are those in the experimental testing models and the exemplary arguments that go along with them. Those standards, if appropriate for their *own* goals, must condemn such question-begging arguments as failing utterly to probe a theory severely. At some point (i.e., with regard to some normal hypothesis), defending a global theory no matter what clashes with the requirements of normal testing. Such defenses can and do occur, but they do not count as warranted—by the strictures of good normal science. Why? Because they come down to a blanket refusal to acknowledge that a hypothesized solution to a normal problem fails, and that betrays an essential requirement of normal science.

Of course, nothing guarantees that actual science obeys the constraints of Kuhnian normal science. In fact, however, Kuhn's account of normal science is descriptively accurate for the bulk of important scientific episodes, episodes from which much has been learned. By and large, when these episodes appear to show otherwise it is due to mistaking as theory testing remarks that record individual biases expressed in hopes and fears, complaints, demands, name-calling, and bullying. Fortunately, the major and minor players in these cases discerned the difference. This is not surprising, given that they had been enjoying a normal science tradition. After all, and this is one of my main points, each such defense is just a hypothesized solution to a normal puzzle (and practitioners are well versed in scrutinizing alleged solutions).

To return to my criticism, let us sketch what happens, according to Kuhn's circularity thesis, when a global theory, T_1, slips into a crisis. Within T_1, which we assume to be a genuine science with a normal tradition and so on, genuine anomalies have been identified. These anomalies, especially if they are quantitative, identify genuine effects that need explaining. These give rise to normal puzzles, that is, normal testing, to scrutinize attempted solutions, to "set the situation right." The criteria of T_1, by dint of its enjoying a normal science tradition, severely constrain attempts to deal with such anomalies. A genuine crisis is afoot when, after considerable effort, T_1 is unable to explain away the anomalies as due either to initial conditions or background hypotheses.

All these points of Kuhn's story have already been told. Notice that it follows that *the experimental testing criteria of T_1 themselves warrant the existence of anomalies and crisis.* Incorporating as they must the general criteria of normal testing, they indicate when an anomaly really is unevadable (i.e., when to put blame elsewhere is tantamount to unwarranted criticism). Do they not, by the same token, indicate that any attempt to save a theory—if that defense depends upon evading the anomaly—violates the very norms upon which enjoying a normal "puzzle-solving" tradition depends? The norms would bar procedures of admitting hypotheses as solutions to puzzles, we saw, if they often would do so erroneously.

Of course, it may take a while until attempted defenses come up against the wall of normal testing strictures. But with a genuine crisis, it seems, that is exactly what happens. Moreover, from Kuhn's demarcation criterion, *it is possible to recognize* (even if not sharply) that a practice is losing its normal puzzle-solving ability. (Astrology is a kind of exemplar of a practice that falls over onto the nonscience line.)

These remarks should not be misunderstood. What normal science must condemn is not saving a global theory in the face of severe anomaly—although that is what Popperian spectacles might have us see. What it must condemn (recalling Kuhn's demarcation) is being prevented from *learning* from normal testing. In any particular case, the obstacles to learning that are condemned are very specific: having to reject experimentally demonstrated effects, contradict known parameter values, change known error distributions of instruments or background factors, and so on. Such moves are not always learning obstacles, but they are when they fly in the face of exemplary experimental models.

Consider what Kuhn calls for when scientists, having split off from global theory T_1 to develop some rival T_2, come knocking on the door of their less adventurous colleagues, who are still muddling through the crisis in T_1. Confronted with rival T_2, which, let's suppose, solves T_1's crisis-provoking problem, crisis scientists in T_1 necessarily defend T_1 circularly. This circularity thesis requires them to do a turnabout and maintain that T_1 will eventually solve this problem, or that the problem was not really very important after all. Once the members of rival T_2 go away (back to their own worlds, presumably), members of T_1 can resume their brooding about the crisis they have identified with their paradigm. Were they to do this, they would indeed be guilty of the unwarranted criticism and mere name-calling Kuhn finds typical of nonsciences. But Kuhn has given no argument to suppose that crisis scientists necessarily do this.

Even if the circularity thesis is rejected, as it should be, many of the more troubling allegations Kuhn raises about global paradigm appraisal persist. These follow, however, not from the high points of normal testing and crisis, but from two additional premises that we should also reject: first, that acceptance (rejection) of a paradigm is the same as or indistinguishable from taking up (stopping work on) its problems, and second, that global theory or paradigm change requires a conversion experience. Taken together, these assumptions render the problem of global theory change as the problem of what causes practitioners working in one paradigm to transfer their allegiance and become converted to another. Seen this way, it is no wonder Kuhn views the answer to be a matter of sociopsychology.

Discrediting a Global Theory versus Stopping Work on It

It would seem that the repeated anomalies that are supposed to bring on crisis provide grounds for thinking that the theory has got it wrong, at least for the anomalous area. Yet Kuhn denies that scientists take even the most well-warranted crisis as grounds to falsify or reject the global theory involved (even as a poor problem solver). For Kuhn, "once it has achieved the status of paradigm, a scientific theory is declared invalid only if an alternate candidate is available to take its place. . . . The decision to reject one paradigm is always simultaneously the decision to accept another" (Kuhn 1962, 77).

This thesis of comparative appraisal, for Kuhn, is not a result of an analysis of warranted epistemic appraisal. It is a consequence of Kuhn's first assumption—equating the acceptance of a global theory with pursuit of its research program. That is why Kuhn denies that scientists can reject a theory without adopting another: "They could not do so and still remain scientists" (1962, 78). Together with his assumption that science has to be done within a single paradigm, the comparative appraisal thesis follows logically. But as we have already said, working within a paradigm—pursuing normal problems and tests—is wholly distinct from according it certain epistemic credentials. A situation that through Kuhnian spectacles appears as "still working within theory T, despite severe crisis" may actually be one where T is discredited (e.g., key normal hypotheses found false), but not yet replaced.[16] Discrediting a theory is not the same as stopping work on it.

Good reasons abound for still working on theory T, despite anomalies. Nor need these reasons go away even if T is replaced. Different types of anomalies must be distinguished. Suppose that the anomaly truly indicts a hypothesis of the theory T, say hypothesis H; that is, "H is in error" passes reliable tests. Consider two cases:

a. Theory T at the moment has got it wrong so far as H goes, but it is the kind of problem we know a fair amount about clearing up. Here the anomaly counts against neither the correctness of T nor the value

---

16. Even when T is replaced, it may very well still enjoy a puzzle-solving tradition. If I am correct, this counts against Kuhn's claim that theories are replaced only when they fail to have a puzzle-solving tradition.

of pursuing T further. Indeed, it is likely to be a fruitful means of modifying or replacing H.

b. In a genuine crisis situation, theory T is found to have got it wrong so far as several key hypotheses go. Here the anomalies discredit the full correctness of T but not the value of continued work on T. Kuhn's idea of "the creative function of crises" should be taken seriously. The anomalies are extremely rich sources of better hypotheses and better normal tests, as well as better theories. After all, whatever passes severe tests, that is, experimental effects—including the anomalies themselves—are things for which any new theory should account: they do not go away.[17]

One reason the thesis about comparative testing appears plausible, even to philosophers who otherwise take issue with Kuhn, is that often the clearest grounds for discrediting a theory arise when a rival, T_2, is at hand. That is because the rival often supplies the experimental grounds for using anomalies to show that T_1 is genuinely in crisis. But the argument, whether it comes in the course of working on T_1 or T_2, must be made out in the experimental testing framework of T_1. More correctly, it must be made out by means of shared canonical models of experiment. Such an argument, where warranted, does not depend on already holding T_2.

An example, to be discussed more fully later, may clarify my point. The two theories are classical thermodynamics and the molecular-kinetic theory. Jean Perrin's experiments, while occurring within the molecular-kinetic account, demonstrated an unevadable anomaly for the classical account by showing that Brownian motion violates a nonstatistical version of the second law of thermodynamics. They did so by showing that Brownian movement exemplifies the type of random phenomenon known from simple games of chance. The canonical model here comes from random walk phenomena. It was well understood and did not belong to any one paradigm. Once the applicability of the canonical model to Brownian motion data was shown, the anomaly, which was quantitative, was unevadable.[18]

---

17. An important task for the experimental program I am promoting would be to identify canonical ways of deliberately learning from anomalies. Contributions from several sources would be relevant. One source would be some of the empirical work in cognitive science, such as Lindley Darden 1991.

18. Those seeking to save a nonstatistical account were not allowed to explain away the anomaly or defend their theory circularly. Once all their attempted explanations were shown wanting—on normal experimental grounds—they had to concede. It is not that they are bound by a sociological convention—doing otherwise violates canons of learning from experiment. I discuss this case in detail in chapter 7.

I do not assert that experimental arguments always exist to guide theory appraisal, but rather deny Kuhn's claim that they never do. Moreover, for experimental arguments to ground theory appraisal, the experimental testing frameworks of the rival large-scale theories need not be identical. It is sufficient for the needed arguments to be made out by appeal to the interparadigmatic canonical experimental models. How can we suppose such a shared understanding? It follows from taking seriously the criteria for good normal scientific practice, criteria that, for Kuhn, must hold for any practice that enjoys a normal scientific tradition. Moreover, the historical record reveals case after case where even the most ardent proponents are forced to relent on the basis of very local but very powerful experimental tests. The Kuhn of normal science can explain this consensus quite naturally; the Kuhn of revolutionary science cannot.

One is justly led to wonder why Kuhn holds to the curious position that the strictures of normal science can compel rejection of hypotheses within a large-scale theory, even to the extent of provoking a crisis, while supposing that when the crisis gets too serious or a rival theory is proposed, the normal practitioner abruptly throws the strictures of normal science out the window and declares (being reduced to aesthetic criteria now) that his or her theory is the most beautiful. One will look in vain for an argument for this position as well as for an argument about why Kuhn takes away what I thought he had promised us—that a crisis compelled by good normal science lets us finally be *warranted* Popperian testers and reject the theory (as being wrong at least so far as its key hypotheses go)—quite apart from stopping work on it. Instead one finds that, when turning his gaze to the problem of large-scale theory appraisal, Kuhn is simply wearing spectacles that necessarily overlook the role of the shared strictures and arguments of normal testing.

Kuhn seeks something that can effect a transfer of allegiance and the gestalt switch that allegedly goes with it. Impressive experimental demonstrations can at most pave the way for this conversion by making a scientist's mind susceptible to the new gestalt. And unlike the gestalt switch of psychology, the scientist cannot switch back and forth to compare global theories.

The Early Innovators

Not everyone switches at the same time, which is one reason why Kuhn supposes that there is no single argument that must rationally convince everyone. We should be glad of this, Kuhn maintains, because such innovations are generally mistaken!

Most judgments that a theory has ceased adequately to support a puzzle-solving tradition prove to be wrong. If everyone agreed in such judgements, no one would be left to show how existing theory could account for the apparent anomaly as it usually does. (P. 248)

But this has a curious consequence for Kuhn. The innovators, with their daring value systems, switch early and proceed to work within the new theory T_2 (never mind how to explain their converting together). Where are they, it might be asked, while they are developing the new theory? Presumably, for Kuhn, they must be within T_2, since working within T_2 is equated with accepting it. It would seem to follow, however, that the early innovators would have to convert back to T_1 when, as happens most of the time pace Kuhn, the innovation is mistaken. Yet this, according to Kuhn, is impossible, or nearly always so.

Whichever paradigm theory they find themselves caught in, the early innovators must still be employing experimental testing tools from the earlier paradigm (in which we include the general pool of canonical models) to test their new hypotheses. Otherwise they could not demonstrate the quantitative experimental successes that Kuhn's own spectacles reveal to be central (if not determinative) in paradigm appraisal. Although their divergent paradigms result in their speaking different languages where translation is at most partial, they can learn how they differ via experiment. Kuhn himself says, "First and foremost, men experiencing communication breakdown can discover by experiment—sometimes by thought-experiment, armchair science— the area within which it occurs" (p. 277).

A Kuhnian may agree with my thesis about shared testing models, yet deny that the experimental arguments provided offer a basis for appraising global theories. Nevertheless, that is still no argument for Kuhn's thesis that global theory change cannot turn on experimental arguments, and, indeed, Kuhn fails to supply one. Rather, his thesis results from assuming that theory change is a conversion experience, that it requires one to "go native," and is complete only when the new theory establishes a grip on one's mind.

We can hold, with Kuhn, that experimental demonstrations and arguments at most allow one to accept a rival theory "intellectually" and yet reach the conclusion opposite from his. Far from downplaying the role of experimental argument, this just shows why scientific appraisal properly turns only on (epistemologically grounded) "intellectual" acceptance and not on psychological conversions. That evidential arguments are incapable of grounding theory change when defined as mind shifts is precisely why mind shifts have nothing to do with grounding theory assessment in science.

It is instructive to consider the new technology of "virtual reality" machines. Fitted with the appropriate helmet and apparatus, one can enter a 3-d world to learn history, medical procedures, and more. In the future a virtual reality program created by the members of one paradigm might well allow scientists from another paradigm to vicariously experience the world through the other's eyes (without the risk of being unable to convert back). This might even prove to heighten the capacity to find solutions to problems set by a research program. However, the ability to tell someone to "get into the machine and see for yourself" will never be an argument, will never substitute for an evidential grounding of the theory thereby "lived in."

2.8 SUMMARY AND CLOSING REMARKS

We began by asking what philosophical mileage could be gotten from exploring Kuhn's contrast of his position with Popper's. How far have we gone and how much of it will be utilized in the project of this book?

On Kuhn's treatment of normal science, we can sensibly construe his comparison with Popper—and it turns out that Kuhn is correct. Normal scientists, in my rereading of Kuhn, have special requirements without which they could not learn from standard tests. They insist on stringent tests, reliable or severe. They could not learn from failed solutions to normal problems if they could always change the question, make alterations, and so on. That is what Kuhn says. That is what having a normal science tradition is all about. But then we have some curious consequences at the level of theory appraisal.

Via the criteria of normal science, Kuhn says, normal science may be led to crisis. It is recognized as crisis because of the stringency of its rules. Suddenly, when confronted with a rival theory, Kuhn says, normal scientists do an about-face, defending their theory and denying it is in crisis. Kuhn gives no argument for supposing this always happens, in fact, my point is that his view of normal science militates against this supposition.

Answering Kuhn does not require us to show that global theory testing is always a function of experimental knowledge, but merely that we deny the Kuhnian view that it cannot be. My solution is based on one thing normal practitioners, even from rival paradigms, have in common (by dint of enjoying a normal testing tradition): they can and do perform the tasks of normal science reliably. That is the thrust of

Kuhn's demarcation criterion. Later we will see how experimental knowledge functions in theory testing.

The problems with Kuhn's account of theory appraisal, however, are not the problems that my approach requires be overcome. I do not seek an account of the comparative testing of rival large-scale theories, as I deny that such a thing occurs (except as understood in an elliptical fashion, to be explained). I do not accept Kuhn's supposition that there are two kinds of empirical scientific activities, normal and revolutionary: there is just normal science, understood as standard testing.[19]

Nevertheless, I will retain several of the key theses I have gleaned from Kuhn's comparison with Popper: Taking Popperian aim at global theories when doing so is not constrained by severe normal testing is a poor strategy for obtaining experimental knowledge. The constraints and norms of normal testing provide the basis for severe tests and informative scientific inquiries. To understand the nature and growth of experimental knowledge, one must look to normal testing.

We can also retain a version of Kuhn's demarcation criterion. The relevant distinction, although it is not intended to be sharp and may well admit of degrees, is between inquiries that are scientific or informative and those that are not. Inquiries are informative to the extent that they enable experimental knowledge, that is, learning from normal science. For Kuhn, in a genuine science, anomalies give rise to research puzzles. In our recasting of Kuhn this translates as, in a genuinely scientific inquiry, anomalies afford opportunities for learning. This learning is tantamount to learning from error, as described in chapter 1 and in what follows. The aim of science is not avoiding anomaly and error but being able to learn from anomaly and error.

Finding things out is a lot like normal science being revisited in the manner discussed in this chapter. Nevertheless, even the idea of normal science as extending and filling in theories should be questioned. Although it is not too far from a description of what scientists generally do, it does not entirely capture knowledge at the forefront—it is still too tied to a theory-dominated way of thinking. In the gathering up of knowledge, it is typical not to know which fields will be called upon to solve problems. There need not even be a stable background theory in place.

Take, for example, recent work on Alzheimer's disease. Clumps of an insoluble substance called beta amyloid have been found in the brains of its victims, which presents problems in its relation to the dis-

---

19. By "empirical scientific activities" I am referring here to experimental activities in the broad sense in which I understand this.

ease and how it builds up in the brain. But what is the background theory or paradigm being extended? It could come from biology or neuroscience, from any one of their specialties—except that recent findings suggest that the solution may come from genetics.

The growth of knowledge, by and large, has to do not with replacing or amending some well-confirmed theory, but with testing specific hypotheses in such a way that there is a good chance of learning something—whatever theory it winds up as part of. Having divorced normal (standard) testing from the Kuhnian dependence upon background paradigms in any sense other than dependence upon an intertheoretic pool of exemplary models of error, it is easy to accommodate a more realistic and less theory-dominated picture of inquiry. In much of day-to-day scientific practice, and in the startling new discoveries we read about, scientists are just trying to find things out.

CHAPTER THREE

The New Experimentalism
and the Bayesian Way

[F]amiliarity with the actual use made of statistical methods in the experimental sciences shows that in the vast majority of cases the work is completed without any statement of mathematical probability being made about the hypothesis or hypotheses under consideration. The simple rejection of a hypothesis, at an assigned level of significance, is of this kind, and is often all that is needed, and all that is proper, for the consideration of a hypothesis in relation to the body of experimental data available.

—R. A. Fisher, *Statistical Methods and Scientific Inference*, p. 40

[T]he job of the average mathematical statistician is to learn from observational data with the help of mathematical tools.

—E. S. Pearson, *The Selected Papers of E. S. Pearson*, p. 275

SINCE LAKATOS, the response to Popper's problems in light of Kuhn has generally been to "go bigger." To get at theory appraisal, empirical testing, and scientific progress requires considering larger units—whole paradigms, research programs, and so on. Some type of holistic move is favored even among the many philosophers who consciously set out to reject or improve upon Kuhn. Whatever else can be said of the variety of holisms that have encroached upon the philosophical landscape, they stand in marked contrast to the logical empiricist approaches to testing that concerned setting out rules of linking bits of evidence (or evidence statements) with hypotheses in a far more localized fashion. For post-Kuhnian holists, observations are paradigm laden or theory laden and testing hypotheses does not occur apart from testing larger units.

The lesson I drew from Kuhn in the previous chapter supports a very different line of approach. The lesson I take for post-Kuhnian philosophy of science is that we need to go smaller, not bigger—to the local tests of "normal science." Having so reworked the activity of normal testing I will now drop the term altogether, except when a reminder of origins seems needed, and instead use our new terms *standard testing* and *error statistics*. But the aims of standard testing are still rather like those that Kuhn sets out for normal science.

I agree with critics of logical empiricism on the inadequacy of a theory of confirmation or testing as a uniform logic relating evidence to hypotheses, that is, the evidential-relationship view. But in contrast to the thrust of holistic models, I take these very problems to show that we need to look to the force of low-level methods of experiment and inference. The fact that theory testing depends on intermediate theories of data, instruments, and experiment, and that data are theory laden, inexact, and "noisy," only underscores the necessity for numerous local experiments, shrewdly interconnected.

The suggestion that aspects of experiment might offer an important though largely untapped resource for addressing key problems in philosophy of science is not new. It underlies the recent surge of interest in experiment by philosophers and historians of science such as Robert Ackermann, Nancy Cartwright, Allan Franklin, Peter Galison, Ronald Giere, and Ian Hacking. Although their agendas, methods, and conclusions differ, there is enough similarity among this new movement to group them together. Appropriating Ackermann's nifty term, I dub them the "New Experimentalists."

Those whom I place under this rubric share the core thesis that focusing on aspects of experiment holds the key to avoiding or solving a number of problems, problems thought to stem from the tendency to view science from theory-dominated stances. In exploring this thesis the New Experimentalists have opened up a new and promising avenue for grappling with key challenges currently facing philosophers of science. Their experimental narratives offer a rich source from which to extricate how reliable data are obtained and used to learn about experimental processes. Still, nothing like a systematic program has been laid out by which to accomplish this. The task requires getting at the structure of experimental activities and at the epistemological rationale for inferences based on such activities.

To my mind, the reason the New Experimentalists have come up short is that the aspects of experiment that have the most to offer in developing such tools are still largely untapped. These aspects cover the designing, modeling, and analyzing of experiments, activities that receive structure by means of statistical methods and arguments.

This is not to say that the experimental narratives do not include the use of statistical methods. In fact, their narratives are replete with applications of statistical techniques for arriving at data, for assessing the fit of data to a model, and for distinguishing real effects from arti-facts (e.g., techniques of data analysis, significance tests, standard er-rors of estimates, and other methods from standard error statistics). What has not been done is to develop these tools into something like an adequate philosophy or epistemology of experiment. What are needed are forward-looking tools for arriving at reliable data and using such data to learn about experimental processes.

In rejecting old-style accounts of confirmation as the wrong way to go to relate data and hypothesis, the New Experimentalists seem to shy away from employing statistical ideas in setting out a general ac-count of experimental inference. Ironically, where there is an attempt to employ formal statistical ideas to give an overarching structure to experiment, some New Experimentalists revert back to the theory-dominated philosophies of decision and inference, particularly Bayes-ian philosophies. The proper role for statistical methods in an adequate epistemology of experiment, however, is not the theory-dominated one of reconstructing episodes of theory confirmation or large-scale theory change. Rather their role is to provide forward-looking, amplia-tive rules for generating, analyzing, and learning from data in a reliable and intersubjective manner. When it comes to these roles, the Bayes-ian Way is the wrong way to go.

My task now is to substantiate all of these claims. In so doing I shall deliberately alternate between discussing the New Experimental-ism and the Bayesian Way. The context of experiment, I believe, pro-vides the needed backdrop against which to show up key distinctions in philosophy of statistics. The New Experimentalist offerings reveal (whether intended or not) the function and rationale of statistical tools from the perspective of actual experimental practice—the very under-standing missing from theory-dominated perspectives on scientific in-ference. This understanding is the basis for both my critique of the Bayesian Way and my defense of standard error statistics.

One other thing about my strategy: In the Bayesian critique I shall bring to the fore some of the statisticians who have contributed to the debates in philosophy of statistics. Their work has received too little attention in recent discussions by Bayesian philosophers of science, which has encouraged the perception that whatever statisticians are doing and saying must be quite distinct from the role of statistics in philosophy of science. Why else would we hear so little in the way of a defense of standard (non-Bayesian) statistics? But in fact statisticians have responded, there is a rich history of their response, and much of what I need to say has been said by them.

I will begin with the New Experimentalism.

3.1 THE NEW EXPERIMENTALISM

Having focused for some time on theory to the near exclusion of experiment, many philosophers and historians of science have now turned
their attention to experimentation, instrumentation, and laboratory
practices.[1] Among a subset of this movement—the New Experimentalists—the hope is to steer a path between the old logical empiricism,
where observations were deemed relatively unproblematic and given
primacy in theory appraisal, and the more pessimistic post-Kuhn-
ians, who see the failure of logical empiricist models of appraisal as
leading to underdetermination and holistic theory change, if not
to outright irrationality. I will begin by outlining what seem to me to
be the three most important themes to emerge from the New Experimentalism.

1. Look to Experimental Practice to Restore to Observation
Its Role as Objective Basis

Kuhn, as we saw in chapter 2, often betrays the presumption that
where an algorithm is unavailable the matter becomes one of sociology. He proposes that problems that are usually put as questions about
the nature of and warrant for theory appraisal be reasked as sociological questions that have "scarcely even been stated before" (Kuhn 1977,
212). Whether intended or not, this has invited sociological studies
into the role that interests and negotiations play in constructing and
interpreting data. Interviews with scientists have provided further grist
for the mills of those who hold that evidence and argument provide
little if any objective constraint.

A theme running through the work of the New Experimentalists
is that to restore the role of empirical data as an objective constraint
and adjudicator in science, we need to study the actual experimental
processes and reasoning that are used to arrive at data. The old-style
accounts of how observation provides an objective basis for appraisal
via confirmation theory or inductive logic should be replaced by an
account that reflects how experimental knowledge is actually arrived
at and how it functions in science.

Peter Galison (1987) rightly objects that "it is unfair to look to experimental arguments for ironclad implications and then, upon finding
that experiments do not have logically impelled conclusions, to ascribe

1. A collection of this work may be found in Achinstein and Hannaway 1985.
For a good selection of interdisciplinary contributions, see Gooding, Pinch, and
Schaffer 1989.

the experimentalists' beliefs entirely to 'interests'" (p. 11). He suggests that we look instead at how experimentalists actually reason.

Similarly, Allan Franklin (1986, 1990) finds in experimental practice the key to combating doubts about the power of empirical evidence in science. He puts forward what he calls an "evidence model" of science—"that when questions of theory choice, confirmation, or refutation are raised they are answered on the basis of valid experimental evidence" (1990, 2)—in contrast to the view that science is merely a social construction.

An evaluation of the New Experimentalists' success must distinguish between their having provided us sticks with which to beat the social constructivists and their having advanced solutions to philosophical problems that persist, even granting that evidence provides an objective constraint in science.

2. Experiment May Have a Life of Its Own

This slogan, from Hacking 1983, 1992a, and 1992b points to several New Experimentalist subthemes, and can be read in three ways, each in keeping with the position I developed in chapter 2.

Topical hypotheses. The first sense, which Hacking (1983, 160) emphasizes, concerns the aims of experiment. In particular, he and others recognize that a major aim of experiment is to learn things without any intention of testing some theory. In a more recent work, Hacking calls the kinds of claims that experiment investigates "topical hypotheses"—like topical creams—in contrast to deeply penetrating theories. Hacking (1992a) claims that

> it is a virtue of recent philosophy of science that it has increasingly come to acknowledge that most of the intellectual work of the theoretical sciences is conducted at [the level of *topical* hypotheses] rather than in the rarefied gas of systematic theory. (P.45)

The New Experimentalists have led in this recognition. Galison (1987) likewise emphasizes that the

> experimentalists' real concern is not with global changes of world view. In the laboratory the scientist wants to find local methods to eliminate or at least quantify backgrounds, to understand where the signal is being lost, and to correct systematic errors. (P. 245)

The parallels with our recasting of Kuhnian normal science in the last chapter are clear.

Theory-independent warrant for data. A second reading of the slogan refers to the justification of experimental evidence—that a theory-
independent warrant is often available. More precisely, the thesis is
that experimental evidence need not be theory laden in any way that
invalidates its various roles in grounding experimental arguments.
Granting that experimental data are not just given unproblematically,
the position is that coming to accept experimental data can be based
on experimental processes and arguments whose reliability is independently demonstrated. Some have especially stressed the independent
grounding afforded by knowledge of instruments; others stress the
weight of certain experimental activities, such as manipulation. The
associated argument, in each case, falls under what I am calling *exemplary* or *canonical arguments* for learning from error.

Experimental knowledge remains. This reading leads directly to the third
gloss of the slogan about experiment having a life of its own. It concerns the continuity and growth of experimental knowledge. In particular, the New Experimentalists observe, experimental knowledge remains despite theory change. Says Galison, "Experimental conclusions
have a stubbornness not easily canceled by theory change" (1987,
259). This cuts against the view that holders of different theories necessarily construe evidence in incommensurable or biased fashions. We
saw in chapter 2 that the criteria of good normal science lead to just the
kind of reliable experimental knowledge that remains through global
theory change. These norms, I argued, themselves belie the position
Kuhn takes on revolutionary science, where everything allegedly
changes.

Continuity at the level of experimental knowledge also has ramifications for the question of scientific progress. It points to a crucial
kind of progress that is overlooked when measures of progress are
sought only in terms of an improvement in theories or other larger
units. Experimental knowledge grows, as do the tools for its acquisition, including instrumentation, manipulation, computation, and most
broadly, argumentation.[2] Giere and Hacking have especially stressed
how this sort of progress is indicated when an entity or process becomes so well understood that it can be used to investigate other objects and processes (e.g., Giere 1988, 140). We can happily accept what

2. Ackermann (1985) and others stress progress through instrumentation, but
there is also progress by means of a whole host of strategies for obtaining experimental knowledge. That is why I take progress through experimental argumentation to be the broadest category of experimental progress.

these authors say about experimental progress while remaining agnostic about what this kind of progress might or might not show about the philosophical doctrine of scientific realism.

3. What Experimentalists Find: Emphasis on Local Discrimination of Error

A third general theme of New Experimentalist work concerns the particular types of tasks that scientists engage in when one turns to the processes of obtaining, modeling, and learning from experimental data: checking instruments, ruling out extraneous factors, getting accuracy estimates, distinguishing real effect from artifact. In short, they are engaged in the manifold local tasks that may be seen as estimating, distinguishing, and ruling out various errors (in our broad sense).

"How do experiments end?" (as in the title of Galison's book) asks "When do experimentalists stake their claim on the reality of an effect? When do they assert that the counter's pulse or the spike in a graph is more than an artifact of the apparatus or environment?" (Galison 1987, 4). The answer, in a nutshell, is only after they have sufficiently ruled out or "subtracted out" various backgrounds that could be responsible for an effect. "As the artistic tale suggests," Galison continues, "the task of removing the background is not ancillary to identifying the foreground—*the two tasks are one and the same*" (p. 256), and the rest of his book explores the vast and often years-long effort to conduct and resolve debates over background.

3.2 What Might an Epistemology of Experiment Be?

Now to build upon the three themes from the New Experimentalist work, which may be listed as follows:

1. Understanding the role of experiment is the key to circumventing doubts about the objectivity of observation.

2. Experiment has a life of its own apart from high level theorizing (pointing to a local yet crucially important type of progress).

3. The cornerstone of experimental knowledge is the ability to discriminate backgrounds: signal from noise, real effect from artifact, and so on.

In pressing these themes, many philosophers of science sense that the New Experimentalists have opened a new and promising avenue within which to grapple with the challenges they face. Less clear is whether the new attention to experiment has paid off in advancing solutions to problems. Nor is it even clear that they have demarcated a program for working out a philosophy or epistemology of experiment.

The New Experimentalist work seems to agree on certain central questions of a philosophy or epistemology of experiment: how to establish well-grounded observational data, how to use data to find out about experimental processes, and how this knowledge bears on revising and appraising hypotheses and theories. Satisfactory answers to these questions would speak to many key problems with which philosophers of science wrestle, but the New Experimentalist work has not yet issued an account of experimental data adequate to the task.

Experimental activities do offer especially powerful grounds for arriving at data and distinguishing real effects from artifacts, but what are these grounds and why are they so powerful? These are core questions of this book and can be answered adequately only one step at a time.

As a first step we can ask, What is the structure of the argument for arriving at this knowledge? My answer is the one sketched in chapter 1: it follows the pattern of *an argument from error* or *learning from error*. The overarching structure of the argument is guided by the following thesis:

> It is learned that an error is absent when (and only to the extent that) a procedure of inquiry (which may include several tests) having a high probability of detecting the error if (and only if[3]) it exists nevertheless fails to do so, but instead produces results that accord well with the *absence* of the error.

Such a procedure of inquiry is highly capable of severely probing for errors—let us call it a *reliable (or highly severe) error probe*. According to the above thesis, we can argue that an error is absent if it fails to be detected by a highly reliable error probe.

Alternatively, the argument from error can be described in terms of a test of a hypothesis, H, that a given error is absent. The evidence indicates the correctness of hypothesis H, when H passes a severe test—one with a high probability of failing H, if H is false. An analogous argument is used to infer the presence of an error.

With this conjecture in hand, let us return to the New Experimentalists. I believe that their offerings are the most interesting and illuminating for the epistemologist of science when they reveal (unwittingly or not) strategies for arriving at especially strong experimental argu-

---

3. The "only if" is already accommodated by the requirement that failing to detect means the result is probable assuming the error is absent. The extreme case would be if the result is entailed by the absence of the error. This thesis is the informal side of the more formal definition of passing a severe test in chapter 6.

ments; and when this is so, I maintain, it is because they are describing ways to arrive at procedures with the capacity to probe severely and learn from errors. The following two examples, from Galison 1987 and Hacking 1983, though sufficient, could easily be multiplied.

Arguments for Real Effects

Galison

> The consistency of different data-analysis procedures can persuade the high-energy physicist that a real effect is present. A similar implicit argument occurs in smaller-scale physics. On the laboratory bench the experimenter can easily vary experimental conditions; when the data remain consistent, the experimentalist believes the effect is no fluke. (1987, 219)

What is the rationale for being thus persuaded that the effect is real? The next sentence contains the clue:

> In both cases, large- and small-scale work, the underlying assumption is the same: under sufficient variation any artifact ought to reveal itself by causing a discrepancy between the different "subexperiments." (Ibid.)

Although several main strategies experimentalists use lie scattered through Galison's narratives, he does not explicitly propose a general epistemological rationale for the inferences reached. The argument from error supplies one.

How do these cases fit the pattern of my argument from error? The evidence is the consistency of results over diverse experiments, and what is learned or inferred is "that it is no fluke." Why does the evidence warrant the no-fluke hypothesis? Because were it a fluke, it would almost surely have been revealed in one of the deliberately varied "subexperiments." Note that it is the entire *procedure* of the various subexperiments that may properly be said to have the probative power—the high probability of detecting an artifact by *not* yielding such consistent results. Never mind just now how to justify the probative power (severity) of the procedure. For the moment, we are just extracting a core type of argument offered by the New Experimentalists. And the pattern of the overall argument is that of my argument from error.

Galison's discussion reveals a further insight: whether it is possible to vary background factors or use data analysis to argue "as if" they are varied, the aim is the same—to argue from the consistency of results to rule out its being due to an artifact.

Hacking. An analysis of Hacking's "argument from coincidence" reveals the same pattern, although Hacking focuses on cases where it is possible to vary backgrounds by way of literal manipulation.

Hacking asks, What convinces someone that an effect is real? Low-powered electron microscopy reveals small dots in red blood platelets, called dense bodies. Are they merely artifacts of the electron microscope?

> One test is obvious: can one see these selfsame bodies using quite different physical techniques? . . . In the fluorescence micrographs there is exactly the same arrangement of grid, general cell structure and of the "bodies" seen in the electron micrograph. It is inferred that the bodies are not an artifact of the electron microscope. . . . It would be a preposterous coincidence if, time and again, two completely different physical processes produced identical visual configurations which were, however, artifacts of the physical processes rather than real structures in the cell. (Hacking 1983, 200–201)

Two things should again be noted: First, the aim is the local one, to distinguish artifacts from real objects or effects—something that can be pursued even without a theory about the entities or effects in question. Second, Hacking's argument from coincidence is an example of sustaining an argument from error. The error of concern is to take as real structure something that is merely an artifact. The evidence is the identical configurations produced by completely different physical processes. Such evidence would be extremely unlikely if the evidence were due to "artifacts of the physical processes rather than real structures in the cell" (ibid.).

As before, much more needs to be said to justify this experimental argument, and Hacking goes on to do that; for example, he stresses that we made the grid, we know all about these grids, and so on. But the present concern is the *pattern* of the argument, and it goes like this: "If you can see the same fundamental features of structure using several different physical systems, you have excellent reason for saying, 'that's real' rather than, 'that's an artifact'" (Hacking 1983, 204).[4] It is not merely the *improbability* of all the instruments and techniques conspiring to make all of the evidence appear as if the effect were real. Rather, to paraphrase Hacking again, it is the fact that it would be akin to invoking a Cartesian demon to suppose such a conspiracy.

Hacking tends to emphasize the special evidential weight afforded by performing certain experimental *activities* rather than what is af-

4. It is important to distinguish carefully between "real" as it is understood here, namely, as genuine or systematic, and as it is understood by various realisms.

forded by certain kinds of *arguments*, saying that "no one actually produces this 'argument from coincidence' in real life" (1983, 201). Not
so. Unless it is obvious that such an argument *could be given*, experimental practitioners produce such arguments all the time. In any
event, as seekers of a philosophy of experiment we need to articulate
the argument if we are to carry out its tasks. The tasks require us to get
at the structure of experimental activities and at the epistemological
rationale for inferences based on the results of such activities. Most
important, understanding the argument is essential in order to justify
inferences where the best one can do is simulate, mimic, or otherwise
argue as if certain experimental activities (e.g., literal manipulation or
variation) had occurred. Experimental arguments, I suggest, often
serve as surrogates for actual experiments; they may be seen as experiments "done on paper."

Other examples from the work of other New Experimentalists
(e.g., conservation of parity), as well as the work of other philosophers
of science, lend themselves to an analogous treatment, but these two
will suffice. In each case, what are needed are tools for arriving at,
communicating, and justifying experimental arguments, and for using
the results of one argument as input into others. Those aspects of experiment that have the most to offer in developing such tools are still
largely untapped, however, which explains, I think, why the New Experimentalists have come up short. These aspects cover the designing,
modeling, and analyzing of experiments—activities that receive structure by means of standard statistical methods and arguments.

Put Your Epistemology of Experiment at the Level of Experiment

In rejecting old-style accounts of confirmation as the wrong way to
go, the New Experimentalists seem dubious about the value of utilizing
statistical ideas to construct a general account of experimental inference. Theories of confirmation, inductive inference, and testing were
born in a theory-dominated philosophy of science, and this is what
they wish to move away from. It is not just that the New Experimentalists want to sidestep the philosophical paradoxes and difficulties that
plagued formal attempts at inductive logics. The complexities and context dependencies of actual experimental practice just seem recalcitrant to the kind of uniform treatment dreamt of by philosophers of
induction. And since it is felt that overlooking these complexities is
precisely what led to many of the problems that the New Experimentalists hope to resolve, it is natural to find them skeptical of the value
of general accounts of scientific inference.

The typical features of what may be called "theory-dominated" accounts of confirmation or testing are these: (1) the philosophical work begins with data or evidence statements already in hand; (2) the account seeks to provide uniform rules for relating evidence (or evidence statements) to any theory or conclusion (or decision) of interest; and (3) as a consequence of (1) and (2), the account functions largely as a way to *reconstruct* a scientific inference or decision, rather than giving us tools scientists actually use or even a way to model the tools actually used.

The New Experimentalists are right to despair of accounts that kick in only after sufficiently sharp statements of evidence and hypotheses are in hand. Galison is right to doubt that it is productive to search for "an after-the-fact reconstruction based on an inductive logic" (1987, 3). Where the New Experimentalists shortchange themselves is in playing down the use of local statistical methods at the experimental level—the very level they exhort us to focus on.[5] The experimental narratives themselves are chock-full of applications of standard statistical methods, methods developed by Fisher, Neyman and Pearson, and others. Despite the alleged commitment to the actual practices of science, however, there is no attempt to explicate these statistical practices on the scientists' own terms. Ironically, where there is an attempt to employ statistical methods to erect an epistemology of experiment, the New Experimentalists revert to the theory-dominated philosophies of decision and inference. A good example is Allan Franklin's appeal to the Bayesian Way in attempting to erect a philosophy of experiment.

The conglomeration of methods and models from standard error statistics, error analysis, experimental design, and cognate methods, I will argue, is the place to look for forward-looking procedures that serve to obtain data in the first place, and that are apt even with only vague preliminary questions in hand. If what I want are tools for discriminating signals from noise, ruling out artifacts, and so on, then I really need tools for doing that. And these tools must be applicable with the kinds of information scientists actually have. At the same time, these tools can provide the needed structure for the practices

5. Hacking's recent work often comes close to what I have in mind, but he is reluctant to worry about the Bayes versus non-Bayes controversy in philosophy of statistics. "It is true that different schools will give you different advice about how to design experiments, but for any given body of data they agree almost everywhere" (Hacking 1992b, 153). When it comes to the use of statistical ideas for a general philosophy of experiment, the divergent recommendations regarding experimental design are crucial. It is as regards the uses of a theory of statistics *for philosophy of science*—the uses that interest me—that the debates in philosophy of statistics matter.

given a central place by the New Experimentalists. Before turning to standard error statistics we need to consider why the Bayesian Way fails to serve the ends in view.

3.3 THE BAYESIAN WAY

> I take a natural and realistic view of science to allow for the acceptance of corrigible statements, both in the form of data and in the form of laws and hypotheses. . . . It is hard to see what motivates the Bayesian who wants to replace the fabric of science, already complicated enough, with a vastly more complicated representation in which each statement of science is accompanied by its probability, for each of us. (Kyburg 1993, 149)

By the "Bayesian Way" I mean the way or ways in which a certain mathematical theory of statistical inference—Bayesian inference—is used in philosophy of science. My criticism is of its uses regarding philosophical problems of scientific inference and hypothesis testing as distinct from its use in certain statistical contexts (where the ingredients it requires, e.g., prior probabilities, are unproblematic or less problematic) and in personal decision-making contexts. It is not that the Bayesian approach is free of problems in these arenas—ongoing controversies are ever present among statisticians and philosophers of statistics. But the problems of central interest to the philosopher of the epistemology of experiment are those that concern the Bayesian Way in philosophy of science, specifically, problems of scientific inference and methodology.[6]

There is another reason to focus on the Bayesian Way in philosophy of science: it is here that the deepest and most philosophically relevant distinctions between Bayesian and non-Bayesian ideas emerge. For a set of well-defined statistical problems, and for given sets of data, Bayesian and non-Bayesian inferences may be found to formally agree—despite differences in interpretation and rationale. When it comes to using statistical methods in philosophy of science,

6. In making this qualification I mean to signal that I recognize the relevance of decision theory to philosophy of science generally. However, the set of philosophical problems for which decision theory is most applicable are distinct from those of scientific inference. It is true that decision theory (Bayesian and non-Bayesian) has also been used to model rational scientific agents. However, I do not find those models useful for my particular purpose, which has to do with identifying rational methods rather than rational agents or rational actions. It would take us too far afield to consider those models here.

differences in experimental design, interpretation, and rationales are all-important.

There are three main ways in which a mathematical theory of probabilistic or statistical inference can be used in philosophy of science:

1. *A way to model scientific inference.* The aim may be to model or represent certain activities in science, such as acquiring data, making inferences or decisions, and confirming or testing hypotheses or theories. The intention may be to capture either actual or rational ways to carry out these activities.

2. *A way to solve problems in philosophy of science.* The aim may be to help solve philosophical problems concerning scientific inference and observation (e.g., objectivity of observation, underdetermination, Duhem's problem).

3. *A way to perform a metamethodological critique.* It can be used to scrutinize methodological principles (according special weight to "novel" facts) or to critique the rationality of scientific episodes (metamethodology).

There are other ways of using a theory of statistics, but the above are the most relevant to the epistemological issues before us. The Bayesian Way has in fact been put to all these uses, and many imagine that it is the only plausible way of using ideas from mathematical statistics to broach these concerns in the philosophy of science. Indeed, its adherents often tout their approach as the only account of inference we will ever need, and some unblushingly declare the Bayesian Way to be the route toward solving all problems of scientific inference and methodology. I do not agree.

Although the Bayesian literature is long and technical, explaining why the Bayesian Way is inadequate for each of the three aims requires little or no technical statistics. Such an explication seems to me to be of pressing importance. Keeping the Bayesian philosophy of science shrouded in mathematical complexity has led to its work going on largely divorced from other approaches in philosophy of science. Philosophers of science who do consult philosophers of statistics get the impression that anything but Bayesian statistics is discredited. Thus important aspects of scientific practice are misunderstood or overlooked by philosophers of science because these practices directly reflect non-Bayesian principles and methods that are widespread in science.

It may seem surprising, given the current climate in philosophy of science, to find philosophers (still) declaring invalid a standard set of experimental methods rather than trying to understand or explain why scientists evidently (still) find them so useful. I think it is surprising. Is there something special about the philosophy of experimental inference that places it outside the newer naturalistic attitudes? By and large, Bayesian statisticians proceed as if there were. Colin Howson and Peter Urbach (1989) charge "that one cannot derive scientifically significant conclusions from the type of information which the Fisher and the Neyman-Pearson theories regard as adequate" (p. 130), despite the fact that for decades scientists and statisticians have made it clear they think otherwise. Nor is their position an isolated case. Howson and Urbach are simply the most recent advocates of the strict Bayesian line of argument worked out by fathers of Bayesianism such as Bruno De Finetti, I. J. Good, Denis Lindley, and L. J. Savage. To their credit, Howson and Urbach attempt to apply the Bayesian Way to current challenges in philosophy of science, and so are useful to our project.

Granted, the majority of Bayesians seem to want to occupy a position less strict than that espoused by Howson and Urbach, although they are not entirely clear about what this means. What is clear is that thus far none of the middle-of-the-road, fallen, or otherwise better-behaved Bayesians have promoted the battery of non-Bayesian methods as the basis for an epistemology of experiment. I hope to encourage a change in that direction. I do not believe that an adequate philosophy of experiment can afford to be at odds with statistical practice in science.

The Focus of My Critique: Bayesian Subjectivism

My critique of Bayesianism in this chapter will focus on the first two ways of using an account of statistical inference in philosophy of science—to model scientific inference and to solve philosophical problems about scientific inference.[7] Simply, the Bayesian tools do not tell us what we want to know in science. What we seek are ampliative rules for generating and analyzing data and for using data to learn about experimental processes in a reliable and intersubjective manner. The kinds of tools needed to do this are crucially different from those the Bayesians supply.

The shortcomings of the Bayesian Way for the first two aims bear directly on its appropriateness for the third aim—using Bayesian prin-

---

7. Let me confess right off that I will give short shrift to many important technical qualifications, historical footnotes, and significant mathematical developments. No doubt some will take me to task for this, and I apologize. For my purposes, I believe, it is of greater importance to get at the main issues in as nontechnical and noncumbersome a manner as possible.

ciples in a metamethodological critique. (Bayesian critiques of non-Bayesian principles and methods will be addressed in later chapters.)

My immediate target is the version or versions of Bayesianism routinely appealed to by philosophers of the Bayesian Way: the standard subjective Bayesian account (with a few exceptions to be noted).[8] To keep the discussion informal I shall proceed concentrically, going once over the main issues, then again more deeply—rotating all the while among the New Experimentalist program, philosophy of statistics, and philosophy of science. In later loops (and later chapters) some of the more formal notions will fall into place. Although proceeding thus means building an argument piecemeal throughout this book, I can make several of my main points now by looking at how the subjective Bayesians, or Personalists, themselves view the task of an account of statistical or inductive inference.

Evidential-Relationship versus Testing Approaches

In delineating approaches to statistical inference, I find it helpful to distinguish between "evidential-relationship" (E-R) approaches and "testing" approaches. E-R approaches grew naturally from what was traditionally thought to be required by a "logic" of confirmation or induction. They commonly seek quantitative measures of the bearing of evidence on hypotheses. What I call testing approaches, in contrast, focus on finding general methods or procedures of testing with certain good properties.

For now, the distinction between E-R and testing approaches may be regarded as simply a way to help put into perspective the different accounts that have been developed. Only later will this descriptive difference be seen to correspond to more fundamental, epistemological ones. A main way to contrast the two approaches is by means of their quantitative measures. The quantities in E-R approaches are probabilities or other measures (of support or credibility) assigned to hypotheses. In contrast, testing approaches do not assign probabilities to hypotheses. The quantities and principles in testing approaches refer only to properties of methods, for example, of testing or of estimation procedures. One example is the probability that a given procedure of testing would reject a null hypothesis erroneously—an error probability. Another is our notion of a severe testing process.

Bayesian inference is an E-R approach, as I am using that term,

8. Many of my remarks here and in chapter 10 also apply to so-called objective Bayesians, e.g., Roger Rosenkrantz. For an excellent critical discussion of objective Bayesianism, see Seidenfeld 1979b.

while testing approaches include non-Bayesian approaches, for example, Popperian corroboration, Fisherian statistics, and Neyman-Pearson statistics. Under the category of a testing approach, I would also include entirely qualitative non-Bayesian approaches, for example, those of Clark Glymour and John Worrall.[9]

In the Bayesian approach the key E-R measure is that of a probability of a hypothesis relative to given data. Computing such probabilities requires starting out with a probability assignment, and a major source of difficulty has been how to construe these *prior probabilities*. One way has been to construe them as "logical probabilities," a second, as subjective probabilities.

Carnapian Bayesians. The pioneer in developing a complete E-R theory based on logical probability is Rudolf Carnap.[10] The Carnapian Bayesian sought to assign priors by deducing them from the logical structure of a particular first order language. The E-R measure was to hold between two statements, one expressing a hypothesis and the other data, sometimes written as $C(h,e)$. The measure was to reflect, in some sense, the "degree of implication" or confirmation that e affords h. Calculating its value, the basis for Carnapian logics of confirmation, was a formal or syntactical matter, much like deductive logic.

Such logics of confirmation, however, were found to suffer from serious difficulties. The languages were far too restricted for most scientific cases, a problem never wholly overcome. Even where applicable, a deeper problem remained: How can a priori assignments of probability be relevant to what can be expected to actually occur, that is, to reliability? How can they provide what Wesley Salmon calls "a guide to life"? There is the further problem, Carnap showed, of having to pick from a continuum of inductive logics. To restrict the field, Carnap was led to articulate several postulates, but these, at best, seemed to rest on what Carnap called "inductive intuition." Salmon remarks:

9. I have hardly completely covered all non-Bayesian approaches. Notable non-Bayesian accounts not discussed are those of Glymour, Kyburg, and Levi. A sect of Bayesians who explicitly consider error probabilities, e.g., Seidenfeld, might seem to be anomalous cases. I regard them as more appropriately placed under the category of testing approaches. I return to this in chapter 10.

10. Carnap 1962, *Logical Foundations of Probability.* See also Carnap's "Replies and Systematic Expositions" in Schilpp's *The Philosophy of Rudolph Carnap* (Schilpp 1963). Wesley Salmon, in many places, clearly and comprehensively discusses the developments of Carnap's work on induction. See, for example, Salmon 1967 and 1988. See also Carnap and Jeffrey 1971.

Carnap has stated that the ultimate justification of the axioms is inductive intuition. I do not consider this answer an adequate basis for a concept of rationality. Indeed, I think that *every* attempt, including those by Jaako Hintikka and his students, to ground the concept of rational degree of belief in logical probability suffers from the same unacceptable apriorism. (Salmon 1988, 13)

Subjective Bayesians. The subjective Bayesian, instead, views prior probabilities as personal degrees of belief on the part of some individual. Subjective Bayesianism is a natural move for inductive logicians still wanting to keep within the general Carnapian (E-R) tradition of what an inductive logic should look like. By replacing logical with subjective probabilities, it provides an evidential-relationship approach to confirmation without the problems of logical probability. The definition and tasks of inductive logic become altered correspondingly. Take Howson and Urbach 1989:

> Inductive logic—which is how we regard the subjective Bayesian theory—is the theory of inference from some exogenously given data and prior distribution of belief to a posterior distribution. (P. 290)

The prior distribution of belief refers to the degrees of belief an agent has in a hypothesis H and its alternatives prior to the data;[11] the posterior (or final) distribution refers to the agent's degree of belief in H after some data or evidence statement is accepted. Inductive inference from evidence is a matter of updating one's degree of belief to yield a posterior degree of belief (via Bayes's theorem).[12]

The Bayesian conception of inductive logic reflects a key feature of theory-dominated philosophies of science: an account of inference begins its work only after sufficiently sharp statements of evidence and hypotheses are in hand. But more is required to get a Bayesian inference off the ground. Also necessary are assignments of degrees of belief to an exhaustive set of hypotheses that could explain the evidence. These are the prior probability assignments. (The full-dress Bayesian requires utilities as well, but I leave this to one side.)

Where do the prior probabilities come from? How does one come to accept the evidence? That the Bayesian approach places no restric-

11. When, as is often the case, the data are known, the prior probability refers to the degree of belief the agent supposes he or she would have if the data were not known. Problems with this occupy us later (chapter 10).
12. Attempts at interval valued probabilities have been proposed but with mixed success. At any rate, nothing in the present discussion is altered by those approaches.

tions on what can serve as hypotheses and evidence, while an important part of its appealing generality, makes it all the more difficult to answer these questions satisfactorily.

Prior Probabilities: Where From?

Many philosophers would agree with Isaac Levi that "strict Bayesians are legitimately challenged to tell us where they get their numbers" (Levi 1982, 387). In particular, it seems they should tell us how to assign prior probabilities. The subjectivist disagrees. The Bayesian subjectivist typically maintains that

> we are under no obligation to legislate concerning the methods people adopt for assigning prior probabilities. These are supposed merely to characterise their beliefs subject to the sole constraint of consistency with the probability calculus. (Howson and Urbach 1989, 273)

Agents presumably are to discover their degrees of belief by introspection, perhaps by considering the odds they might give if presented with (and required to take?) a series of bets.

But would not such personal opinions be highly unstable, varying not just from person to person, but from moment to moment? That they would, subjectivists accept and expect.

In their classic paper, Edwards, Lindman, and Savage (1963) tell us that the probability of a hypothesis H, $P(H)$ "might be illustrated by the sentence: 'The probability for you, now, that Russia will use a booster rocket bigger than our planned Saturn booster within the next year is .8'" (p. 198). Throughout the introductory text by Richard Savage (L. J.'s brother) "my probability" is quite deliberately used instead of "probability."

Quantitatively expressing the degree of belief "for you now" is quite outside what Bayesian inference officially supplies. Bayesian inference takes it as a given that agents have degrees of belief and assumes that these are expressible as probabilities; its work is to offer a way of fitting your beliefs together coherently. In particular, your beliefs prior to the data should cohere with those posterior to the data by Bayes's theorem (whether you do it by conditionalization or by changing your prior probability assignment[13]).

---

13. Not all Bayesians hold the posterior to result from conditionality only. It might be due to a change in prior probability assignment for reasons other than new evidence e. Those who violate conditionality have the Bayesian approach doing even less work—it only tells you to be coherent. Nothing in our discussion turns on this qualification, however.

(Much of the technical work by Bayesian philosophers concerns so-called Dutch Book arguments, which come in various forms. These arguments purport to show that if we are rational, we will be coherent in the Bayesian sense. The basic argument is that if it is given that beliefs are expressible as probabilities, then, assuming you must accept every bet you are offered, if your beliefs do not conform to the probability calculus, you are being incoherent and will lose money for sure. In as much as these givens hardly seem to describe the situation in science, as many have argued,[14] we need not accept what such arguments purport to show.)

Personal Consistency versus Scientific Prediction

Bayes's theorem, to be stated shortly, follows from the probability calculus and is unquestioned by critics. What is questioned by critics is the relevance of a certain use of this theorem, namely, for scientific inference. Their question for the subjective Bayesian is whether scientists have prior degrees of belief in the hypotheses they investigate and whether, even if they do, it is desirable to have them figure centrally in learning from data in science. In science, it seems, we want to know what the data are saying, quite apart from the opinions we start out with. In trading logical probabilities for measures of belief, the problem of relevance to real world predictions remains.

Leonard "L. J." Savage, a founder of modern subjective Bayesianism, makes it very clear throughout his work that the theory of personal probability "is *a code of consistency for the person applying it, not a system of predictions about the world around him*" (Savage 1972, 59; emphasis added). Fittingly, Savage employs the term "personalism" to describe subjective Bayesianism.

But is a personal code of consistency, requiring the quantification of personal opinions, however vague or ill formed, an appropriate basis for scientific inference? Most of the founders of modern statistical theory—Fisher, Neyman, Pearson, and others—said no. Pearson (of Neyman and Pearson) put his rejection this way:

> It seems to me that . . . [even with no additional knowledge] I might quote at intervals widely different Bayesian probabilities for the same set of states, simply because I should be attempting what would be for me impossible and resorting to guesswork. It is difficult to see how the matter could be put to experimental test. (Pearson 1966e, 278)

Stating his position by means of a question, as he was wont to do, Pearson asks:

14. For an excellent recent discussion, see Baccus, Kyburg, and Thalos 1990.

Can it really lead to my own clear thinking to put at the very foundation of the mathematical structure used in acquiring knowledge, functions about whose form I have often such imprecise ideas? (Pearson 1966e, 279)

Fisher expressed his rejection of the Bayesian approach far more vehemently (which is not to say that he favored the one erected by Neyman and Pearson, but more on that later). Bayesians, Fisher declared,

> seem forced to regard mathematical probability, not as an objective quantity measured by observable frequencies, but as measuring merely psychological tendencies, theorems respecting which are useless for scientific purposes. (Fisher 1947, 6–7)

As is evident from this chapter's epigraph, Fisher denied the need for posterior probabilities of hypotheses in science in the first place.

In an earlier generation (late nineteenth century), C. S. Peirce, anticipating the later, non-Bayesian statisticians, similarly criticized the use of subjective probabilities in his day. Considering Peirce will clarify Fisher's claim that Bayesians "seem forced to regard" probability as subjective degrees of belief.

Why the Evidential-Relationship Philosophy Leads to Subjectivism

Peirce, whom I shall look at more closely in chapter 12, is well aware that probabilities of hypotheses are calculable by the doctrine of "inverse probability" (Bayes's theorem). However, Peirce explains,

> this depends upon knowing antecedent probabilities. If these antecedent probabilities were solid statistical facts, like those upon which the insurance business rests, the ordinary precepts and practice [of inverse probability] would be sound. But they are not and cannot be statistical facts. What is the antecedent probability that matter should be composed of atoms? Can we take statistics of a multitude of different universes? . . . All that is attainable are subjective probabilities. (Peirce 2.777)[15]

And subjective probabilities, Peirce continues, "are the source of most of the errors into which man falls, and of all the worst of them" (ibid.).

By "solid statistical facts" Peirce means that they have some clear stochastic or frequentist interpretation. (I discuss my gloss on frequentist statistics in chapter 5.) It makes sense to talk of the relative

---

15. All Peirce references are to C. S. Peirce, *Collected Papers*. References are cited by volume and paragraph number. For example, Peirce 2.777 refers to volume 2, paragraph 777.

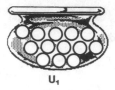

U_1

FIGURE 3.1.   A single-universe context.

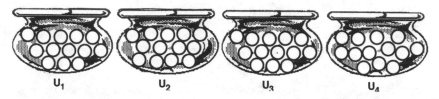

U_1 U_2 U_3 U_4

FIGURE 3.2.   A multiple-universe context (universes as plenty as blackberries).

frequency of events such as "heads" in a population of coin-tossing experiments but not of the relative frequency of the truth of a hypothesis such as matter is composed of atoms. The probability of a hypothesis would make sense, Peirce goes on to say, only

> if universes were as plenty as blackberries, if we could put a quantity of them in a bag, shake them well up, draw out a sample and examine them to see what proportion of them had one arrangement and what proportion another. (2.684)

Single- versus Multiple-Universe Contexts. Figures 3.1 and 3.2 illustrate the distinction between the situation Peirce regards scientists as being in and one where universes are "as plenty as blackberries." The first situation (fig. 3.1) involves just one universe or one urn. It contains, let us suppose, some fixed proportion of white balls.

The second situation (fig. 3.2) involves some (possibly infinite) number of urns, U_1, U_2, . . . , each with some proportion of white balls. Here universes, represented as urns, are "as plenty as blackberries." Consider a hypothesis about the single-universe context, hypothesis *H*: the proportion of white balls (in this one universe U_1) equals .6. *H* asserts that in a random selection from that urn, the probability of a white ball equals .6. Because there is just this one urn, hypothesis *H* either is or is not true about *it*. *H* is true either 100 percent of the time or 0 percent of the time. The only probabilities that could be assigned to *H* itself are the trivial ones, 1 and 0.

Now consider the second situation, where there are many universes or urns. Hypothesis *H* may be true in some and not in others. If we can imagine reaching in and drawing out one of the universes, like selecting well-shaken blackberries from a bag, then it makes sense to talk about the probability that we will select a universe (urn) in which *H* is true. (A specific example to come.) But the second context is not our situation in science, says Peirce. Since our scientific hypotheses refer to just this one universe that we are in, like it or not, probabilities of such hypotheses cannot be regarded as "solid statistical facts."

By interpreting probabilities as subjective degrees of belief in hypotheses, however, it becomes meaningful to talk about nontrivial assignments of probabilities to hypotheses—even about this one universe. Until a hypothesis is known to be true or false by an agent, the agent may be supposed to have some quantitative assessment, between 0 and 1, of the strength of credibility one feels toward the hypothesis. Believing a certain "Big Bang" hypothesis to be very credible, for instance, you might assign it a degree of belief of .9.

We can now understand why the desire for a posterior probability measure, coupled with a single-universe context (as well as the rejection of logical probabilities), "seems to force" the subjective interpretation of the probability calculus, as Fisher alleged. Fisher, Peirce, Neyman, and Pearson, as well as contemporary frequentists, view the attempt to model quantitatively the strengths of opinion and their changes as useless for science. Thus except for contexts appropriately modeled as in figure 3.2—multiple urn experiments—they view theories of statistics appropriate for the second case as inappropriate for scientific inquiry into our one universe.[16]

An Illustration: A Multiple-Universe Context and Bayes's Theorem

It will be useful to have a very simple example of a context that *can* be modeled as a multiple-universe or multiple-urn context in which Bayes's theorem can be applied. It will also help clarify the notion of conditional probability.

a. Consider a game of chance, *rouge et noire:* You bet on either black or red, (randomly) select a card from the deck, and win if it is of a suit with your color. Let the possible outcomes be either "win" or "lose." Suppose that the probability of a win given that *rouge et noire* is played equals .5. We can abbreviate this sentence using probability notation:

$$P(\text{win} \mid rouge\ et\ noire) = \frac{1}{2}.$$

16. Some may regard Fisher's "fiducial probabilities" as falling outside this delineation, and they may be correct. This is no doubt bound up with the reason that such probabilities lead to inconsistencies.

For our purposes, conditional probability need not be technically explored. Grasp it by reading whatever comes after the "given bar" (|) as announcing the specific type of experiment, condition, or hypothesis to which you are restricted in considering the probability of the outcome of interest. With *rouge et noire* we are asserting that "the probability of winning, given that the experiment is a *rouge et noire* experiment, is one-half."

b. Now consider a second game, that of betting on 1 of 36 numbers in roulette (assume that no 0 or 00 outcomes are on the wheel). Let the probability of a win, given that the second game is played, equal $\frac{1}{36}$. We can write this as

$P(\text{win} \mid \text{single-number game}) = \frac{1}{36}.$

c. Now consider a third game, a sort of second-order game. A fair coin is tossed to decide whether to play the first or second game above. Say that "heads" results in *rouge et noir* being played, "tails," in the single-number roulette game. Then, with probability $\frac{1}{2}$, *rouge et noire* is played, and with probability $\frac{1}{2}$ the single-number game is played. Games *a* and *b* are like blackberries that we shake up in a bag and draw from. (Never mind why anyone would play this game!)

What has happened in the third game is that the game to be played *is itself* an outcome of a game of chance. That is, there are two outcomes: *"rouge et noire* is played" and "single-number roulette is played"—where it is given that these are the only two possibilities. We can write these as two "hypotheses":

H_1: *rouge et noire* is played.

H_2: single-number roulette is played.

Notice that here we have stipulated that the truth of these two hypotheses is determined by the outcome of a game of chance. Each hypothesis is true with probability $\frac{1}{2}$. That is, the context has two blackberries, H_1 and H_2, and each has equal chance of being drawn from the bag. Thus, we have two (perfectly objective) *unconditional* probabilities:

$P(H_1)$ (i.e., the probability that *rouge et noire* is played)

and

$P(H_2)$ (i.e., the probability that single-number roulette is played).

Further, we know the values of these two unconditional probabilities, because we have stipulated that they are each $\frac{1}{2}$.

$P(H_1) = P(H_2) = \frac{1}{2}.$

d. Now imagine that you are told the following: a woman who has gone through the game in (*c*), and played whatever game it selected for her, has won. What might be inferred about whether she won through playing *rouge et noire* (H_1 is true) or through single-number roulette (H_2 is true)? The prior (unconditional) probability in each is ½; but with this new information—the result was a win—we can update this probability and calculate the probability that the game played was *rouge et noire* given that it yielded a win. That is, we can calculate the (posterior) conditional probability

$P(H_1 \mid \text{win})$.

(We can likewise calculate $P[H_2 \mid \text{win}]$, but let us just do the first.)

The formula for this updating is *Bayes's theorem,* and in this case even one who insists on objective probabilities can use it. It just follows from the definition of conditional probability.[17]

$$P(H_1 \mid \text{win}) = \frac{P(\text{win} \mid H_1)\, P(H_1)}{P(\text{win} \mid H_1)P(H_1) + P(\text{win} \mid H_2)\, P(H_2)}.$$

Here, the needed probabilities for the computation are given. The prior probabilities are given, and from (*a*) and (*b*) we have

$P(\text{win} \mid H_1) = $ ½, and
$P(\text{win} \mid H_2) = $ ¹⁄₃₆.

The reader may want to calculate $P(H_1 \mid \text{win})$. The answer is ¹⁸⁄₁₉. So the evidence of a win gives a Bayesian "confirmation" of hypothesis H_1: the posterior probability exceeds the prior probability.

Bayes's Theorem

We can generalize this result. For an exhaustive set of disjoint hypotheses, $H_1, H_2, \ldots, H_n$, whose probabilities are not zero, and outcome *e* where $P(e) > 0$:

17. In general,

$$P(A \mid B) = \frac{P(A \text{ and } B)}{P(B)}.$$

So we have

$$P(H_1 \mid \text{win}) = \frac{P(H_1 \text{ and win})}{P(\text{win})},$$

and

$$P(\text{win}) = P(\text{win and } H_1) + P(\text{win and } H_2).$$

$$P(H_1 \mid e) = \frac{P(e \mid H_1)\, P(H_1)}{P(e \mid H_1)\, P(H_1) + P(e \mid H_2)\, P(H_2) + \ldots + P(e \mid H_n)\, P(H_n)}.$$

A Role for Opinions?

In a case like the above illustration, the truth of a hypothesis can be seen as an outcome of an experimental process, and it makes sense to talk about the probability of that outcome in the usual frequentist sense. Since it makes sense to talk about the probability of the outcome, it makes sense, in this special kind of case, to talk about the probability of the hypothesis being true. In such cases there is no philosophical problem with calculating the posterior probabilities using Bayes's theorem. *Except* for such contexts, however, the prior probabilities of the hypotheses are problematic. Given that logical probabilities will not do, the only thing left is subjective probabilities. For many, these are unwelcome in scientific inquiry. Not to subjectivists.

Subjectivists or personalists, by contrast, seem only too happy to announce, as Savage (1964) puts it, that

> the Bayesian outlook reinstates opinion in statistics—in the guise of the personal probabilities of events. (P. 178)

and that

> the concept of personal probability . . . seems to those of us who have worked with it an excellent model for the concept of opinion. (P. 182)

But whether personal probability succeeds well or badly in modeling opinion—something that is itself open to question—is beside the point for those who, like Peirce, Fisher, Neyman, and Pearson, see this kind of reliance on opinion as entirely wrongheaded for scientific inference. Knowledge of the world, many think, is best promoted by *excluding* so far as possible personal opinions, preferences, and biases. The arguments of Peirce in his day, and of Fisher, and Neyman and Pearson in theirs, remain the major grounds for rejecting the Bayesian approach in science.

Bayesians will object that it is impossible to exclude opinions, that at least the Bayesian brings them out rather than "sweeping them under the carpet," to paraphrase I. J. Good (1976). The charge of subjectivity leveled at non-Bayesian statistics will occupy us later (e.g., in chapters 5 and 11). I will argue that the type of arbitrariness in non-Bayesian error statistics is very different from that of subjective probabilities. The subjectivity of personalism, as Henry Kyburg (1993) has aptly put it, is particularly pernicious.

The Pernicious Subjectivity of Prior Probabilities

That scientists regularly start out with differing opinions in hypotheses is something the subjective Bayesian accepts and expects. Consequently, Bayesian consistency instructs agents to reach different posterior degrees of belief, even on the very same experimental evidence. What should be done in the face of such disagreement? Is there a way to tell who is right? Denis Lindley, also a father of modern Bayesianism, has this to say:

> I am often asked if the method gives the *right* answer: or, more particularly, how do you know if you have got the *right* prior. My reply is that I don't know what is meant by "right" in this context. The Bayesian theory is about *coherence*, not about right or wrong. (Lindley 1976, 359)

It is understandable that Lindley wonders what "right" can mean in the personalist context, for there is no reason to suppose that there is a correct degree of belief to hold. My opinions are my opinions and your opinions are yours. Without some way to criticize prior degrees of belief, it is hard to see how any criticism of your opinion can be warranted. If "right" lacks meaning, how can I say that you are in error? This leads to Kyburg's charge:

> This is almost a touchstone of objectivity: the possibility of error. There is no way I can be in error in my prior distribution for μ— unless I make a logical error—. . . . It is that very fact that makes this prior distribution perniciously subjective. It represents an assumption that has consequences, but cannot be corrected by criticism or further evidence. (Kyburg 1993, 147)

Of course one can change it. Kyburg's point is that even when my degree of belief changes on new evidence, it in no way shows my previous degree of belief to have been mistaken.

The subjectivity of the subjective Bayesian Way presents a major obstacle to its serving as an adequate model for scientific practice. Being right may be meaningless for a personalist, but in scientific contexts being right, avoiding specific errors, is generally well understood. Even where uncertainty exists, this understanding at least guides practitioners toward making progress in settling disagreements. And it guides them toward doing something right now, with the kind of evidence they can realistically obtain.

Swamping Out of Priors

The problem of accounting for consensus is not alleviated by the often heard promise that with sufficient additional evidence differences in prior probability are washed away. For one thing, these "washout theorems" assume that agents assign nonzero priors to the same set of hypotheses as well as agree on the other entries in the Bayesian algorithm. For another, they assume statistical hypotheses, while the Bayesian Way is intended to hold for any type of hypothesis. While some of these assumptions may be relaxed, the results about convergence are far less impressive.

Many excellent critical discussions of these points can be found in the literature.[18] Committed Bayesians will direct me to so-and-so's new theorem that extends convergence results. But these results, however mathematically interesting, are of no help with our problem. The real problem is not that convergence results hold only for very special circumstances; even where they hold they are beside the point. The possibility of eventual convergence of belief is irrelevant to the day-to-day problem of evaluating the evidential bearing of data in science.

Imagine two scientists reporting degrees of belief in H of .9 and .1, respectively. Would they find it helpful to know that with some amount of additional data their degree of belief assignments would differ by no more than a given amount? Would they not instead be inclined to dismiss reports of degrees of belief as irrelevant for evaluating evidence in science?[19]

John Earman (1992), despite his valiant efforts to combat the problems of the Bayesian Way, despairs of grounding objectivity via washout theorems:

> Scientists often agree that a particular bit of evidence supports one theory better than another or that a particular theory is better supported by one experimental finding than another. . . . What happens in the long or the short run when additional pieces of evidence are added is irrelevant to the explanation of shared judgments about the evidential value of present evidence. (P. 149)

What, then, explains the consensus about present evidence? Is the choice really, as Earman's title states, "Bayes or Bust"? I see this as a false choice. Science is not a bust. Yet scientists regularly settle or at

18. See, for example, Earman 1992 and Kyburg 1993. In the case where hypotheses are statistical and outcomes are independent and identically distributed, it is unexceptional that convergence can be expected. It is hard to imagine any theory of statistical inference not having such an asymptotic result for that special case (it follows from the laws of large numbers, chapter 5).

19. What is more ,the tables can be turned on the washout claims. As Kyburg (1993, 146) shows, for any body of evidence there are prior probabilities in a hypothesis H that, while nonextreme, will result in the two scientists having posterior probabilities in H that *differ* by as much as one wants.

least make progress with disputes about the import of evidence, and they do so with arguments and analyses based on non-Bayesian principles. The question of how to understand the evidence, in the jargon of chapter 2's Kuhnian analysis, regularly gives rise to a "normal research problem." It is tackled by reliable testing of low-level hypotheses about error.

Making Subjectivism Unimpeachably Objective

Howson and Urbach, staunch defenders of the subjective Bayesian faith, are unfazed by the limited value of the washout theorems, declaring them unnecessary to counter the charge of subjectivity in the first place. The charge, they claim, "is quite misconceived. It arises from a widespread failure to see the subjective Bayesian theory for what it is, a theory of inference. And as such, it is unimpeachably objective: though its subject matter, degrees of belief, is subjective, the rules of consistency imposed on them are not at all" (Howson and Urbach 1989, 290).

Howson and Urbach press an analogy with deductive logic. Just as deductive logic concerns theories of valid inferences from premises to conclusions where the truth of the premises is unknown, inductive logic concerns inferences from premises to some quantitative measure on the conclusion where the truth of the premises is unknown.

When Bayesians talk this way, they reveal just how deeply disparate their view of inductive inference is from what is sought by an account of ampliative inference or experimental learning. Although most Bayesians would not go as far as Howson and Urbach in calling the Bayesian approach "unimpeachably objective," all seem to endorse their analogy between inductive and deductive logic. As Kyburg (1993) has put it, neo-Bayesianism is "yet another effort to convert induction to deduction" (p. 150) in the form of a deductive calculus of probabilities.

A Fundamental Difference in Aims

This fundamental difference in their views of what an account of scientific inference should do has played too little of a role in the Bayes–non-Bayes controversy. Once we recognize that there is a big difference between the goals of a "deductive inductive" inference and what we seek from an ampliative account, we can agree to disagree with Bayesians on the goals of an account of scientific inference. This recognition has two consequences:

First it explains why Bayesian criticisms of non-Bayesian (standard error) statistics cut no ice with non-Bayesians. Such criticisms tend to show only that the latter fail to pass muster on Bayesian grounds. (Examples will occupy us later.) It is true that standard error statistics is "incoherent" according to the Bayesian definition. But Bayesian coherence is of no moment to error statisticians. At a 1970 conference on the foundations of statistics at the University of Waterloo, the statistician Irwin Bross put it bluntly:

> I want to take this opportunity to flatly repudiate the Principle of Coherence which, as I see it, has very little relevance to the statistical inference that is used in the sciences. . . . While we do want to be coherent in ordinary language, it is not necessary for us to be coherent in a jargon that we don't want to use anyway—say the jargon of L. J. Savage or Professor Lindley. (Bross 1971, 448)

This is not to say that all Bayesian criticisms of error statistics may be just dismissed; I will return to them later.

The second consequence of recognizing the difference in aims is more constructive. Conceding the limited scope of the Bayesian algorithm might free the Bayesian to concede that additional methods are needed, if only to fill out the Bayesian account. We will pursue this possibility as we proceed.

Can Bayesians Accept Evidence?

Perhaps the most obvious place for supplementary methods concerns the data or evidence. For just as with arriving at prior probabilities, the Bayesian response when asked about the grounds for accepting data is that it is not their job:

> The Bayesian theory we are proposing is a theory of inference from data; we say nothing about whether it is correct to accept the data. . . . The Bayesian theory of support is a theory of how the *acceptance as true of some evidential statement* affects your belief in some hypothesis. How you came to accept the truth of the evidence, and whether you are correct in accepting it as true, are matters which, from the point of view of the theory, are simply irrelevant. (Howson and Urbach 1989, 272; emphasis added)

The idea that we begin from the "acceptance as true of some evidential statement" is problematic for two reasons.[20] First, the Bayesian Way is to hold for any evidence and hypothesis, not just those associated with a specific statistical model. So the evidential statement e will

20. Nor does Richard Jeffrey's (1965) approach, in which the evidence need not be accepted as certain but is allowed to be merely probable, help us. See, for example, the discussion in Kyburg 1974, 118–22.

often be the type of claim that a theory of ampliative inference should help us to assess—not require us to begin with as given. In applying the Bayesian Way to classic episodes of hypothesis appraisal, for example, statements that are called upon to serve as evidence *e* include "Brownian motion exists" and "The estimated deflection of light is such and such."

This leads to the second problem with beginning with accepting the evidence statement as true—one that is more troubling. The standard Bayesian philosophy, after all, eschews acceptance of hypotheses (unless they have probability one), preferring instead to assign them degrees of belief. But why should it be all right to accept a statement when it plays the role of evidence and not when it plays the role of the hypothesis inferred? Now there are Bayesian accounts of the acceptance of hypotheses, but they do not help with our problem of accepting the evidence to get a Bayesian inference going.

Patrick Maher (1993a) proposes "a conception of acceptance that does not make acceptance irrational from a Bayesian perspective" (p. 154). Maher argues (1993a, 1993b), contrary to the general position of Bayesian philosophers of science, that a Bayesian account requires a theory of acceptance to be applicable to the history of science. For the history of science records the acceptance of claims, not scientists' probability assignments. Maher (1993b) argues that Bayesian philosophers of science, for example, Dorling, Franklin, and Howson, "operate with a tacit theory of acceptance" (p. 163)—one that identifies acceptance with high probability. Maher argues that a more adequate Bayesian theory of acceptance is a decision-theoretic one, where acceptance is a function both of probabilities of hypotheses and (cognitive) utilities. While Maher is to be credited for pointing up the shortcomings in Bayesian analyses of scientific episodes, his approach, as with all Bayesian decision approaches, only adds to the ingredients needed to get a Bayesian inference going.[21]

To return to the problem of accepting evidence claims, a solution could possibly be found by supplementing Bayesian algorithms with some separate—non-Bayesian—account for accepting evidential claims. Although the problems of arriving at prior probabilities and setting out alternative hypotheses (and their likelihoods) would still persist, such a supplement might at least offer reliable grounds for accepting the evidence. Still, this tactic would have at least one serious

---

21. It would take me too far afield to discuss the various decision-theoretic accounts, Bayesian and non-Bayesian, in this work. Maher (1993b) provides a good overview from the point of view of theories of acceptance.

drawback for Bayesians: the need for a supplementary account of evidence would belie one of the main selling points of the Bayesian approach—that it provides a single, unified account of scientific inference.

All You Need Is Bayes

Subjective Bayesians, especially leaders in the field, are remarkable for their ability to champion the Bayesian Way as the one true way (its many variants notwithstanding). If we take these Bayesians at their word, it appears that they view the Bayesian approach as the only account of inference (and perhaps decision) that we shall ever need. To solve the fundamental problems of inference and methodology, we need only to continue working out the details of the Bayesian paradigm. Consider Lindley's reasoning at the Waterloo conference:

> Now any decision that depends on the data that is being used in making the inference only requires from the data the posterior distribution. Consequently the problem of inference is effectively solved by stating the posterior distribution. This is the reason why I feel that the basic problem of inference is solved. (Lindley 1971, 436)

Even if it were true that stating the posterior solves the problem of inference, it would not follow that the Bayesian Way solves the problem of inference because it does not give one *the* posterior distribution. It gives, at best, a posterior distribution for a given agent reporting *that agent's* degree of belief at a given time. If we restate Lindley's claim to read that the basic problem of inference is solved by stating a given agent's degree of belief in a hypothesis, then I think we must conclude that Lindley's view of the basic problem of inference differs sharply from what scientists view it to be. Learning of an agent's posterior degree of belief, a scientist, it seems to me, would be interested only in what the evidence was for that belief and whether it was warranted. This calls for intersubjective tools for assessing the evidence and for adjudicating disagreements about hypothesis appraisal. This, subjective posteriors do not provide.

M. S. Bartlett, also a prominent statistician attending the Waterloo conference, had this response for Lindley:

> What does professor Lindley mean when he says that "the proof of the pudding is in the eating"? If he has done the cooking it is not surprising if he finds the pudding palatable, but what is his reply if we say that we do not. If the Bayesian allows some general investigation to check the frequency of errors committed . . . this might be

set up; but if the criterion is inner coherency, then to me this is not acceptable. (Bartlett 1971, 447).

Bartlett puts his finger on a central point over which the error statistician is at loggerheads with the Bayesian: the former's insistence on checking "the frequency of errors" or on *error probabilities*.[22] The centrality of the notion of error probabilities to non-Bayesian statisticians is why it is apt to call them error statisticians. To get a rough and ready idea of the error frequency check for which Bartlett is asking, imagine that the Bayesian agent reports a posterior degree of belief of .9 in hypothesis *H*. Bartlett, an error statistician, would require some way of checking how often such a high assignment would be expected to occur even if *H* is false. What for Bartlett would be necessary for the palatability of the Bayesian posterior, however, would, for the Bayesian, be quite irrelevant. Bayesian principles, as will be seen, conflict with error probability principles (chapter 10). Moreover, error probabilities call for an objective (frequentist[23]) notion of probability—while Bayesians, at least strict (or, to use Savage's [1964] term, "radical") ones, declare subjective probabilities to be all we need.

> I will confess . . . that I and some other Bayesians hold this [personal probability] to be the only valid concept of probability and, therefore, the only one needed in statistics, physics, or other applications of the idea. (Savage 1964, 183)

Scientists, as we shall shortly see, beg to differ. Scientific practice does not support the position that "all you need is Bayes," but the position held by the founders of non-Bayesian methods in the epigraphs to this chapter: when it comes to science, subjective Bayesianism is not needed at all.

Giere: Scientists Are Not Bayesian Agents

In reality, scientists do not proceed to appraise claims by explicit application of Bayesian methods. They do not, for example, report results by reporting their posterior probability assignments to one hypothesis compared with others—even dyed-in-the-wool Bayesians apparently grant this. Followers of the Bayesian Way do not seem very

22. Those who have investigated the error probabilities of Bayesian methods have found them to be problematic. See, for example, Giere 1969 and Kempthorne and Folks 1971, 304–7. I return to this in chapter 10.

23. Some, it is true (e.g., Giere), prefer propensities. I will not take up the problems with propensity definitions, but will argue for the appropriateness of frequentist statistics.

disturbed by this. One retort is that they are modeling only the ideally rational scientist, not the actual one. This type of defense is not very comfortable in the present climate where aprioristic philosophy of science is unwelcome. More modern Bayesians take a different tack. They view the Bayesian approach as a way to reconstruct actual scientific episodes and/or to model scientific judgments at some intuitive level, although it is not clear what the latter might mean.

Ronald Giere argues that empirical studies refute the claim that typical scientists are intuitive Bayesians and thereby count against the value of Bayesian reconstructions of science. Giere, a major (non-Bayesian) player in the philosophy of statistics debates of the 1970s, now declares that "we need not pursue this debate any further, for there is now overwhelming empirical evidence that no Bayesian model fits the thoughts or actions of real scientists" (Giere 1988, 149). But I do not think the debate is settled. To see why not we need to ask, What are these empirical studies?

The empirical studies refer to experiments conducted since the 1960s to assess how well people obey Bayes's theorem. These experiments, such as those performed by Daniel Kahneman, Paul Slovic, and Amos Tversky (1982), reveal substantial deviations from the Bayesian model even in simple cases where the prior probabilities are given, and even with statistically sophisticated subjects.

> Human beings are not naturally Bayesian information processors. And even considerable familiarity with probabilistic models seems not generally sufficient to overcome the natural judgment mechanisms, whatever they might be. (Giere 1988, 153)

Apparent success at Bayesian reconstructions of historical cases, Giere concludes, is mistaken or irrelevant.

> Scientists, as a matter of empirical fact, are not Bayesian agents. Reconstructions of actual scientific episodes along Bayesian lines can at most show that a Bayesian agent would have reached similar conclusions to those in fact reached by actual scientists. Any such reconstruction provides no explanation of what actually happened. (Giere 1988, 157)

Although I agree with the upshot of Giere's remarks, I do not claim that the probability experiments are what vitiate the Bayesian reconstructions. While interesting in their own right, these experiments seem to be the wrong place to look to test whether Bayes's theorem is a good model for scientific inference. Why? Because in these experiments the problem is *set up* to be one in which the task is calculating probabilities (whether of a posterior or of a conjunction of claims, or whatever). The experiments refer to classic games of chance or other setups where the needed probabilities are either given or assumed. The probabilities, moreover, refer to objective frequency calculations, not degrees of belief. Even with fully representative subjects, the results are at most relevant to how well people's intuitive judgments of probability obey the calculus of probabilities. They say nothing about whether scientists are engaged in attempting to assign probabilities to the hypotheses about which they inquire. If, as I, and error statisticians, urge, scientific inference is not a matter of assigning probabilities to hypotheses in the manner in which we would assign probabilities to outcomes of games of chance, then it is irrelevant whether subjects' judgments in these contexts accord well or badly with the probability calculus.

The finding of the probability experiments, that humans violate the probability calculus—*when asked to carry out a probability problem*—also says nothing about the methodology actually used in appraising scientific claims. For this we have to look at the kinds of experimental tools and arguments scientists use. One hardly does justice to inferences in science by describing them merely as violations of the Bayesian account. What one finds is a systematic pattern of statistical reasoning—but of the non-Bayesian sort I call standard error statistics. Familiar applications include the typical methods we hear about every day in reports of polling results and of studies on new drugs, cancer-causing substances, and the like. Contrary to what Giere had hoped, the debates concerning these methods still need to be pursued. Only now they should be pursued by considering actual experimental practice.

Where to look? Most classical cases of theory change are too fossilized to help much, unless detailed accounts of data analysis are available. Two such examples (Brownian motion and eclipse experiments) will be considered later. A rich source of examples of standard error statistics in experimental inference is the New Experimentalist narratives that we began discussing earlier. Our discussion now picks up where we left off in section 3.2.

3.4 THE NEW EXPERIMENTALISTS: EXPERIMENTAL PRACTICE IS NON-BAYESIAN

Regardless of what one thinks of the Bayesian Way's ability to reconstruct or model learning, looking at the tools actually used in the building up of knowledge reveals a use of probabilistic ideas quite unlike that of an after-trial sum-up of a theory's probable truth. I share the view of Oscar Kempthorne, a student of R. A. Fisher:

> It seems then that a use of "probability" as in "the probability that the theory of relativity is correct" does not really enter at all into the building up of knowledge. (Kempthorne and Folks 1971, 505)

One way to capture how probability considerations are used, I propose, is as tools for sustaining experimental arguments even in the absence of literal control and manipulation. The statistical ideas, as I see them, embody much of what has been learned about how limited information and errors lead us astray, as discussed in chapter 1. Using what has been learned about these mistakes, we have erected a conglomeration of interrelated tools that are good at practically forcing mistakes to show themselves, so to speak.

Galison: Neutral Currents

Let us now turn to the tools used in the trenches and brought to light in experimentalist work. Galison's 1987 work is especially congenial, and all references to him in this section refer to that work. Although Galison is not trying to draw any lessons for statistical philosophy and perhaps *because* he is not, his efforts to get at the arguments used to distinguish genuine effect from artifact effectively reveal the important roles served by error statistics. As Galison remarks, a key characteristic of twentieth-century experimental physics is "how much of the burden of experimental demonstration has shifted to data analysis" (p. 151) to distinguish signal from background. The increasingly central role played by data analysis makes the pronouncements of R. A. Fisher and E. S. Pearson (in the epigraphs to this chapter) as relevant today as in their own time.

I shall follow a portion of Galison's discussion of the discovery of neutral currents, thought to be one of the most significant in twentieth-century physics. By the end of the 1960s, Galison tells us, the "collective wisdom" was that there were no neutral currents. Bubble chamber evidence from many experiments indicated that neutral currents either did not exist or were well suppressed (pp. 164, 174). Soon after, however, from 1971 to 1974, "photographs . . . that at first appeared to be mere curiosities came to be seen as powerful evidence for" their existence (p. 135).

This episode, lasting from 1971 to 1974, occupies one-third of Galison's book, but my focus will be on the one analysis for which he provides the most detailed data. Abstracted from the whole story, this part cannot elucidate either the full theory at stake or the sociological context, but it can answer Galison's key question: "How did the experimentalists themselves come to believe that neutral currents existed? What persuaded them that they were looking at a real effect and not at an artifact of the machine or the environment?" (p. 136).

Here is the gist of their experimental analysis: Neutral currents are described as those neutrino events without muons. Experimental outcomes are described as muonless or muonful events, and the recorded result is the ratio of the number of muonless and muonful events. (This ratio is an example of what is meant by a statistic—a function of the outcome.) The main thing is that the more muonless events recorded, the more the result favors neutral currents. The worry is that recorded muonless events are due, not to neutral currents, but to inadequacies of the detection apparatus.

Experiments were conducted in collaboration by researchers from Harvard, Wisconsin, Pennsylvania, and Fermilab, the HWPF group. They recorded 54 muonless events and 56 muonful events, giving a ratio of 54/56. The question is, Does this provide evidence for the existence of neutral currents?

> For Rubbia [from Harvard] there was no question about the statistical significance of the effect . . . Rubbia emphasized that "the important question in my opinion is whether neutral currents exist or not. . . . The evidence we have is a 6-standard-deviation-effect." (P. 220)

The "important question" revolved around the question of the statistical significance of the effect. I will refer to it as the *significance question*. It is this:

> Given the assumption that the pre-Glashow-Weinberg-Salam theory of weak interactions is valid (no neutral currents), then what is the probability that HWPF would have an experiment with as many recorded muonless events as they did? (P. 220)

Three points need to be addressed: How might the probability in the significance question be interpreted? Why would one want to know it? and How might one get it?

Interpreting the Significance Question: What is being asked when one asks for the probability that the HWPF group would have an experiment with as many recorded muonless events as they did, given no neutral currents? In statistical language the question is, How (statistically) significant is the number of excess muonless events? The general concept of statistical significance will be taken up later (e.g., in chapter 5). Here I want to informally consider how it might be interpreted.

The experimental result, we said, was the recorded ratio of muonless to muonful events, namely, 54/56. The significance question, then, is, What is the probability that the HWPF group would get as many as (or more than) 54 muonless events, given the hypothesis that there are no neutral currents? The probability, notice, is not a probability of the hypothesis, it is the probability of a certain kind of experimental outcome or event. The event is that of recording *as large* a ratio of muonless to muonful events as the HWPF group did in this one experiment. It refers not only to this one experimental result, but to a set of results—54 *or more* muonless events. Wanted is the probability of the occurrence of this event given that there are no neutral currents. One way to cash out what is wanted is this: how often, in a series of experiments such as the one done by the HWPF group, would as many (or more) muonless events be expected to occur, given that there are no neutral currents?

But there is only one actual experimental result to be assessed, not a series of experiments. True, the series of experiments here is a kind of hypothetical construct. What we need to get at is why it is perceived as so useful to introduce this hypothetical construct into the data analysis.

What Is the Value of Answering the Significance Question? The quick answer is that it is an effective way of distinguishing real effects from artifacts. Were the experiment so well controlled that the only reason for failing to detect a muon is that the event is a genuine muonless one, then artifacts would not be a problem and this statistical construct would not be needed. But artifacts are a problem. From the start a good deal of attention was focused on the backgrounds that might fake neutral currents (p. 177). As is standard, one wants to assess the maximum amount of the effect for which such backgrounds are likely to be responsible and then "subtract them out" in some way. In this case, a major problem was escaping muons. "From the beginning of the HWPF neutral-current search, the principal worry was that a muon could escape detection in the muon spectrometer by exiting at a wide angle. The event would therefore look like a neutral-current event in which no muon was ever produced" (p. 217, fig. 4.40).

The problem, then, is to rule out a certain error: construing as a genuine muonless event one where the muon simply never made it to the spectrometer, and thus went undetected. To relate this problem to the significance question, let us introduce some abbreviations. If we let hypothesis *H* be

   H: neutral currents are responsible for (at least some of) the results,

then, *within this piece of data analysis,* the falsity of *H* is the artifact explanation:

H is false (the artifact explanation): recorded muonless events are due not to neutral currents, but to wide-angle muons escaping detection.

Our significance question becomes

What is the probability of a ratio (of muonless to muonful events) as great as 54/56, given that *H* is false?

The answer is the *statistical significance level* of the result.[24]

Returning to the relevance of knowing this probability, suppose it was found to be high. That is, suppose that as many or even more muonless events would occur frequently, say more often than not, even if *H* is false (and it is simply an artifact). What is being supposed is that a result as or even more favorable to *H* than the actual HWPF result is fairly common due not to neutral currents, but to wide-angle muons escaping detection. In that case, the HWPF result clearly does *not* provide grounds to rule out wide-angle muons as the source. Were one to take such a result as grounds for *H*, and for ruling out the artifact explanation, one would be wrong more often than not. That is, the probability of erroneously finding grounds for *H* would exceed .5. This would be a very unreliable way to proceed. Therefore, a result with a high significance level is an unreliable way to affirm *H*. Hence, results are not taken to indicate *H* unless the significance level is very low.

Suppose now that the significance level of the result is very low, say .01 or .001. This means that it is extremely improbable for so many muonless events to occur, if *H* were false and the HWPF group were really only observing the result of muons escaping. Since escaping muons could practically never be responsible for so many muonless events, their occurrence in the experiment is taken as good grounds for rejecting the artifact explanation. That is because, following an argument from error, the procedure is a highly reliable probe of the artifact explanation. This was the case in the HWPF experiment, although the significance level in that case was actually considerably smaller.

This result by itself is not grounds for *H*. Other experiments addressing this and other artifacts are needed. All I am showing, just now, is the relevance of answering the significance question for ruling out an artifact. But how do you get the probability needed for this answer?

How Is the Significance Question Answered? The reasoning just described does not require a precise value of the probability. It is enough to know that it is or is not extremely low. But how does one arrive at even a ballpark figure? The answer comes from the use of various canonical statistical analyses, but to apply them (even qualitatively) requires in-

24. Here the "null hypothesis" is that *H* is false (i.e., not-*H*).

formation about how the artifact in question could be responsible for certain experimental results. Statistical analyses are rather magical, but they do not come from thin air. They send the researcher back for domain-specific information. Let us see what the HWPF group did.

The data used in the HWPF paper are as follows (p. 220):

Visible muon events	56
No visible muon events	54
Calculated muonless events	24
Excess	30
Statistical significant deviation	5.1

The first two entries just record the HWPF result. What about the third entry, the calculated number of muonless events? This entry refers to the number calculated or expected to occur because of escaping muons. Where does that calculation come from? It comes from separate work deliberately carried out to find out how an event can wind up being recorded "muonless," not because no muon was produced (as would be the case in neutral currents), but because the muon never made it to the detection instrument.

The group from Harvard, for example, created a computer simulation to model statistically how muons could escape detection by the spectrometer by exiting at a wide angle. This is an example of what is called a "Monte Carlo" program.

> By comparing the number of muons expected not to reach the muon spectrometer with the number of measured muonless events, they could determine if there was a statistically significant excess of neutral candidates. (P. 217)

In short, the Monte Carlo simulation afforded a way (not the only way) of answering the significance question.

The reason probability arises in this part of the analysis is not because the hypothesis about neutral currents is a statistical one, much less because it quantifies credibility in H or in not-H. Probabilistic considerations are deliberately *introduced* into the data analysis because they offer a way to model the expected effect of the artifact (escaping muons). Statistical considerations—we might call them "manipulations on paper" (or on computer)—afford a way to subtract out background factors that cannot literally be controlled for. In several places, Galison brings out what I have in mind:

> One way to recapture the lost ability to manipulate the big machines has been to simulate their behavior on a computer. In a sense the computer simulation allows the experimentalist to see, at least

through the eye of the central processor, *what would happen* if a larger spark chamber were on the floor, if a shield were thicker, or if the multiton concrete walls were removed. (P. 265; emphasis added)

The Monte Carlo program can do even more. It can simulate situations that *could never exist in nature. . . .* Such altered universes do work for the experimentalist. One part of the Gargamelle demonstration functioned this way: suppose the world had only charged-current neutrino interactions. How many neutral-current *candidates* would there be? Where (statistically) would they be in the chamber? (Ibid.)

Returning to the specific analysis, it was calculated that 24 muonless events would be expected in the HWPF experiment due to escaping muons. This gives the number expected to be misinterpreted as genuinely muonless.

They wanted to know how likely it was that the observed ratio of muonless to muon-ful events (54/56) would fall within the statistical spread of the calculated ratio (24/56), due entirely to wide-angle muons. (P. 220)

They wanted to "display the probability" (as the report put it) that the difference between the number of observed and expected muonless events was merely an ordinary chance fluctuation. The difference between the ratio observed and the ratio expected (due to the artifact) is 54/56 − 24/56 = 0.536. How improbable is such a difference even if the HWPF group were experimenting on a process where the artifact explanation was true (i.e., where recorded muonless events were due to escaping muons)? This is "the significance question" again, and finally we can answer it.

What would it be like if the HWPF study actually was an experiment on a process where the artifact explanation is true? The simulation lets us model the relevant features of what it would be like: it would be like experimenting on (or sampling from) a process that generates ratios (of m events to m-less events) where the average (and the most likely) ratio is 24/56. (This corresponds to the hypothetical sequence of experiments we spoke of.) This value is just an average, however, so some experiments would yield greater ratios, others smaller ratios. Most experiments would yield ratios close to the average (24/56); the vast majority would be within two standard deviations of it. The statistical model tells us how probable different observed ratios are, given that the average ratio is 24/56.[25] In other words, the

25. Of course, this would be correct only if the tests were at least approximately independent.

statistical model tells us what it would be like to experiment on a process where the artifact explanation is true; namely, certain outcomes (observed ratios) would occur with certain probabilities. In short, information about "what it would be like" is given by "displaying" an *experimental distribution.*

Putting an observed difference between recorded and expected ratios in standard deviation units allows one to use a chart to read off the corresponding probability. The standard deviation (generally only estimated) gives just that—a standard unit of deviation that allows the same standard scale to be used with lots of different problems (with similar error distributions). Any difference exceeding two or more standard deviation units corresponds to one that is improbably large (occurring less than 3 percent of the time).

Approximating the standard deviation of the observed ratio shows the observed difference to be 5.1 standard deviations.[26] This observed difference is so improbable as to be off the charts; so, clearly, by significance test reasoning, the observed difference indicates that the artifact explanation is untenable. It is practically impossible for so many muonless events to have been recorded, had they been due to the artifact of wide angle muons. The procedure is a reliable artifact probe.

This analysis is just one small part of a series of experimental arguments that took years to build up.[27] Each involved this kind of statistical data analysis to distinguish real effects or signals from artifacts and to rule out key errors piecemeal. They are put together to form the experimental arguments that showed the experiment could end. I would be seriously misunderstood if I were taken as suggesting that the substantive inference is settled on the basis of a single such analysis. Nothing could be further from my intent.

As Galison points out, by analyzing the HWPF data in a different manner, in effect, by posing a different question, the same data were seen to yield a different level of statistical significance—still highly significant. The error statistics approach does not mandate one best approach in each case. Its principal value is that it allows different analyses to be understood and scrutinized. Galison's excellent narration of this episode reveals a hodgepodge of different results both on the same and different data, by the same and different researchers at different

---

26. The standard deviation is estimated using the recorded result and a standard statistical model. It equals $\dfrac{24}{56}\sqrt{\dfrac{1}{24} + \dfrac{1}{56}} = 0.105$ (Galison 1987, 220–21).

27. The recent inference to the identification of so-called top quarks followed an analogous pattern.

times in different labs. This calls for just the kinds of tools contained in a tool kit of error statistics.

Some Contrasts with the Bayesian Model

The Bayesian model is neater, but it does not fit the actual procedure of inquiry. The Bayesian model requires the researchers to start out with their degrees of belief in neutral currents and then update them via Bayes's theorem. It also requires assessing their strength of belief in all the other hypotheses that might explain some experimental result, such as the artifact explanation of escaping muons. The researchers did not do this.

As Galison shows, different institutes at different times came up with different estimates of parameters of interest. No one is surprised by this, and, more importantly, the researchers can use these reports as the basis for criticism and further work. Imagine, in contrast, different institutes reporting their various posterior degrees of belief in H, neutral currents. Suppose one institute reports that the degree of belief in H is low. Lindley says all the information resides in an agent's posterior probability. But it is not clear what other institutes could make of this. For one thing, one would not know whether it was due to a highly discrepant result or a small prior degree of belief. A two-standard-deviation difference and a ten-standard-deviation difference indicate different things to the practitioner, but they could both very well yield an identical (low) posterior. The posterior would not have provided the information the researchers actually used to learn things such as what to do next, where the source of error is likely to lie, how to combine it with other results, or how well the data accord with the model.

Bayesians, at least officially, reject the use of significance tests and other error probability methods. (Indeed, it is hard to see how one can be a consistent Bayesian and *not* reject them. See chapter 10.) Howson and Urbach (1989), following the Bayesian fathers cited earlier, maintain that "the support enjoyed by [error statistics methods] . . . among statisticians is unwarranted" (p. 198). They declare that one of the staples of the experimenter's tool kit for assessing "goodness of fit" for a model to data (the chi-square test) "should be discarded" (p. 136)! Their criticisms, to be taken up later, stem from the fact that error statistics methods aim to perform a very different role from the one envisaged in the Bayesian model of inductive inference.

Error probabilities are not final evidential-relation measures in hypotheses. However, error probabilities of the experiment from which a claim is arrived at perform a much valued service in experiments. They provide for an objective communication of the evidence and for debate over the reasons a given claim was reached. They indicate what experiments have been performed and the process by which the estimate or result came about. They can be checked by experimenting with a different type of test. Scientists obviously find such information valuable.[28] Their value is as part of an iterative and messy series of small-scale probes using a hodgepodge of ready-to-use and easy-to-check methods. (Much like ready-to-wear [versus designer] clothes, these "off the shelf" methods do not require collecting vast resources before you can get going with them.) They will not appeal to ultra neatnicks.

By working with the data and arguments of specific cases, however, it is possible to see how the messiness of a host of piecemeal analyses gives way to rather neat strategies. The ingredients, for at least several important cases, I maintain, are already available in the works of the New Experimentalists. This is so even for Allan Franklin's work, despite his appeal to the Bayesian Way in his proposed epistemology of experiment, for his extensive examples reveal page after page of error statistics. Separate from these experimental narratives, Franklin attempts to give a Bayesian gloss to the experimental strategies he so aptly reveals. In doing so, the actual epistemological rationale of those strategies gets lost.

Scientists Are Bayesians in Disguise (and Artists Paint by Number)

Even where the use of error-statistical methods is of indisputable value, the ardent Bayesian still withholds credit from them. What the Bayesian would have us believe is that the methods used are really disguised attempts to apply Bayes's theorem, and the Bayesian will happily show you the priors that would give the same result. We may grant that experimental inferences, once complete, may be reconstructed so as to be seen as applications of Bayesian methods—even though that would be stretching it in many cases. My point is that the inferences actually made are applications of standard non-Bayesian methods. That an after-the-fact Bayesian reconstruction is possible provides no reason to think that if the researchers had started out only with Bayesian tools they would have reached the result they did. The point may be made with an analogy. Imagine the following conversation:

Paint-by-number artist to Leonardo Da Vinci: I can show that the *Mona Lisa* may be seen as the result of following a certain paint-by-number kit that

---

28. This is what endears these methods to practitioners. See, for example, Lucien LeCam 1977 and B. Efron 1986.

I can devise. Whether you know it or not you are really a painter by number.

Da Vinci: But you devised your paint-by-number *Mona Lisa* only by starting with my painting, and I assure you I did not create it by means of a paint-by-number algorithm. Your ability to do this in no way shows that the paint-by-number method is a good way to produce new art. If I were required to have a paint-by-number algorithm before beginning to paint, I would not have arrived at my beautiful *Mona Lisa*.

CHAPTER FOUR

Duhem, Kuhn, and Bayes

DEFENDERS of the Bayesian Way can and do argue that even if scientists are not conscious or unconscious Bayesians, reconstructing scientific inference in Bayesian terms is of value in solving key problems of philosophy of science. In this chapter I will consider how Bayesian reconstructions have been used to grapple with Duhem's problem, and to bridge the logical empiricist approach to confirmation with the historicist approach promoted by Kuhn. In both cases I will argue that if the goal is solving rather than reconstructing problems, then the Bayesian Way comes up short.

4.1 THE BAYESIAN WAY OUT OF THE DUHEM PROBLEM

The problem for which the Bayesian Way is most often touted as scoring an impressive success is the Duhem problem—the problem of which of a group of hypotheses used to derive a prediction should be rejected when experiment disagrees with that prediction. Although I will argue that the Bayesian Way out of Duhem's problem is really no way out at all, my aim is not primarily negative. Rather, my hope is to lay the groundwork for a satisfactory non-Bayesian approach to the problem based on error statistics.

Some philosophers of science dismiss the Duhem problem as the product of old-fashioned (hypothetico-deductive) philosophy of science and therefore not really an issue for New Experimentalists. What Duhem's problem shows, strictly speaking, is that logic alone permits an anomalous result to be blamed not on the primary hypothesis being tested, but on the host of auxiliary principles and hypotheses involved in testing. And we know formal logic is not all we have at our disposal. But the problem that still remains is to show that there are good grounds for localizing the bearing of evidence. If an inference account cannot at least make headway toward showing which assignment of error is warranted, it cannot be seen to have gotten around the Duhem problem in its modern guise.

Lakatos, we saw, attempted to improve on Popper in the light of

Duhem's problem as brought home by Kuhn. For Lakatos, anomalies are blamed on suitable auxiliary hypotheses, hard-core theories remaining protected. But he conceded that any hard-core theory can be defended "progressively" this way. Bayesians believe that they have a more adequate solution to Duhem's problem, that "the questions left unanswered by Lakatos are answered with the help of Bayes's theorem" (Howson and Urbach 1989, 96). They and other Bayesians appeal to the Bayesian strategy of Jon Dorling (1979), which I will outline shortly. In the section "The Duhem Problem Solved by Bayesian Means" (p. 96), Howson and Urbach declare just that. Let us see what they mean.

The Duhem Problem Solved by Bayesian Means

When Bayesians say they can solve Duhem's problem, what they mean is this: Give me a case in which an anomaly is taken to refute a hypothesis H out of a group of hypotheses used to derive the prediction, and I'll show you how certain prior probability assignments can justify doing so. The "justification" is that H gets a low (or lower) posterior probability than the other hypotheses. As with the general Bayesian Way of explaining a scientific episode, solving Duhem comes down to a homework assignment—not to say a necessarily easy one—of how various assumptions and priors allow the scientific inference reached to be in accord with that reached via Bayes's theorem.

In addition to accounting for specific episodes, the Bayesian Way can be used to derive a set of general statements of the probabilistic relationships that would have to hold for one or another parceling out of the blame. These equations are neat, and the algorithms they offer for solving such homework problems are interesting. What they do not provide, however, is a solution to Duhem's problem. Duhem's problem, as Howson and Urbach themselves say, is to determine "which of the several distinct theories involved in deriving a false prediction should be regarded as the false element" (Howson and Urbach 1989, 94). The possibility of a degree of belief reconstruction does not help to pinpoint which element ought to be regarded as false.

From all we have already seen, we might expect the subjective Bayesian to retort that I am misunderstanding the subjectivist account. For the subjective Bayesian, the hypotheses an agent ought to consider disconfirmed *are* the ones with low posterior probabilities, and these follow deductively from the agent's prior degrees of belief (and other subjective probabilities), which agents are assumed to have. That is what a subjectivist means by an inference being rational. Dorling (1979), to his credit, admits as much. He says that adopting a personalist reconstruction "automatically" yields a resolution of Duhem, but quite correctly stresses "that it is the adoption of a *personalist* Bayesianism which yields this way out of the Duhem problem" (p. 178). The question that remains is whether to adopt the Bayesian Way out is really to have a way out of the problem. Not, I claim, if the problem is understood normatively. What the Bayesian Way offers, at best, is a way of reconstructing given inferences probabilistically. The Duhem problem, if it is not simply defined away, just returns as the problem of justifying the correctness of the probabilities in the Bayesian equations.

Since Dorling's work is credited as the exemplar for the Bayesian solution to Duhem, I will take it as my example too.

Dorling's Homework Problem

Dorling considers a situation where despite the fact that an anomalous result e' occurs, the blame is placed on an auxiliary hypothesis A while the credibility placed on theory T is barely diminished. In Dorling's simplified problem, only one auxiliary hypothesis A is considered (I am replacing his H with A).

In the historical case considered here, Dorling (1979, 178) takes T to be "the relevant part of solidly established Newtonian theory which Adams and Laplace used" to compute e, the predicted secular acceleration of the moon, which conflicted with the observed result e'. The auxiliary, A, is the hypothesis that the effects of tidal friction are not of a sufficient order of magnitude to affect appreciably the lunar acceleration.

Dorling's homework problem is to provide probability assignments so that, in accordance with the episode, an agent's credibility in theory T is little diminished by the anomaly e', while the credibility in auxiliary A is greatly diminished. We can sidestep the numerical gymnastics to get a feel for one type of context where the agent faults A. Afterwards I will give a numerical algorithm (calculated at the end of section 4.1).

Theory T and auxiliary A entail e, but e' is observed. When might e' blame A far more than T? Here's one scenario sketched in terms that I intend to be neutral between accounts of inference. Suppose (1) there is a lot of evidence for theory T, whereas (2) there is hardly more evidence for the truth of auxiliary hypothesis A than for its falsity. Suppose, further, that (3) unless A is false, there is no other way to explain e'. This is a rough account, it seems to me, of a situation where e' indicates (or is best explained by) A being in error.[1]

---

1. A more extreme situation would give a very low prior probability to hypothesis A. Dorling is trying to describe a case where it is not so obvious how things come out.

A Bayesian rendering may be effected by inserting "agent x believes that" prior to assertions 1, 2, and 3. We then have a description of a circumstance where the agent believes or decides that A is discredited by e'. Nothing is said about whether the assignments are warranted, or, more importantly, how a scientist should go about determining where the error really lies. Assigning the probabilities differently puts blame elsewhere, and the Bayesian "solution" is not a solution for adjudicating such assignments.

The Numerical Solution to the Homework Assignment

The numerical "solution" that corresponds to what I described above is this: The scientist's degree of belief is such that a high degree of belief is initially accorded to T (e.g., $P(T) = .9$); in any case, it is substantially more probable than A, which is considered only slightly more probable than not (e.g., $P(A) = .6$). These numbers are introduced by the personalist, Dorling explains, as approximate descriptions of the belief state of a particular scientist at the time. Let us see how we might describe the agent's beliefs so that the third and key assumption is cashed out probabilistically.

First, imagine the agent considering the possibility that auxiliary hypothesis A is true:

The agent contemplates auxiliary A *being true.* Clearly, T could not also be true (since together they counterpredict e'). But might not some rival to T explain e'? Here is where the key assumption enters. The agent believes there to be no plausible rival that predicts e'. That is to say, the agent sees no rival that, in his or her opinion, has any plausibility, that would make anomaly e' expected. In subjective probability terms, this becomes

  a. The probability of e', given that A holds and T is false, is very small. Let this very small value be ε.

Since the anomaly e' has been observed, it might seem that the agent would assign it a probability of 1. Doing so would have serious ramifications (i.e., this is the "old-evidence problem"). To avoid assigning degree of belief 1 to e', Bayesian agents need to imagine how strongly they *would have believed* in the occurrence of anomaly e' *before* it was observed—no mean feat. But never mind the difficulties in assigning such probabilities just now (see chapter 10). The Bayesian assumes that the agent can and does make the key assumption that, in the agent's view, the e' observed is extremely improbable if A is true. Now consider the agent's beliefs assuming that auxiliary A is false.

The agent contemplates auxiliary A *being false.* In contrast, if auxiliary A were false, the agent would find e' much more likely than if T were false and A true. In fact, Dorling imagines that scientists assign a probability to e', given that A is false, 50 times as high as that in (a), whether or not T is true. That is, $P(e' \mid A \text{ is false}) = 50\varepsilon$. We have

    $b.$ i. The probability of e', given that T holds and A is false, is 50ε.

         ii. The probability of e', given that T is false and A is false, is 50ε.

Of course, (i) and (ii) need not be exactly equal, but what they must yield together is a probability of e' given A is false many times that in (a). A further assumption, it should be noted, is that T and A are independent.

Together, (a) and (b) describe a situation where the outcome e' is believed to be far more likely if A is false than if A is true. This yields assumption 3. The result is that the posterior probability of T remains rather high, that is, .897, while the posterior of A becomes very low, dropping from .6 to .003.

This gives one algorithm—Dorling's—for how evidence can yield a Bayesian disconfirmation of auxiliary A, despite A's being deemed reasonably plausible at the start. Nonquantitatively put, the algorithm for solving the homework problem is this: Start with a suitably high degree of belief in T as compared with A, believe no plausible rival to T exists that would make you expect the anomalous result, and hold that the falsity of A renders e' many times more expected than does any plausible rival to T.

Reconstructing versus Solving Duhem

Dorling's homework problem can be done in reverse. Scientists who assign the above degrees of belief, but with A substituted for T, reach the opposite conclusion about T and A. So being able to give a Bayesian retelling does not, by itself, say which apportionment of blame is warranted.

Bayesians may retort that the probabilities stipulated in their reconstruction are plausible descriptions of the beliefs actually held at the time, and others are not. That may well be, though it is largely due to the special way in which they describe the prediction. I leave that to one side. For my own part, I have no idea about the odds "a typical non-Newtonian would have been willing to place [on] a bet on the correct quantitative value of the effect, in advance even of its qualitative discovery" (Dorling 1979, 182). (Something like this is the contortion required to get around assigning e' a probability of 1.)

Nor is it easy to justify the prior probability assignments needed to solve the homework problem, in particular, that theory T is given a prior probability of .9. The "tempered personalism" of Abner Shimony (e.g., 1970) advises that fairly low prior probabilities be assigned to hypotheses being considered, to leave a fairly high probability for their denial—for the "catchall" of other hypotheses not yet considered. The Dorling assignment leaves only .1 for the catchall hypothesis.

A Highly Qualified Success? If Bayesian reconstructions fail to count as solving Duhem, it seems fair to ask what value such reconstructions might have. Bayesians apparently find them useful. John Earman, for example, shares my position that the Bayesian Way is no solution to Duhem. While calling it a "highly qualified success for Bayesianism," Earman finds that "the apparatus provides for an illuminating representation of the Quine and Duhem problem" (Earman 1992, 85). For my part, I find the problem stated by Duhem (1954) clear enough— how to determine the error responsible:

> The only thing the experiment teaches us is that among the propositions used to predict the phenomenon . . . there is at least one error; but where this error lies is just what it does not tell us. The physicist may declare that this error is contained in exactly the proposition he wishes to refute, but is he sure it is not in another proposition? If he is, he accepts implicitly the accuracy of all the other propositions he has used, and the validity of his conclusion is as great as the validity of his confidence. (Duhem 1954, 185)

This last clause can be put in Bayesian terms by replacing "the validity of his confidence" with "the validity of his prior and other degree of belief assignments." But I do not see how attaching a degree of belief phrase to the claims in Duhem's statement helps to illuminate the matter. Indeed, attaching probabilities to statements only complicates things.

But there is something that might be said about the Bayesian reconstructions that may explain why philosophers find them appealing to begin with. A purely syntactical theory of confirmation along the lines of a hypothetico-deductive account seems to lack a way to account for differential assignments of blame for an anomaly.[2] Two different cases may go over as the same syntactical configuration, even though our intuitions tell us that in one case the primary hypothesis is discredited while in the other the auxiliary is. The complaint against syntactical approaches is correct. But this shows only that syntax alone won't do and that substantive background knowledge is needed. What

---

2. Even Glymour's bootstrapping version, Earman argues, seems to have no way to solve it.

it does not show, and what I have been urging we should deny, is that the background should come in by way of subjective degrees of belief.

A Sign of Being a Correct Account? Howson and Urbach follow Dorling's treatment in their own example, giving assignments very similar to, though less striking than, Dorling's. The ability of the Bayesian model to accord with actual cases of attributing blame, they conclude, shows that "Bayes's Theorem provides a model to account for the kind of scientific reasoning that gave rise to the Duhem problem" (Howson and Urbach 1989, 101). If this just means that there are Bayesian reconstructions of the sort we have been considering, then we can agree. However, Howson and Urbach go on to claim that the ability to give a Bayesian reconstruction of cases shows "that the Bayesian model is essentially correct"! (p. 101). But merely being able to offer reconstructions of episodes says nothing about the Bayesian model's correctness.[3]

If the name of the game is reconstruction, it is quite simple to offer a non-Bayesian one. How would our error-testing model reconstruct an episode where an auxiliary A is blamed, rather than theory T? We would want to distinguish between two cases: (*a*) the case where there are positive grounds for attributing the error to auxiliary A, and (*b*) the case where there are simply inadequate grounds for saying an error in A is absent. In coming out with a posterior probability of .003 in A, Dorling is describing the case as (*a*), yet anomaly e' itself seems at best to warrant regarding it as in case (*b*)—where there is simply not enough information to attribute blame to T.

An error-statistical description of the episode might go like this: Theory T is not shown to be in error as a result of anomaly e' unless the evidence warrants ruling out the possibility that an error in auxiliary hypothesis A is responsible. Evidence does not warrant ruling out an error in auxiliary A unless A has been shown to pass a sufficiently severe test. But the assumption of lukewarm evidence for A (reconstructed by Dorling as its having a prior probability of .6) would be taken as denying that A had passed a severe test. This explains why e' was not taken to discredit T. To take e' as grounds for condemning T would be to follow a very unreliable procedure. To reconstruct an episode as a case of (*a*), in contrast, would require there to be positive grounds to consider A false, and its falsity to blame for the anomaly. In that case what must have passed a severe test is hypothesis "not-A":

3. Similar criticisms of the Bayesian solution to Duhem are raised by Worrall (1993).

that the extraneous factor (tidal friction) is responsible for the anomalous effect (lunar acceleration).

My error statistics reconstruction enjoys several advantages over the Bayesian one: First, it does not suppose that for any anomaly there is some inference to be reached about where to lay the blame. The description in (b) may be all that would be allowed until positive grounds for fingering an auxiliary were obtained. Second, the question whether there are grounds for an error in A does not turn on opinions in T and there is no need to imagine having a prior in all the other possible theories (i.e., the so-called catchall). This second reason leads to a third, which is what allows us to go beyond mere reconstruction: unlike the probabilities needed for the Bayesian reconstruction, philosophers do not have to invent the components we need in depicting the scientific inference, nor work with make-believe calculations (e.g., imagining the odds scientists would place if they did not already know the evidence).

There seems to be no suggestion, even by Bayesians, that scientists actually apply Bayes's theorem in reaching their conclusion. Most important, the Bayesian description fails to capture how Duhemian problems are actually grappled with before they are solved. Adjudicating disputes with a measure of objectivity calls for methods that can actually help to determine whether given auxiliaries are responsible for the anomaly. Scientists do not succeed in justifying a claim that an anomaly is due to an auxiliary hypothesis by showing how their degrees of subjective belief brought them there. Were they to attempt to do so, they undoubtedly would be told to go out and muster evidence for their claim, and in so doing, it is to non-Bayesian methods that they would turn.

What's Belief Got to Do with It?

Howson and Urbach (1989) state, without argument, that "by contrast [with the Bayesian model], non-probabilistic theories seem to lack entirely the resources that could deal with Duhem's problem" (p. 101) where "non-probabilistic theories" include the error statistics methods of Fisher and Neyman and Pearson. In truth, these methods contain just the resources that are needed and regularly relied upon to solve real-life Duhemian problems.

A major virtue of the error statistics approach is that the issue of whether the primary or auxiliary hypothesis is discredited is not based on the relative credence accorded to each. The experiment is supposed to find out about these hypotheses; it would only bias things to make interpreting the evidence depend on antecedent opinions. After all, in

Dorling's examples, and I agree that the assumption is plausible, theory *T* is assumed to be *independent* of auxiliary *A*. There is no reason to suppose that assessing auxiliary *A* should depend at all on one's opinion about *T*. What is called for are separate researches to detect whether specific auxiliaries are responsible for observed anomalies.

Let me allude to an example to be considered later (chapter 8). When one of the results of the 1919 eclipse experiments on the deflection of light disagreed with Einstein's prediction, there was a lengthy debate about whether the anomaly should be attributed to certain distortions of the mirror, to Einstein's theory, or to something else. The debate over where to lay the blame was engaged in by scientists with very different opinions about Einstein's theory. Such attitudes were no part of the arguments deemed relevant for the question at hand. The relevant argument, put forth by Sir Arthur Eddington (and others), turned on a rather esoteric piece of data analysis showing (holdouts notwithstanding) that the mirror distortion was implicated.

Eddington believed in the correctness of Einstein's account, but nobody cared how strongly Eddington believed in Einstein. Quite the contrary—it only made those who favored a Newtonian explanation that much more suspicious of Eddington's suggestion that the faulty mirror, not Einstein's account, was to blame. Being an ardent proponent of either of the two rivals entered the debate: it explained the lengths to which players in the debate were willing to go to scrutinize each other's arguments. But ardor did not enter into the *evidential appraisal* of the hypotheses involved.

The argument to blame an auxiliary such as the mirror is the flip side of the argument to rule out an artifact. Here the anomalous effect may be shown to go away when there is no distortion of the lens. Additional positive arguments that the lens was the culprit were given, but I will save those for later.

Ronald Giere (1988) suggests a "technological fix for the Duhem-Quine problem" (p. 138), observing that often auxiliary hypotheses are embodied in instruments, and "Scientists' knowledge of the technology used in experimentation is far more reliable than their knowledge of the subject matter of their experiments" (p. 139). My position for solving Duhem extends this technological fix to include any experimental tool. It is the reliability of experimental knowledge in general, the repertoire of errors and strategies for getting around them, that allows checking auxiliaries, and for doing so quite apart from the primary subject matter of experiments.

When it comes to finding out which auxiliaries ought to be blamed, and to adjudicating disputes about such matters, error statistics provides forward-looking methods to turn to. I do not claim that scientists will always be able to probe the needed errors successfully. My claim is that scientists do regularly tackle and often enough solve Duhemian problems, and that they do so by means of error statistical reasoning. Once we have set out the ingredients of an experimental framework (in chapter 5) we will see more clearly how an inquiry may be broken down so that each hypothesis is a local assertion about a particular error. There, and again in later chapters (e.g., chapters 6 and 13) we will return to Duhem's problem.

In the following subsection, I summarize the calculations that yield the results of Dorling's homework problem.

Calculations for the Homework Problem:
  BACKGROUND ASSUMPTIONS:
    Hypotheses A and T entail e, but e' is observed: $P(e' \mid A \text{ and } T) = 0$. A and T are statistically independent
  ASSUMED PRIOR PROBABILITIES
    $P(T) = .9$, $P(A) = .6$.
  ASSUMED LIKELIHOODS:
    a. $P(e' \mid A \text{ and } {\sim}T) = \varepsilon$ (very small number, e.g., .001).
    b. i. $P(e' \mid {\sim}A \text{ and } T) = 50\varepsilon$
       ii. $P(e' \mid {\sim}A \text{ and } {\sim}T) = 50\varepsilon$

Bayes's theorem:

$$P(T \mid e') = \frac{P(e' \mid T)\, P(T)}{P(e')}$$

From the above we get the following:

$$P(e') = P(e' \mid T)P(T) + P(e' \mid {\sim}T)P({\sim}T).$$

$$\begin{aligned} P(e' \mid T) &= P(e' \mid A \text{ and } T)\, P(A) + P(e' \mid {\sim}A \text{ and } T)P({\sim}A) \\ &= \quad 0 \qquad\qquad + 50\varepsilon(.4) \\ &= 20\varepsilon. \end{aligned}$$

$$\begin{aligned} P(e' \mid {\sim}T) &= P(e' \mid A \text{ and } {\sim}T)P(A) + P(e' \mid {\sim}A \text{ and } {\sim}T)P({\sim}A) \\ &= \varepsilon(.6) \qquad\qquad + 50\varepsilon(.4) \\ &= 20.6\varepsilon. \end{aligned}$$

So

$$\begin{aligned} P(e') &= 20\varepsilon(.9) + 2.06\varepsilon \\ &= 20.06\varepsilon. \end{aligned}$$

The posterior probability of T can now be calculated:

$$P(e') = \frac{20\varepsilon(.9)}{20.06\varepsilon}$$
$$= 0.897.$$

Next we can calculate the posterior probability of A: By Bayes's theorem: $P(A \mid e') = \dfrac{P(e' \mid A)\, P(A)}{P(e')}$. Since

$$
\begin{aligned}
P(e' \mid A) &= P(e' \mid A \text{ and } T)P(T) + P(e' \mid A \text{ and } {\sim}T)P({\sim}T) \\
&= \quad\quad 0 \quad + \quad\quad\quad \varepsilon(.1) \\
&= .1\varepsilon.
\end{aligned}
$$

We get

$$P(A \mid e') = \frac{.06\varepsilon}{20.06\varepsilon}$$
$$= .003.$$

4.2 THOMAS KUHN MEETS THOMAS BAYES, INTRODUCTIONS BY WESLEY SALMON

I have thus far confined my criticism to the standard subjectivist Bayesian approach. There have been attempts to constrain the prior probabilities but with very limited success, especially when it comes to the Bayesian Way in philosophy of science. To discuss them here would require introducing technical ideas beyond the scope of our discussion. There is, however, one line of approach, developed by Wesley Salmon, that will tie together and illuminate a number of the themes I have taken up. My focus will be on his paper "Rationality and Objectivity in Science, *or* Tom Kuhn Meets Tom Bayes" (Salmon 1990).

As with the discussion in the previous section, Salmon's discussion is an attempt to employ the Bayesian Way to solve a philosophical problem, this time to answer Kuhn's challenge as to the existence of an empirical logic for science. Reflecting on the deep division between the logical empiricists and those who adopt the "historical approach," a division owing much to Kuhn's *Structure of Scientific Revolutions*, Salmon (1990) proposes "that a bridge could be built between the differing views of Kuhn and Hempel if Bayes's theorem were invoked to explicate the concept of scientific confirmation" (p. 175). The idea came home to Salmon, he tells us, during an American Philosophical Association (Eastern Division 1983) symposium on Carl Hempel, in which

Kuhn and Hempel shared the platform.[4] "At the time it seemed to me that this maneuver could remove a large part of the dispute between standard logical empiricism and the historical approach to philosophy of science" on the fundamental issue of confirmation (p. 175).

Granting that observation and experiment, together with hypothetico-deductive reasoning, fail adequately to account for theory choice, Salmon argues that the Bayesian Way can accommodate the additional factors Kuhn seems to think are required. In building his bridge, Salmon often refers to Kuhn's (1977) "Objectivity, Value Judgment, and Theory Choice." It is a fitting reference: in that paper Kuhn himself is trying to build bridges with the more traditional philosophy of science, aiming to thwart charges that he has rendered theory choice irrational.

Deliberately employing traditional terminology, Kuhn attempts to assuage his critics. He assures us that he agrees entirely that the standard criteria—accuracy, consistency, scope, simplicity, and fruitfulness—play a vital role in choosing between an established theory and a rival (p. 322). But as noted in chapter 2, Kuhn charges that these criteria underdetermine theory choice: they are imprecise, differently interpreted and differently weighed by different scientists. Taken together, they may contradict each other—one theory being most accurate, say, while another is most consistent with background knowledge. Hence theory appraisals may disagree even when agents ostensibly follow the same shared criteria. They function, Kuhn says, more like values than rules.

Here's where one leg of Salmon's bridge enters. The shared criteria of theory choice, Salmon proposes, can be cashed out, at least partly, in terms of prior probabilities. The conflicting appraisals that Kuhn might describe as resulting from different interpretations and weightings of the shared values, a Bayesian could describe as resulting from different assignments of prior probabilities. We have at least a partial bridge linking Bayes and Kuhn, but would a logical empiricist want to cross it?

Logical empiricists, it seems, would need to get around the Kuhnian position that the shared criteria are never sufficient to ground the choice between an accepted theory and a competitor, that consensus, if it occurs, always requires an appeal to idiosyncratic, personal factors beyond the shared ones. They would need to counter Kuhn's charge that in choosing between rival theories "scientists behave like philosophers," engaging in what I called "mere critical discourse" in chapter 2.

---

4. See "Symposium: The Philosophy of Carl G. Hempel," *Journal of Philosophy* 80, no. 10 (October 1983):555–72. Salmon's contribution is Salmon 1983.

Interestingly, Kuhn's single reference to a Bayesian approach is to combat criticism of his position. For the sake of argument, Kuhn says, suppose that scientists deploy some Bayesian algorithm to compute the posterior probabilities of rival theories on evidence and suppose that we could describe their choice between these theories as based on this Bayesian calculation (Kuhn 1977, 328). "Nevertheless," Kuhn holds that "the algorithms of individuals are all ultimately different by virtue of the subjective considerations with which each must complete the objective criteria before any computations can be done" (p. 329). So sharing Bayes's theorem does not count as a "shared algorithm" for Kuhn. Kuhn views his (logical empiricist) critic as arguing that since scientists often reach agreement in theory choice, the subjective elements are eventually eliminated from the decision process and the Bayesian posteriors converge to an objective choice. Such an argument, Kuhn says, is a non sequitur. In Kuhn's view, the variable priors lead different scientists to different theory choices, and agreement, if it does occur, results from sociopsychological factors, if not from unreasoned leaps of faith. Agreement, in other words, might just as well be taken as evidence of the further role of subjective and sociopsychological factors, rather than of their eventual elimination.

But perhaps building a logical empiricist bridge out of Bayesian bricks would not require solving this subjectivity problem. Perhaps Salmon's point is that by redescribing Kuhn's account in Bayesian terms, Kuhn's account need not be seen as denying science a logic based on empirical evidence. It can have a logic based on Bayes's theorem. It seems to me that much of the current appeal of the Bayesian Way reflects this kind of move: while allowing plenty of room for "extrascientific" factors, Bayes's theorem ensures at least some role for empirical evidence. It gives a formal model, we just saw, for reconstructing (after the fact) a given assignment of blame for an anomaly, and it may well allow for reconstructing Kuhnian theory choice. Putting aside for the moment whether a bridge from Bayes to Kuhn holds us above the water, let us see how far such a bridge would need to go.

Right away an important point of incongruity arises. While Kuhn talks of theory acceptance, the Bayesian talks only of probabilifying a theory—something Kuhn eschews. For the context of Kuhnian normal science, where problems are "solved" or not, this incongruity is too serious to remedy. But Salmon is talking about theory choice or theory preference, and here there seem to be ways of reconciling Bayes and Kuhn (provided radical incommensurabilities are put to one side), although Salmon does not say which he has in mind. One possibility would be to supplement the Bayesian posterior probability assessment with rules for acceptance or preference (e.g., accept or prefer a theory if its posterior probability is sufficiently higher than that of its rivals).

A second possibility would be to utilize the full-blown Bayesian decision theory. Here, averaging probabilities and utilities allows calculating the average or expected utility of a decision. The Bayesian rule is to choose the action that *maximizes expected utility.* Choosing a theory would then be represented in Bayesian terms as adopting the theory that the agent feels maximizes expected utility. If it is remembered that, according to Kuhn, choosing a theory means deciding to work within its paradigm, this second possibility seems more apt than the first. The utility calculation would provide a convenient place to locate the variety of values—those shared as well as those of "individual personality and biography"—that Kuhn sees as the basis for theory choice.

Even this way of embedding Kuhn in a Bayesian model would not quite reach the position Kuhn holds. In alluding to the Bayesian model, Kuhn (1977) concedes that he is tempering his position somewhat, putting to one side the problems of radically theory-laden evidence and incommensurability. Strictly speaking, comparing the expected utilities of choosing between theories describes a kind of comparison that Kuhn deems impossible for choosing between incommensurables. It is doubtful that a genuine Kuhnian conversion is captured as the result of a Bayesian conditionalization. Still, the reality of radical incommensurability has hardly been demonstrated. So let us grant that the subjective Bayesian Way, with the addition of some rule of acceptance such as that offered by Bayesian decision theory, affords a fairly good bridge between Bayes and a slightly-tempered Kuhn. Note also that the Kuhnian problems of subjectivity and relativism are rather well modeled—though not solved—by the corresponding Bayesian problems. The charge that Kuhn is unable to account for how scientists adjudicate disputes and often reach consensus seems analogous to the charge we put to the subjective Bayesian position. (For a good discussion linking Kuhn and Bayes, see Earman 1992, 192–93.)

But this is not Salmon's bridge. Our bridge pretty much reaches Kuhn, but the toll it exacts from the logical empiricist agenda seems too dear for philosophers of that school to want to cross it. Salmon's bridge is intended to be free of the kinds of personal interests that Kuhn allows, and as such it does not go as far as reaching Kuhn's philosophy of science. But that is not a mark against Salmon's approach, quite the opposite. A bridge that really winds up in Kuhnian territory is a bridge too far: a utility calculation opens theory choice to all manner of interests and practical values. It seems the last thing that would appeal to those wishing to retain the core of a logical empiricist philosophy. (It opens too wide a corridor for the enemy!) So let us look at Salmon's bridge as a possible link, not between a tempered Kuhn and Bayes, but between logical empiricism and a tempered Bayesianism. Before the last brick is in place, I shall question whether the bridge does not actually bypass Bayesianism altogether.

4.3 SALMON'S COMPARATIVE APPROACH AND A BAYESIAN BYPASS

Salmon endorses the Kuhnian position that theory choice, particularly among mature sciences, is always a matter of choosing between rivals. Kuhn's reason, however, is that he regards rejecting a theory or paradigm in which one had been working without accepting a replacement as tantamount to dropping out of science. Salmon's reason is that using Bayes's theorem comparatively helps cancel out what he takes to be the most troubling probability: the probability of the evidence e given not-T ("the catchall"). (Salmon, like me, prefers the term hypotheses to theories, but uses T in this discussion because Kuhn does. I shall follow Salmon in allowing either to be used.)

Because of some misinterpretations that will take center stage later, let us be clear here on the probability of evidence e on the catchall hypothesis.[5] Evidence e describes some outcome or information, and not-T, the catchall, refers to the disjunction of all possible hypotheses other than T, including those not even thought of, that might predict or be relevant to e. This probability is not generally meaningful for a frequentist, but is necessary for Bayes's theorem.[6] Let us call it the *Bayesian catchall factor* (with evidence e):[7]

Define the *Bayesian catchall factor* (in assessing T with evidence e) as

$P(e \mid \text{not-}T)$.

Salmon, a frequentist at heart, rejects the use of the Bayesian catchall factor.

> What is the likelihood of any given piece of evidence with respect to the catchall? This question strikes me as utterly intractable; to answer it we would have to predict the future course of the history of science. (Salmon 1991, 329)

5. To my knowledge, it was L. J. Savage who originated the term *catchall*.
6. See chapter 6.
7. I take this term from that of the Bayes factor, which is the ratio of the Bayesian catchall factor and $P(e \mid T)$.

This recognition is a credit to Salmon, but since the Bayesian catchall factor is vital to the general Bayesian calculation of a posterior probability, his rejecting it seems almost a renunciation of the Bayesian Way. The central role of the Bayesian catchall factor is brought out in writing Bayes's theorem as follows:

$$P(T \mid e) = \frac{P(e \mid T)\, P(T)}{P(e \mid T)\, P(T) + \textbf{P(e} \mid \textbf{not-T)}\, P(\text{not-}T).}$$

Clearly, the lower the value of the Bayesian catchall factor, the higher the posterior probability in T, because the lower its value, the less the denominator in Bayes's theorem exceeds the numerator. The subjectivist "solution" to Duhem turned on the agent assigning a very small value to the Bayesian catchall factor (where the evidence was the anomalous result e'), because that allowed the posterior of T to remain high despite the anomaly. Subjective Bayesians accept, as a justification for this probability assignment, that agents believe there to be no plausible rival to T that they feel would make them expect the anomaly e'. This is not good enough for Salmon.

In order to get around such a subjective assignment (and avoid needing to predict the future course of science), Salmon says we should restrict the Bayesian Way to looking at the ratio of the posteriors of two theories T_1 and T_2: In the ratio of the posteriors of the two theories, we get a canceling out of the Bayesian catchall factors (the probability of e on the catchall).[8] Let us see what the resulting comparative assessment looks like. Since the aim is no longer to bridge Kuhn, we can follow Salmon in talking freely about either theories or hypotheses. Salmon's Bayesian algorithm for theory preference is as follows (to keep things streamlined, I drop the explicit statement of the background variable B):

Salmon's Bayesian algorithm for theory preference (1990, 192):

Prefer T_1 to T_2 whenever $P(T_1 \mid e)/P(T_2 \mid e)$ exceeds 1, where:

8. This is because

$$P(T_i \mid e) = \frac{P(e \mid T_i)P(T_i)}{P(e \mid T_i)P(T_i) + P(e \mid {\sim}T_i)P({\sim}T_i).}$$

Note that the denominator equals $P(e)$. Since that is so for the posterior of T_1 as well as for T_2, the result of calculating the ratio is to cancel $P(e)$, and thereby cancel the probabilities of e on the catchalls.

$$\frac{P(T_1 \mid e)}{P(T_2 \mid e)} = \frac{P(T_1)\; P(e \mid T_1)}{P(T_2)\; P(e \mid T_2)}.$$

To start with the simplest case, suppose that both theories T_1 and T_2 entail e.[9] Then $P(e \mid T_1)$ and $P(e \mid T_2)$ are both 1. These two probabilities are the *likelihoods* of T_1 and T_2, respectively.[10] Salmon's rule for this special case becomes:

> *Special case:* Salmon's rule for relative preference (where each of T_1 and T_2 entails e):
>
> Prefer T_1 to T_2 whenever $P(T_1)$ exceeds $P(T_2)$.

Thus, in this special case, the relative preference is unchanged by evidence e. You prefer T_1 to T_2 just in case your prior probability in T_1 exceeds that of T_2 (or vice versa). Note that this is a general Bayesian result that we will want to come back to. In neutral terms, it says that if evidence is entailed by two hypotheses, then *that evidence* cannot speak any more for one hypothesis than another—according to the Bayesian algorithm.[11] If their appraisal differs, it must be due to some difference in prior probability assignments to the hypotheses. This will not be true on the error statistics model.

To return to Salmon's analysis, he proposes that where theories do not entail the evidence, the agent consider auxiliary hypotheses (A_1 and A_2) that, when coinjoined with each theory (T_1 and T_2, respectively), would entail the evidence. That is, the conjunction of T_1 and A_1 entails e, and the conjunction of T_2 and A_2 entails e. This allows, once again, the needed likelihoods to equal 1, and so to drop out. The relative appraisal of T_1 and T_2 then equals the ratio of the prior probabilities of the conjunctions of T_1 and A_1, and T_2 and A_2. We are to prefer that conjunction (of theory and auxiliary) that has the higher prior probability.[12] In short, in Salmon's comparative analysis the weight is taken from the likelihoods and placed on the priors, making the appraisal even more dependent upon the priors than the noncomparative Bayesian approach.

9. While this case is very special, Salmon proposes that it be made the standard case by conjoining suitable auxiliaries to the hypotheses. I will come back to this in a moment.

10. Note that likelihoods of hypotheses are *not* probabilities. For example, the sum of the likelihoods of a set of mutually exclusive, collectively exhaustive hypotheses need not equal 1.

11. This follows from the likelihood principle to be discussed in later chapters.

12. For simplicity, we could just replace T_1, T_2, in the statement of the special case, with the corresponding conjunctions T_1 and A_1, and T_2 and A_2, respectively.

Problems with the Comparative Bayesian Approach

Bayesians will have their own problems with such a comparative Bayesian approach. How, asks Earman (1992, 172), can we plug in probabilities to perform the usual Bayesian decision theory? But Earman is reluctant to throw stones, confessing that "as a fallen Bayesian, I am in no position to chide others for acts of apostasy" (p. 171). Earman, with good reason, thinks that Salmon has brought himself to the brink of renouncing the Bayesian Way. Pursuing Salmon's view a bit further will show that he may be relieved of the yoke altogether.

For my part, the main problem with the comparative approach is that we cannot apply it until we have accumulated sufficient knowledge, by some non-Bayesian means, to arrive at the prior probability assignments (whether to theories or theories conjoined with auxiliaries). Why by some non-Bayesian means? Couldn't prior probability assessments of theories and auxiliaries themselves be the result of applying Bayes's theorem? They could, but only by requiring a *reintroduction* of the corresponding assignments to the Bayesian catchall factors—the very thing Salmon is at pains to avoid. The problems of adjudicating conflicting assessments, predicting the future of science, and so on, remain.

Could not the prior probability assignments be attained by some more hard-nosed assessment? Here is where Salmon's view becomes most interesting. While he grants that assessments of prior probabilities, or, as he prefers, plausibilities, are going to be relative to agents, Salmon demands that the priors be constrained to reflect objective considerations.

> The frightening thing about pure unadulterated personalism is that nothing prevents prior probabilities (and other probabilities as well) from being determined by all sorts of idiosyncratic and objectively irrelevant considerations (Salmon 1990, 183)

such as the agent's mood, political disagreements with or prejudices toward the scientists who first advanced the hypothesis, and so on.

> What we want to demand is that the investigator make every effort to bring all of his or her *relevant* experience in evaluating hypotheses to bear on the question of whether the hypothesis under consideration is of a type likely to succeed, and to leave aside emotional irrelevancies. (P. 183)

Ever the frequentist, Salmon proposes that prior probabilities "can be understood as our best estimates of the frequencies with which certain kinds of hypotheses succeed" (p. 187).[13] They may be seen as personalistic so long as the agent is guided by "the aim of bringing to bear all his or her experience that is relevant to the success or failure of hypotheses similar to that being considered." According to Salmon, "On the basis of their training and experience, scientists are qualified to make such judgments" (p. 182).

But are they? How are we to understand the probability Salmon is after? The context may be seen as a single-universe one. The members of this universe are hypotheses similar to the hypothesis H being considered, presumably from the population of existing hypotheses. To assign the prior probability to hypothesis H, I imagine one asks oneself, What proportion of the hypotheses in this population are (or have been) successful? Assuming that H is a random sample from the universe of hypotheses similar to H, this proportion equals the probability of interest. Similar to hypothesis H in what respects? Successful in what ways? For how long? The reference class problem becomes acute.

I admit that this attempt at a frequentist prior (also found in Hans Reichenbach) has a strong appeal. My hunch is that its appeal stems from unconsciously equating this frequency with an entirely different one, and it is this different one that is really of interest and, at the same time, is really obtainable.

Let us imagine that one had an answer to Salmon's question: what is the relative frequency with which hypotheses relevantly similar to H are successful (in some sense)? Say the answer is that 60 percent of them are. If H can be seen as a fair sample from this population, we could assign H a probability of .6. Would it be of much help to know this? I do not see how. I want to know how often *this* hypothesis will succeed.

What might an error statistician mean by the probability that this hypothesis H will succeed? As always, for a frequentist, probability applies to outcomes of a type of experiment. (They are sometimes called "generic" outcomes.) The universe or population here consists of possible or hypothetical experiments, each involving an application of hypothesis H. Success is some characteristic of experimental outcomes. For example, if H is a hypothesized value of a parameter μ, a successful outcome might be an outcome that is within a specified margin of error of μ. The probability of success construed this way is just the ordinary

13. This was also Reichenbach's view. He did not consider that enough was known at present to calculate such a probability, but thought that it might be achievable in the future.

probability of the occurrence of certain experimental outcomes. (Further discussion occurs in chapter 5.) Indeed, for the error theorist, the only kinds of things to which probabilities apply are things that can be modeled as experimental outcomes. Knowledge of H's probable success is knowledge of the probability distributions associated with applying H in specific types of experiments. Such knowledge captures the spirit of what C. S. Peirce would call the "experimental purport" of hypothesis H.

Two Meanings of the Probability That a Hypothesis Is Successful. Let us have a picture of our two probabilities. Both can be represented as one-urn models. In Salmon's urn are the members of the population of hypotheses similar to H. These hypotheses are to be characterized as successful or not, in some way that would need to be specified. The probability of interest concerns the relative frequency with which hypotheses in this urn are successful. This number is taken as the probability that H is successful.

In my urn are members of a population of outcomes (a sample space) of an experiment. Each outcome is defined as successful or not according to whether it is close to what H predicts relative to a certain experiment (for simplicity, omit degrees of closeness). The probability of interest concerns the relative frequency with which outcomes in this urn are successful. Hypothesis H can be construed as asserting about this population of outcomes that with high probability they will be successful (e.g., specifiably close to what H predicts). The logic of standard statistical inference can be pictured as selecting one outcome, randomly, from this "urn of outcomes" and using it to learn whether what H asserts is correct.

Take one kind of hypothesis already discussed, that a given effect is real or systematic and not artifactual. In particular, take Hacking's hypothesis H (discussed in chapter 3).

H: dense bodies are real structures in blood cells, not artifacts. We have no idea what proportion of hypotheses similar to H are true, nor do we have a clue as to how to find out, nor what we would do if we did. In actuality, our interest is not in a probabilistic assignment to H, but in whether H is the case. We need not have infallible knowledge about H to learn about the correctness of H.

We ask: what does the correctness of H say about certain experimental results, ideally, those we can investigate? One thing Hacking's H says is that dense bodies will be detected even using radically different physical techniques, or at least that they will be detected with high reliability. Experimenting on dense bodies, in other words, will *not* be

like experimenting on an artifact of the experimental apparatus. Or so
H asserts. This leads to pinpointing a corresponding notion of success.
We may regard an application of *H* as successful when it is specifiably
far from what would be expected in experimenting on an artifact.
More formally, a successful outcome can be identified as one that a
specified experimental test would take as failing the artifact explanation. That *H* is frequently successful, then, asserts that the artifact explanation, that is, not-*H*, would very frequently be rejected by a certain
statistical test.

The knowledge of the quality of the evidence in hand allows assessing whether there are good grounds for the correctness of *H* (in the
particular respects indicated in defining success). In particular, knowledge that *H* passes extremely severe tests is a good indication that *H* is
correct, that it will often be successful. Why? Because were the dense
bodies an artifact, it would almost certainly not have produced the
kinds of identical configurations that were seen, despite using several
radically different processes.

What are needed, in my view, then, are arguments that *H* is correct, that experimental outcomes will very frequently be in accordance
with what *H* predicts—that *H* will very frequently succeed, *in this sense.*
These are the arguments for achieving experimental knowledge
(knowledge of experimental distributions). We obtain such experimental knowledge by making use of probabilities—not of hypotheses
but probabilistic characteristics of experimental testing methods (e.g.,
their reliability or severity). Where possible, these probabilities are arrived at by means of standard error probabilities (e.g., significance levels). In those cases, what I have in mind is well put in my favorite
passage from Fisher:

> In relation to the test of significance, we may say that a phenomenon
> is experimentally demonstrable when we know how to conduct an
> experiment which will rarely fail to give us a statistically significant
> result. (Fisher 1947, 14)

A Natural Bridge between Salmon and Error Statistics

Whereas Salmon construes "the probability of *H*'s success" as the
relative frequency with which hypotheses similar to *H* are successful,
the error statistician proposes that it be given a very different frequentist construal. For the error statistician, "the probability of *H*'s success" or, more aptly, *H*'s reliability, is viewed as an elliptical way of
referring to the relative frequency with which *H* is expected to succeed
in specifically defined experimental applications. There is no license to

use the latter frequentist notion in applying Bayes's theorem. Nevertheless, it may be used in Salmon's comparative approach (where the likelihoods drop out), and doing so yields a very natural bridge connecting his approach to that of error statistics.

To see in a simple way what this natural bridge looks like, let the two hypotheses H_1 and H_2 entail evidence e (it would be adequate to have them merely fit e to some degree). Then, on Salmon's comparative Bayesian approach, H_1 is to be preferred to H_2 just in case the prior probability assessment of H_1 exceeds H_2. The assignment of the prior probabilities must not contain irrelevant subjective factors, says Salmon, but must be restricted to assessing whether the hypotheses are likely to be successful. Hypothesis H_1 is to be preferred to H_2 just in case H_1 is accorded a higher probability of success than H_2. Now let us substitute my error statistical construal of probable success (for some specified measure of "successful outcome"). Evaluating H's probable success (or H's reliability) means evaluating the relative frequency with which applications of H would yield results in accordance with (i.e., specifiably close to) what H asserts. As complex as this task sounds, it is just the kind of information afforded by experimental knowledge of H. The task one commonly sets for oneself is far less technically put. The task, informally, is to consider the extent to which specific obstacles to H's success have been ruled out. Here is where the kind of background knowledge I think Salmon has in mind enters. What training and experience give the experimenter is knowledge of the specific ways in which hypotheses can be in error, and knowledge of whether the evidence is so far sufficient to rule out those errors.

To put my point another way, Salmon's comparative approach requires only the two prior probabilities or plausibilities to be considered, effectively wiping out the rest of the Bayesian calculation. The focus is on ways of assessing the plausibilities of the hypotheses or theories themselves. However, Salmon's approach gives no specific directions for evaluating the plausibilities or probable success of the hypotheses. Interpreting probable success in the way I recommend allows one to work out these directions. Salmon's comparative appraisal of H_1 against a rival H_2 would become: prefer H_1 to H_2 just to the extent that the evidence gives a better indication of H_1's likely success than H_2's.

Further, the kinds of evidence and arguments relevant to judge H's success, in my sense, seem quite congenial to what Salmon suggests should go into a plausibility assessment. In one example Salmon refers explicitly to the way in which standard (non-Bayesian) significance tests may be used to give plausibility to hypotheses (Salmon 1990, 182). In particular, a statistically significant association between saccharin and bladder cancer in rats, he says, lends plausibility to the hypothesis H that saccharin in diet drinks increases the risk of bladder cancer in humans. Provided that errors of extrapolating from rats to humans and from high to low doses are satisfactorily ruled out, a statistically significant association may well provide evidence that H has been shown, *that H will be successful in our sense.* This success may be cashed out in different ways, because the truth of H has different implications. One implication of the correctness of H here is that were populations to consume such and such amount of saccharin the incidence of bladder cancer would be higher than if they did not. My point is that such experiments are evidence for the correctness of H in this sense. Such experiments do not provide the number Salmon claims to be after, the probability that hypotheses similar to the saccharin hypothesis are successful. So even if that probability were wanted (I claim it is not), that is not what the saccharin experiments provide.[14]

By allowing for this error statistical gloss of "H's probable success," the reader should not be misled into viewing our account as aiming to assign some quantitative measure to hypotheses—the reverse is true. My task here was to erect a bridge between an approach like Salmon's and the testing account I call error statistics. By demonstrating that the

14. In the case of the saccharin hypothesis, it might look as if Salmon's frequentist probability is obtained. That, I think, is because of a tendency to slide from one kind of probability statement to another. Consider hypotheses of the form x causes cancer in humans. They are all similar to H: saccharin causes cancer in humans. But what should be included in the reference set for getting Salmon's probability? Might x be anything at all? If so, then only a very tiny proportion would be successful hypotheses. That would not help in assessing the plausibility of H. I suggest that the only way this probability makes sense is if hypotheses "similar to H" refers to hypotheses similarly grounded or tested. In trying to specify the reference set in the case of the saccharin hypothesis we might restrict it to those causal hypotheses (of the required form) that have been shown to hold about as well as H. So it would consist of causal claims where a statistically significant correlation is found in various animal species, where certain dosage levels are used, where certain extrapolation models are applied (to go from animal doses to human doses as well as from rats to humans), where various other errors in identifying carcinogens are ruled out, and so on. Notice how these lead to a severity assessment.

A relative frequency question of interest that can be answered, at least qualitatively, is this: What is the relative frequency with which hypotheses of this sort (x causes cancer in humans) pass experimental tests $E_1, \ldots, E_n$ as well as H does, and yet do not succeed (turn out to be incorrect)? One minus this gives the severity of the test H passes. The question boils down to asking after the severity of the test H passes (where, as is common, several separate tests are taken together).

It does not matter that the hypotheses here differ. Error probabilities of procedures hold for applications to the same or *different* hypotheses. Neyman (e.g., 1977) often discussed the mistake in thinking they hold only for the former.

role Salmon gives to plausibility assessments is better accomplished by an assessment of the reliability of the tests hypotheses pass, I mean to show that the latter is all that is needed.

There are plenty of advantages to the testing account of scientific inference. First, by leading to accepting hypotheses as approximately correct, as well indicated, or as likely to be successful—rather than trying to assign some quantity of support or probability to hypotheses—it accords with the way scientists (and the rest of us) talk. Second, reporting the quality of the tests performed provides a way of communicating the evidence (summarizing the status of the problem to date) that is intersubjectively testable. A researcher might say, for example, that the saccharin rat study gives good grounds for holding that there is a causal connection with cancer in rats, but deny that the corresponding hypothesis about humans has been severely tested. This indicates what further errors would need to be ruled out (e.g., certain dose-response models are wrong).

Now it is open to a Bayesian to claim that the kinds of arguments and evidence that I might say give excellent grounds for the correctness of H, for accepting H, or for considering H to have passed a severe test can be taken as warranting a high prior probability assignment in H. For example, "there are excellent grounds for H" may be construed as "H has high prior probability" (say, around .9). (That Bayesians implicitly do this in their retrospective reconstructions of episodes is what gives their prior probability assessments their reasonableness.) Used in a purely comparative approach such as Salmon's, it might do no harm. However, there is nothing Bayesian left in this comparative approach! It is, instead, a quantitative sum-up of the quality of *non-Bayesian* tests passed by one hypothesis compared with those passed by another. (Whether such a non-Bayesian assessment of Bayesian priors could even be made to obey the probability calculus is not clear.)

To call such an approach Bayesian, even restricting it to comparisons, would be misleading. It is not just that the quantitative sum-up of H's warrant is not arrived at via Bayes's theorem. It is, as critics of error statistics are happy to demonstrate, that the principles of testing used in the non-Bayesian methods conflict with Bayesian principles.[15] (I will have much more to say about this later, e.g., in chapter 10.) The Bayesian Way supposes, for any hypothesis one wishes to consider,

15. To anticipate a little, the Bayesian Way follows the likelihood principle, which conflicts with error-probability principles. Quoting Savage: "Practically none of the 'nice properties' respect the likelihood principle" (1964, 184). The "nice properties" refer to error characteristics of standard statistical procedures, such as unbiasedness and significance levels. I return to this in chapter 10.

that a Bayesian prior is available for an agent, and that an inference can be made. In general, however, there are not going to be sufficient (non-Bayesian) grounds to assign even a rough number to such hypotheses. We are back to the problem of making it too difficult to get started when, as is commonly the case, one needs a forward-looking method to begin learning something.

Bayesian Heretics, Fallen and Disgruntled Bayesians

The Bayesian landscape is littered with Bayesians who variably describe themselves or are described by others as fallen, heretical, tempered, nonstrict, or whatnot. Many Bayesians in this category came to the Bayesian Way in the movements led by Carnap and Reichenbach. Assigning probabilities to hypotheses was a natural way of avoiding the rigidities of a hypothetico-deductive approach. Inadequacies in the two main objective ways philosophers tried to define the prior probabilities—Carnapian logical or Reichenbachian frequentist—have left some in limbo: wanting to avoid the excesses of personalism and not sure how non-Bayesian statistics can help. Those Bayesians do not see themselves as falling under the subjectivist position that I criticized earlier. I invite them to try out the natural bridge proffered above, to see where it may lead.

What is really being linked by this bridge? Might it be said to link the cornerstone of logical empiricism on the one hand and the centerpiece of the New Experimentalism on the other? Such a bridge, as I see it, would link the (logical empiricist) view that the key to solving problems in philosophy of science is an inductive-statistical account of hypothesis appraisal with the view that the key to solving problems in philosophy of science is an understanding of the nature and role of experiment in scientific practice. It provides a way to *model* Kuhn's view of science—*where he is correct*—as well as a way to "*solve* Kuhn" where he challenges the objectivity and rationality of science.

In this chapter and the last I have brought out the main shortcomings of appeals to the Bayesian Way in modeling scientific inference and in solving problems about evidence and inference. Understanding these shortcomings also puts us in a better position to see what would be required of any theory of statistics that purports to take a leading role in an adequate philosophy of experiment. For one thing, we need an account that explicitly incorporates the intermediate theories of data, instruments, and experiment that are required to obtain experimental evidence in the first place. For another, the account must enable us to address the question of whether auxiliary hypotheses or experimental assumptions are responsible for observed anomalies from a hypothesis *H*, quite apart from how credible we regard hypothesis *H*. In other words, we need to be able to split off from some primary inquiry or test those questions about how well run the experiment was, or how well its assumptions were satisfied. Let us now turn to an experimental framework that will lend itself to these requirements.

CHAPTER FIVE

Models of Experimental Inquiry

The conception of chance enters into the very first steps of scientific activity, in virtue of the fact that no observation is absolutely correct.

—Max Born, *Natural Philosophy of Cause and Chance,* p. 47

5.1 OVERVIEW

IN THE LAST TWO CHAPTERS I have argued that an adequate account of experimental testing must not begin at the point where data and hypotheses are given, but rather must explicitly incorporate the intermediate theories of data, instruments, and experiment that are required to obtain experimental evidence in the first place. The account must also find places to house the many piecemeal local experiments that are needed to link raw data to substantive experimental questions. One might describe what is needed as the homes within which experiments "live lives of their own," lives connected to high-level theorizing, but with their own models, parameters, and theories. To this end, I propose we view experimental inquiry in terms of a series of conceptual representations or models ranging from the primary scientific hypotheses or questions that are being addressed in an inquiry to the nitty-gritty details of the generation and analysis of data. For each experimental inquiry we can delineate three types of models: models of primary scientific hypotheses, models of data, and models of experiment that link the others by means of test procedures. Figure 5.1 gives a schematic representation:

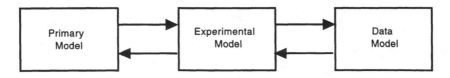

FIGURE 5.1. Framework of models.

128

Precisely how to break down a given experimental inquiry into these models is not a cut and dried affair—most inquiries will call for several of each. The main thing is to have at our disposal a framework that permits us to delineate the relatively complex steps from raw data to scientific hypotheses, and to systematically pose the questions that arise at each step. Let us begin with the key questions that would be posed.

1. Primary scientific hypotheses or questions. A substantive scientific inquiry is to be broken down into one or more local or "topical" hypotheses, each of which corresponds to a distinct *primary question* or *primary problem*. In a comprehensive research situation, several different piecemeal tests of hypotheses are likely to correspond to a single primary question. Typically, primary problems take the form of estimating quantities of a model or theory, or of testing hypothesized values of these quantities. They may also concern the form or equation linking theoretical variables and quantities. These problems often correspond to questions framed in terms of one or more canonical errors: about parameter values, causes, accidental effects, and assumptions involved in testing other errors.

2. Experimental models. The experimental models serve as the key linkage models connecting the primary model to the data, and conversely. Two general questions are how to relate primary questions to (canonical) questions about the particular type of experiment at hand, and how to relate the data to these experimental questions.

3. Data models. Modeled data, not raw data, are linked to the experimental models. Accordingly, two questions arise, one before the data are collected—"before-trial"—and one after the data are in hand—"after-trial." The before-trial question is how to generate and model raw data so as to put them in the canonical form needed to address the questions in the experimental model. The after-trial question is how to check whether actual data satisfy various assumptions of experimental models.

Table 5.1 summarizes these questions. Clearly, the answers to these questions depend on what one wants to learn in a given inquiry as well as on the methodological strategies and tools that are available for investigating the primary questions. Different principles of experimental methodology relate to one or more of these questions, and methodological rules should be assessed in relation to these tasks.

This framework of models does double duty for my account: it provides a means for spelling out the relationship between data and hypotheses—one that organizes but does not oversimplify actual sci-

TABLE 5.1

PRIMARY MODELS
How to break down a substantive inquiry into one or more local questions that can be probed reliably

EXPERIMENTAL MODELS
How to relate primary questions to (canonical) questions about the particular type of experiment at hand

How to relate the data to these experimental questions

DATA MODELS
How to generate and model raw data so as to put them in canonical form

How to check if the actual data generation satisfies various assumptions of experimental models

entific episodes—and it organizes the key tasks of a philosophy of experiment.

The idea of appealing to something like the above framework of models is hardly new. A formal analog is found in standard statistical inference. I am not the first to notice this. Patrick Suppes was. In his short but seminal paper "Models of Data," he remarks that

> it is a fundamental contribution of modern mathematical statistics to have recognized the explicit need of a model in analyzing the significance of experimental data. (Suppes 1969, 33)

Because it is always a model of data that links experiment to theories, Suppes proposes "that exact analysis of the relation between empirical theories and relevant data calls for a hierarchy of models" of different types (p. 33): models of data, models of experiment, and models of theories. The Suppean idea that understanding the relationship between theory and experiment calls for investigating the linkages between models of data and other models in this hierarchy, while in marked contrast with the more traditional approaches of the time, struck a chord when I was first introduced to statistics in graduate school. It still does.

More recently, Suppes's framework has gained currency from developments on the semantic view of theories (Suppes's view is one variant), as noted by Fred Suppe (e.g., 1977, 1989) and Bas van Fraassen (1989).[1] Proponents of the semantic view of theories convincingly argue that it promotes a more realistic picture of the relationship between theories and experiment than the so-called received

1. A full list of contributors to the semantic approach and an in-depth discussion of the specific developments can be found in Fred Suppe's work. A good part of the attention has been to employ the semantic view to promote realism (e.g.,

view. The last section of Suppes's "Models of Data" merits quoting at length:

> One of the besetting sins of philosophers of science is to overly sim-
> plify the structure of science. Philosophers who write about the repre-
> sentation of scientific theories as logical calculi then go on to say that
> a theory is given empirical meaning by providing interpretations or
> coordinating definitions for some of the primitive or defined
> terms. . . . What I have attempted to argue is that a whole hierarchy
> of models stands between the model of the basic theory and the com-
> plete experimental experience. . . . For each level . . . there is a theory
> in its own right. Theory at one level is given empirical meaning by
> making formal connections with theory at a lower level. . . . Once the
> empirical data are put in canonical form (at the level of model of data
> . . .), every question of systematic evaluation that arises is a formal
> one. . . . It is precisely the fundamental problem of scientific method
> to state the principles of scientific methodology that are to be used to
> answer these questions—questions of measurement, of goodness of
> fit, of parameter estimation, of identifiability, and the like. (Suppes
> 1969, 34)

In saying that these principles are formal, Suppes means "that they have as their subject matter set-theoretical models and their compari-son" (p. 34). A goal of Suppes's program is to articulate the formal, set-theoretical relationships between the models involved—something he masterfully pursued.

However, "the principles of scientific methodology that are to be used to answer these questions" are open to controversy, and to flesh them out and assess their epistemological rationale requires going be-yond the set-theoretical relationships between models. In any case, I am much less concerned with the possibility of specifying the set-theoretic relationships of these models than in using them to embark upon the kind of philosophy of experiment I have been sketching. My interest is in how the various models of inquiry offer a framework for canonical models of error, methodological rules, and theories of statis-tical testing. Except for the level of data models, Suppes's "Data Mod-els" is fairly sketchy; much filling in is in order. Since my chief interest is in the relationships between the models, particularly between the data models and the primary models, my emphasis will be on these linkages. The list of questions in table 5.1 is a guide to adapting Suppes's hierarchy of models to our purposes.

Giere 1988, Suppe 1989) or antirealism (e.g., van Fraassen 1989). The semantic
view has also been adopted and developed by a number of philosophers of biology
(e.g., Beatty 1980, Lloyd 1988, and Thompson 1988).

Before turning to that adaptation, here is a summary of the ground to be covered in this chapter. In section 5.2, I will discuss the details of Suppes's hierarchy of models and then adapt that framework to the present project. In section 5.3, I will use the hierarchy of models developed in section 5.2 to trace the main tasks of an epistemology of experiment. In section 5.4, I will illustrate a canonical inquiry into an error concerning "chance effects." And in section 5.5, I will attempt to demystify and justify the roles of probability models and relative frequency notions in experimental reasoning. The reader hungry for a full-blown scientific example may wish to turn to chapter 7 following section 3 (or before).

5.2 A SUPPEAN HIERARCHY OF MODELS

The top of Suppes's hierarchy is occupied by the primary model being evaluated (his example is a linear response model from learning theory). Beneath this are models of experiment and models of data, then a level of experimental design and a level of data generation. At the bottom are extraneous ceteris paribus conditions. Associated with each type of model is a theory in its own right: the primary theory of interest, theory of experiment, theory of data, and finally, theory of experimental design and data analysis.

The Suppean hierarchy begins with a problem: Experimental data are to be used to assess values of theoretical quantities, despite the fact that the data constitute only a finite sample of outputs from the experimental system, the accuracy, precision, and reliability of which are limited by various distortions introduced by intermediary processes of observation and measurement. Data are finite and discrete, while primary hypotheses may refer to an infinite number of cases and involve continuous quantities such as weight and temperature. The problem is how to link the detailed raw data to primary theoretical hypotheses.

Let us draw from an example to be considered in more detail later (chapter 8). The substantive theory is Einstein's theory of gravitation. A single question that may be split off for testing might concern a parameter λ, the mean deflection of light at the limb of the sun. Jumping down to the other end of the experimental spectrum is the description of the experimental apparatus, and the means by which raw data were obtained and analyzed. The following is a snippet from the celebrated account of the expedition to Sobral, Brazil, during the eclipse of 1919:

> When the crescent disappeared the word "go" was called and a metronome was started by Dr. Leocadio, who called out every tenth beat

during totality, and the exposure times were recorded in terms of these beats. It beat 320 times in 310 seconds. . . . The plates remained in their holders until development, which was carried out in convenient batches during the night hours of the following days, being completed by June 5. . . . On June 7 . . . we left Sobral for Fortaleza, returning on July 9 for the purpose of securing comparison plates of the eclipse field. Before our departure we dismounted the mirrors and driving clocks which were brought into the house to avoid exposure to dust. The telescopes and coelostats were left *in situ*. (Dyson, Eddington, and Davidson 1920, 299–301)

A wide gap exists between the nitty-gritty details of the data gathering experience and the primary theoretical model of the inquiry. The result of this gap, in Suppes's terminology, is that "a possible realization [model] of the theory cannot be a possible realization [model] of experimental data" (Suppes 1969, 26). Two additional models are therefore introduced: experimental models and data models. It is through these "linkage models" that experimental methodology links data and theories.

Nothing turns on whether we choose to array the models of inquiry into a hierarchy or into something like the flow chart in figure 5.1. Our present purposes are best served by Suppes's hierarchical design. The idea is that as one descends the hierarchy, one gets closer to the data and the actual details of the experimental experience. In addition, hypotheses at one level are checked as well as applied by making connections to the level below. I will consider each link in turn.

Experimental Models

The experimental model, the first step down from the primary hypothesis or theory, serves two chief functions. First, an experimental model provides a kind of experimental analog of the salient features of the primary model. In this way the primary question or hypothesis may be articulated with reference to the kind of experiment being contemplated, the number of trials, the particular experimental characteristics to be recorded (experimental statistics). If the primary question is to test some hypothesis H, the job of the experimental model is to say, possibly with the aid of other auxiliary hypotheses, what is expected or entailed by the truth of H, with respect to the kind of experiment of interest. But this stated expectation, at the level of the experimental model, still retains the main idealizations of the primary hypothesis (e.g., frictionless planes, perfect control). It is not yet a statement of the actual experimental outcome. If the primary hypothesis is Einstein's law of gravitation, and the contemplated test is on deflection of starlight, an experimental prediction might hypothesize what the mean deflection would be for light right at the edge or "limb" of the sun, even though no actual experiment will observe this.

This first function of the experimental model, then, involves specifying the key features of the experiment, and stating the primary question (or questions) with respect to it.

Experimental models, as I see them, serve a second function: to specify analytical techniques for linking experimental data to the questions of the experimental model. Whereas the first function addresses the links between the primary hypotheses and experimental hypotheses (or questions), the second function concerns links between the experimental hypotheses and experimental data or data models. In housing this second function within experimental models I seem to be going beyond Suppes's scheme, but not in any way that prevents us from keeping to the broad outlines of his conception.

What do the linkages of the second kind look like? Suppose the primary question is a test of a hypothesis. Because of the many sources of approximation and error that enter into arriving at the data, the data would rarely if ever be expected to agree exactly with theoretical predictions. As such, the link between data model and experimental hypothesis or question may often be modeled statistically, whether or not the primary theory or hypothesis is statistical. This statistical link can be modeled in two ways: the experimental prediction can itself be framed as a statistical hypothesis, or the statistical considerations may be seen to be introduced by the test (in its using a statistical test rule).

To refer to the example of the predicted deflection of light, the first way of modeling the relationship between the data model and experimental hypothesis would be to regard the prediction as including a margin of error (e.g., the Einstein prediction might be that "the mean deflection of light is 1.75" plus or minus some experimental error"). The second way would regard the predicted deflection as 1.75", but then classify the data as in accordance with that prediction so long as it did not differ by more than some margin of experimental error. Either way, the primary hypothesis may be seen to entail something about the probability with which certain experimental results would occur. This is an *experimental distribution*—and it is in terms of the experimental distribution (in the experimental model) that the primary question is asked and answered. More exactly, when the primary question is tackled it is by means of a hypothesis about an experimental distribution (i.e., a statistical hypothesis).

In specifying a test, the experimental model also includes some experimental *measure of fit or distance* by which to compare the experimental hypothesis and the data model. One example is a significance level (discussed in chapter 3 and more fully later in this chapter); another would be the number of probable errors (between a hypothesized and an observed quantity). Commonly, a test's job would be to specify which distances to count as "in accordance with" a given hypothesis, and which not to count. In an estimation problem, the experimental model supplies a rule for using the modeled data to arrive at an inference about the value of a parameter being estimated. Generally, the inferred estimate is of a range or interval of values.

According to the present account, the overarching criterion in performing the analytical function of an experimental model is that of reliability or severity. The concern is that estimators are not sufficiently reliable or that tests are not sufficiently probative to test severely hypotheses of interest. Principles of statistical inference enter at the level of the experimental model to address these worries. (Since test problems can be converted to estimation problems, severity serves to cover both.)

I follow Suppes in distinguishing experimental models from experimental *design*, which occurs two levels lower in the hierarchy. This distinction reflects statistical practice and will be explained more fully after we have considered data models.

Data Models

Data models, not raw data, are linked to the questions about the primary theory, and a great deal of work is required to generate the raw data and massage them into the canonical form required to actually apply the analytic tools in the experimental model. What should be included in data models? The overriding constraint is the need for data models that permit the statistical assessment of fit (between prediction and actual data) just described. A common practice in statistics is to seek data models that parallel the relevant features and quantities of the primary hypothesis for which they are to provide information. For instance, if the primary question concerns the mean value of a quantity μ, a characteristic of the data might be the corresponding sample mean abbreviated as $\overline{X}$, read "x-bar."

Let us return to the account of the photographs discussed from Sobral. Of the eight photographs taken during the eclipse, seven gave measurable images of these stars. These measurements of apparent star positions did not yet yield data on light *displacement*. "The problem was to determine how the apparent positions of the stars, affected by the sun's gravitational field, compared with the normal positions on a photograph taken when the sun was out of the way" (Eddington [1920] 1987, 115).

> Plates of the same field taken under nearly similar conditions . . .
> were taken on July 14, 15, 17 and 18 . . . for comparison with the
> eclipse plates. . . . The micrometer at the Royal Observatory is not
> suitable for the direct comparison of plates of this size. It was there-
> fore decided to measure each plate by placing, film to film upon it,
> another photograph of the same region reversed by being taken
> through the glass [the scale plate]. . . . The measures, both direct and
> reversed [through 180 degrees], were made by two measurers (Mr.
> Davidson and Mr. Furner), and the means taken. (Dyson, Eddington,
> and Davidson 1920, 302)

This is followed by a page with a vast array of measurements recording
the positions of the seven stars in the eight eclipse plates and in the
seven comparison plates. These measures provide the raw data on displacement, but simply looking at this vast array researchers could not
see what the "observed" mean deflection was. They needed a technique of data analysis that would let them *infer* this observation. Con-
sulting their "experimental tool kits," as it were, they employed a standard statistical technique known from other astronomical problems,
and arrived at an estimated mean deflection. This estimated deflection
gave them the needed modeled data. (I discuss this further in chapter 8.)

This estimated deflection is more accurate than the individual
measurements by themselves. In this sense, we come out of the statistical analysis with more than we put in! The estimate we come out with
is actually a *hypothesis* (in the theory of the data) that says something
about what deflection value *would have resulted* had more accurate and
precise measurements been made.

This hypothesis about the data, however, holds only if certain assumptions are met. This brings us to the key function of data models.
The statistical prediction in the experimental model relates to canoni-
cally modeled data, that is, to data models. But in order for the comparison offered by the statistical link in the experimental model to go
through, the assumptions of the experimental model must hold sufficiently in the actual experiment. There are two ways to do this. One is
to design an experiment that conforms sufficiently to the assumptions.
The second is to find a way (through corrections or whatever) of arriving at a desired data statement that effectively subtracts out (or renders
harmless) any violation of assumptions. This second way suffices because it supplies a statement of what the data would have been were
the assumptions actually met. So another "would have been" enters.
Often one can take one's pick: one can either utilize theories of experimental design to get data that pretty well conform to assumptions, or
utilize theories of data analysis to arrive at cleaned-up data. Whether

it is through literal manipulation or "would have been" reasoning, the upshot, in Suppes's terminology, is to turn the actual data into a model of the data.

In complex cases, as with the eclipse experiments, both careful design and extensive use of after-trial corrections are needed. With respect to the Sobral plates alluded to above, the researchers reported:

> The probable error of the result judging from the accordance of the separate determinations is about 6 per cent. It is desirable to consider carefully the possibility of systematic error. The eclipse and comparison photographs were taken under precisely similar instrumental conditions, but there is the difference that the eclipse photographs were taken on the day of May 29, and the comparison photographs on nights between July 14 and July 18. A very satisfactory feature of the photographs is the essential similarity of the star images on the two sets of photographs. (Dyson, Eddington, and Davidson 1920, 306)

The eclipse researchers' determination of this "essential similarity" used a standard technique of statistical analysis. Such standard statistical tests (e.g., tests of goodness of fit) are used to check assumptions of experimental data models, once the results are in (i.e., after-the-trial).

The task of checking assumptions may be regarded as a secondary experimental question in relation to the primary one. (Or if one chooses, this task may be set out as a distinct primary question.) Generally, the modeled data intended for the primary question must be re-modeled to ask this secondary question. In the eclipse experiment, the data for the primary question condensed the data from individual stars into an estimated mean deflection. When it comes to the secondary question, the relevant data are the uncondensed data reporting the observed positions of the individual seven stars in the eclipse plates. These remodeled data were then compared with the positions observed in the comparison plates. "The satisfactory accordance of the eclipse and comparison plates is shown by a study of the plate constants" (Dyson, Eddington, and Davidson 1920, 306).

In contrast to these results, when it came to checking experimental assumptions on a second set of photographs from Sobral, no such satisfactory accordance could be shown. Instead, this second set revealed violations of key experimental assumptions, and on these grounds was discounted. The validity of discounting this second set was of particular interest because it appeared to yield an estimate in accord with the "Newtonian" rather than the Einstein prediction. (More on this in chapter 8.)

The extensive battery of low-level tests of experimental assumptions, while typically overlooked in philosophical discussions, occupies center stage in an approach that properly recognizes the importance of data models. (These tests deal with the fourth type of error in my list: violations of experimental assumptions.) For Suppes, a complete description of models of data should include a delineation of the statistical tests that hypothesized data models would have to pass. The secondary tests generally demand less stringent standards than the primary ones—thanks to certain robustness arguments. In Suppes's learning theory example, one of the checks concerns "stationarity," whether the conditional relative frequency of a reinforcement is constant over the trials. An assertion like "with respect to stationarity the data are satisfactory" is generally all that is required to rule out each of the ways data could fail to be a model of data for an experiment. Typically, the tests of assumptions are applications of the same kinds of statistical tests that are used in testing primary claims, so discussing these does double duty. In this book I will only be able to discuss tests of assumptions of this type. There is also a large battery of tests designed specifically to test assumptions (e.g., runs tests), which would certainly need to be included in a full-blown experimental account.

Checking assumptions does not always call for running explicit statistical tests. In many experimental contexts this can be accomplished by formal or informal checks of instruments used in data generation. An aberrant instrumental measurement can often be discerned right off. Nor am I saying that in practice elaborate checks are always needed to affirm that the experimental assumptions are satisfactorily met. The theory of experimental design and data generation may be so well understood and well checked that they are already known to produce data that would pass the tests (at the level of the data model). (Or if not, it may often be argued, the error is almost sure to come to light before any damage is done.) This is the key value of standard statistical mechanisms for data generation—which I will return to in section 5.5.

Admittedly, I am only scratching the surface here. A full-blown philosophy of experiment calls for much more work in explicating the formal tests and the informal reasonings that go into checking experimental assumptions. This work would take place at the lowest levels of the hierarchy, which I will combine into one.

Experimental Design, Data Generation, and Ceteris Paribus

An important first step has already been taken by some of the New Experimentalists. They have given us experimental narratives chock-full of discussions of laboratory equipment and instrumental devices. It is the relationship between these tools and the overall experimental reasoning that has not been explored in any systematic way. This can be remedied, I believe, by considering how the various methods for planning and carrying out the generation of data—the theory of the experimental design—relate to satisfying the requirements of the canonical data models.

To some extent, the difference between the typical experimental narrative and what I am calling for is a matter of emphasis. For example, an important consideration under experimental design would be various protocols for instrumentation—something already found in many narratives—but the emphasis should be on how these protocols bear on experimental assumptions, for example, assessing experimental errors, making any needed adjustments, circumventing pitfalls. Information, say, about how the eclipse and scale plates were held together, about temperature, about exposure times, and so on would be relevant because these factors were vital in reaching probable errors and in answering charges about the influence of extraneous factors. In a very different sort of inquiry, considerations of experimental design would include procedures for assigning subjects to treatments, because that bears directly on an assumption of random selection in the data model.

In short, what we need to extract from the experimental experience under the rubric of "experimental design" are all of the considerations in the data generation that relate explicitly to the recorded data, that is, to some feature of the data models. But there are also numerous and sundry factors that are not incorporated in any of the models of data for assessing the primary hypothesis. These are the leftovers that Suppes places under the heading of the ceteris paribus conditions of an experiment. "Here is placed every intuitive consideration of experimental design that involves no formal statistics" (Suppes 1969, 32)—for example, the control of bad odors, the time of day, noises, the sex of investigator—where these are not incorporated in models of data. Suppes creates a distinct level for these ceteris paribus conditions—the idea being that they are assumed to hold or be controlled and are not explicitly checked. For two reasons I think it is better to include them with experimental design: First, even those features of data generation that do relate explicitly to data models need not always be checked by formal statistics. Second, features assumed to be irrelevant or controlled may at a later stage turn out to require explicit scrutiny. For example, the sex of an experimenter in a study on human subjects may not be explicitly incorporated in a data model, and thereby placed under ceteris paribus, yet a suspected influence of gender on a subject's response may later be studied as a possible obstacle to satisfying the experimental model assumptions.

Experimental design and ceteris paribus both concern possible flaws

TABLE 5.2 Hierarchy of Models of Inquiry

PRIMARY MODELS

Break down inquiry into questions that can be addressed by canonical models for
testing hypotheses and estimating values of parameters in equations and theories

Test hypotheses by applying procedures of testing and estimation to models of data

Typical form of errors: Hypothesis is false: a discrepancy of ∂ exists, the difference
between modeled data and hypothesis exceeds ε, form of model or equation is
incorrect, H fails to solve a problem adequately

MODELS OF EXPERIMENTS

Break down questions into tests of experimental hypotheses, select relevant canonical
models of error for performing primary tests

Specify experiments: choice of experimental model, sample size, experimental
variables, and test statistics

Specify analytical methods to answer questions framed in terms of the experiment:
choice of testing or estimating procedure, specification of a measure of fit and of test
characteristics (error probabilities), e.g., significance level

Specification errors: Experimental specifications fail to provide a sufficiently
informative or severe test of primary hypotheses, unable to assess error
probabilities

MODELS OF DATA

Put raw data into a canonical form to apply analytical methods and run hypothesis tests

Test whether assumptions of the experimental model hold for the actual data (remodel
data, run statistical tests for independence and for experimental control), test for
robustness

Typical errors: Violations of experimental assumptions

EXPERIMENTAL DESIGN, DATA GENERATION, AND CETERIS PARIBUS CONDITIONS

Planning and executing data generation procedures:

Introduce statistical considerations via simulations and manipulations on paper or on
computer

Apply systematic procedures for producing data satisfying the assumptions of the
experimental data model, e.g., randomized treatment-control studies

Insure the adequate control of extraneous factors or estimate their influence to
subtract them out in the analysis

in the *control* of the experiment—auxiliary factors that might be confused with those being studied. An adequate account of an experimental
inquiry requires explicating all the considerations, formal and informal,
that bear on the assumptions of the data from the experiment.

Table 5.2 is my version of Suppes 1969, page 31, table 1. It gives a rough idea of how the key models of inquiry organize the problems and errors in experimental inquiry. It corresponds to the list of questions delineated in table 5.1. While I am filling out Suppes's delineation a good deal, I do not think it differs substantially from what he has in mind. (See also table 7.2.)

Now for an example. Although I have already drawn from the eclipse experiments on the deflection of light, they are too complex to fill out now, when what we want are the bare bones of the hierarchy. A simpler example with a lower level theory is needed. The following example has the additional merit of illustrating a canonical type of inquiry when causal knowledge is sought—one based on a standard treatment-control study.

Example 5.1: Birth Control Pills

The following passage comes from a review of the evidence from studies conducted in the United States in the 1960s and 1970s on the relationship between oral contraceptives and blood-clotting disease:

> A study from Puerto Rico by Fuertes, et al., which was a randomized controlled trial of oral contraceptives, has reported *no increased incidence of thromboembolic disease in persons taking the oral contraceptives.* . . . This experiment is often cited as evidence that the case-control studies from the United States and from England . . . may be finding a spurious association. (Stolley 1978, 123)

The primary analysis in a large study by Fuertes et al. (1971) was to test whether taking birth control pills causes an increased incidence of a blood-clotting disorder in the population of women, or whether the increases observed in some studies are really spurious. I discuss this example in some detail in Mayo 1985b, but our present purpose requires only a sketch.

Primary Question. We can state the primary hypothesis as a question: is there an increased risk of developing a blood-clotting disorder among women who use the pill (for a specified time)? Would it be an error, our question asks, to suppose that the risk of a clotting disorder is *no greater* among pill takers than among non–pill takers? We are entitled to infer that a genuine increased risk exists only to the extent that we can rule out this error. The hypothesis that the risk is no greater among pill users ("the treated group") than among nonusers ("the control group") is the *test* or *null hypothesis* here. That is,

> H_0: the incidence of clotting disorders in women treated (with the pill) *does not* exceed the incidence among the controls.

If μ_T is the average or mean incidence of clotting disorders among pill users and μ_C that among nonusers, the question of interest may be framed as concerning a parameter Δ, the difference in average incidence rates (i.e., $\Delta = \mu_T - \mu_C$). The question in this study concerned the error of supposing H_0 (no increased risk) when in fact there is a positive increase in risk. The assertion of a positive increase in risk is the alternative hypothesis to the null hypothesis H_0. Whereas H_0 asserts that $\Delta = 0$, the alternative, which we can abbreviate as H', asserts that $\Delta > 0$. So we can represent the primary question as a test of H_0 against H'. That is,

Primary question: test H_0: $\Delta = 0$ versus H': $\Delta > 0$.

Experimental Model. An experiment is specified to include two groups of women, say, with 5,000 in each group. What does our test hypothesis H_0 predict about the results of this study? An idea from statistics is to represent the experiment as observing an experimental *statistic,* a function of the experiment, that parallels the population parameter Δ—the difference in the population means. The standard statistic chosen is the difference in the observed rates of a blood-clotting disease in the two groups. Let $\overline{X}_T$ represent the observed proportion with a clotting disease in the treated group, and $\overline{X}_C$ the observed proportion with the disease in the control group. Then the test statistic is $\overline{X}_T - \overline{X}_C$. This difference in the proportions with clotting disease may be called the *risk increase RI*. That is,

Test statistic (risk increase) RI: $\overline{X}_T - \overline{X}_C$.

Suppose that H_0 is true, and that the mean (or "expected") risk of clotting disease is the same in both treated and nontreated women. Then hypothesis H_0 predicts that if the two groups of women had the same chance of suffering a blood-clotting disorder at the start of the experiment, and if the only relevant difference introduced over a period of years is that one group is given birth-control pills, then the two groups would still have about the same chance of a blood-clotting disorder. However, even if H_0 is true (and the treated and untreated groups have the same chance of a blood-clotting disorder), the actual data, that is, the value of *RI*, is not expected to be exactly 0. Some experiments might yield an *RI* of 0, others will exceed 0, others will be less than 0. However, *on the average, RI* would be 0. That is, if we imagine the hypothetical population of experiments that consist in observing *RI*, under the assumptions of this experiment, the average value will be 0 if H_0 is true.

We can say more about what it would be like in a hypothetical series of experiments on a population where H_0 is true and RI calculated. We can say about each possible value of statistic RI how probable that outcome would be under H_0. A statement of the probability associated with each value of a statistic is what I call its *experimental distribution*. (I prefer this to its official name, the *sampling distribution*.[2]) The truth of H_0 (together with experimental assumptions) entails the experimental distribution of statistic RI.

Given the predicted experimental distribution we can define a standard measure of fit or distance between H_0 and statistic RI. A natural measure would be the difference between the experimental risk increase RI and the expected or predicted increase according to H_0 (namely, 0). That is,

Distance from H_0 = observed RI − expected RI (i.e., 0).

The further away from 0 the risk increase RI is observed to be, the worse is the fit between the data and what is expected if H_0 is true. For each possible fit, for each possible *distance* between RI and 0, the experimental distribution of RI gives us the probability of a worse fit, given that hypothesis H_0 is true. This is the *significance level* of the difference.

The test may simply report the significance level of the difference observed, or it may specify a cutoff beyond which the fit is so poor as to lead to rejecting H_0 and passing alternative H', that $\Delta > 0$. (A difference that reaches this cutoff is also said to be *statistically significant*.) A central concern is that the experimental test provide a severe test of whatever hypothesis passes. In a test such as this one, where the effect of interest is likely to be small, a common mistake is that the sample size is not large enough to have detected an increased risk even if there is one. The possibility of this mistake needs to be investigated in the eventuality that the observed RI is not statistically significantly different from 0. I shall return to the Fuertes study (section 6.5) in tackling the problem of insufficiently severe tests.

Data Model. Approximately 10,000 women were randomly assigned to either the treated group or the control group, each with approximately 5,000 women. The recorded data, the data model, is the value of RI obtained once the experiment has been run and the results observed.

2. My reason for this preference is that experimental distributions (distributions of experimental statistics) are utilized in science in contexts not ordinarily construed as sampling from a population.

We can abbreviate the observed value of RI as RI_{obs}. Its actual value in this study was 0.0002.

Data model: the observed value of the risk increase: RI_{obs} (e.g., 0.0002).

In order for RI_{obs} to be used to test H, it must satisfy adequately the assumptions of a *comparative* or *treatment-control random experiment*.

Separate tests for each such assumption may be carried out. To ensure that the treated and control women were sufficiently homogeneous with respect to the chance of suffering a blood-clotting disorder, the researchers tested a series of hypotheses that were *themselves* examples of null or "no effect" hypotheses: that the two groups do not differ significantly with respect to year of admission, age, number of pregnancies, last school year completed, income, years of marriage, and other factors thought to be relevant to clotting disorders. They found no significant difference between the treated and control groups on any of these factors. They found, in short, that the assumption of no difference on each of these factors was sufficiently met in their experiment.

Experimental Design, Data Generation, and Ceteris Paribus. Experimental design includes all of the pretrial methods aimed at producing a sample that satisfies the assumptions of a comparative-randomized study, for example, randomized assignment of subjects into treated and control groups, use of placebos, blindness of subjects and doctors. Notice that these methods correspond to the goal of satisfying the experimental assumptions—the assumptions that are tested by the series of significance tests at the level of the data models. After-trial corrections (e.g., eliminating certain subjects) may also be called for.

Then there is the array of extraneous factors assumed to be either irrelevant to the effect of interest or satisfactorily controlled. The correctness of this assumption itself can, in principle, come up for questioning after-the-trial. In this study extraneous factors would include those not likely to have been recorded, such as hair color, patient height, astrological sign.

5.3 THE HIERARCHY OF MODELS AND THE TASKS OF AN EPISTEMOLOGY OF EXPERIMENT

This framework of models of experimental inquiry does double duty for my account: it provides a means for setting out the relationship between data and hypotheses that organizes but does not oversimplify the complexities of actual experimental inquiries, and it provides a framework to address systematically the key tasks of an epistemology of experiment thus far sketched out. The sampling that follows, which shows some of the ways it does this, is also a preview of some of the problems yet to be taken up in this book as well as the many tasks that still lay ahead for this experimental program.

Vague Statements of Data or Evidence Will Not Do

Recognizing the need for each step in the hierarchy should ward off the central ambiguities and oversimplifications of approaches that try to articulate some general relationship between data and hypothesis. In particular, it checks the tendency to blur the different types of data that arise as evidence in a substantive scientific appraisal. The framework being proposed not only requires that references to "data" be situated within a specific inquiry, but also that they be attached to a particular level and model of that inquiry.

Take the example of the use of observations and measurements from a single eclipse expedition to address questions about Einstein's law of gravitation. The "data" may refer to *at least* this many different things: (*a*) the raw measurements of the positions of a group of stars before and after the eclipse; (*b*) the estimate, resulting from data analysis (least squares averaging, error corrections) of the deflection of light observed for the actual stars measured (which were at least two times the solar radius away from the sun);[3] (*c*) the estimate, using the data in (*b*), of the deflection that would have been observed at the limb of the sun, together with its standard errors; or (*d*) the estimate, using results in (*c*), of the deflection due to the sun's gravitational field. These different data statements roughly correspond to the results of the theory of experimental design, the data models, and [(*c*) and (*d*)] to two experimental models.[4]

It follows that epistemologists of experiment need to be able to cope with real data and methods for their analysis, not only to have anything genuinely relevant to say about how to conduct inquiries, but also to understand historical episodes in science.

Canonical Experimental Questions

Breaking down a substantive inquiry into the framework of a hierarchy of models is needed to render the analysis tractable, but not just

---

3. This can also be stated counterfactually as the deflection that would have been observed had repeated measurements been made. It too is an inference. Note that this is an example of how a more reliable observation can arise from inferences built on less reliable ones.

4. Basically, statement *c* corresponds to an estimation of the mean deflection of light at the limb of the sun, statement *d* to a test (or estimation) of the mean deflection of light (at the limb of the sun) that is due to the sun's gravitational field.

any delineation will do. Hark back to our recasting of Kuhn's demarcation of good scientific inquiries: in a good inquiry it must be possible
to learn from failures. As such, we said, the criteria of good normal
science (good standard testing) direct us to become appropriately small
in our questioning. There is nothing about this framework that requires starting from the scientific hypotheses and working one's way
down to the data. A familiar situation is to start with some data model
and then ask, of what experimental hypothesis could these data be
regarded as having provided a good test? In an appropriately specified
experimental model, the answer feeds up to the primary hypothesis as
well. But it may turn out that only some more restricted hypothesis
has been well tested.

There is a variant on this strategy that needs emphasis. Here, one
starts out by surveying the feasible experimental models and the questions they allow to be posed, and then designs the primary hypotheses
so that these questions are applicable. Suppose, for example, that the
only feasible experimental question would be to ask, *retrospectively,*
whether women with blood-clotting disorders had had exposure to
oral contraceptives. Then one might choose to pose a primary question
about the existence of such an association, rather than about a causal
connection (where a prospective study is wanted).

These considerations lead me to introduce the idea of *canonical experimental questions.* One set of (sufficient but not necessary) conditions
that would make questions standard or canonical is that (1) they correspond to questions of interest in a large variety of inquiries, (2) they
are associated with standard experimental design techniques known
to make it likely that assumptions are sufficiently met, (3) they are
open to applying standard statistical tests (e.g., goodness-of-fit tests) to
check systematically experimental assumptions, and (4) there are well-
understood methods of estimation and testing that link data models
with answers to the canonical question.

Clearly, an adequate experimental inquiry will often have to pose
several different primary questions whose answers are inputs into
other inquiries. But the complexity of the story pays off in terms of
understanding actual inquiries and disputes. It is also the key to grappling with a number of philosophical problems.

The Three Decisions

In chapter 1 we discussed Popper's "three decisions" and the problems each raised for his account. They were: falsifying (or, more generally, testing) statistical claims, arriving at test statements, and ruling
out alternative auxiliary factors. These correspond most nearly to the

problems addressed by theories of experiment (or experimental testing), data, and experimental design, respectively.

For example, the problem of articulating what counts as a genuine
or severe test—a problem Popper never satisfactorily solved—will be
addressed within a theory of testing based upon standard error statistics (in the next chapter). Considerations of severity are the pivot point
around which the tasks at the levels of data models and experimental
design will turn. Where the need for reliable test statements and "falsifying hypotheses" were embarrassments for Popper, and where auxiliaries posed an insuperable threat, tools for achieving reliable (modeled) data and adequate experimental control are seat-of-the-pants
techniques of data analysis and experimental design.

Duhemian Problems Revisited

Suppes emphasizes, quite correctly, that checking whether the actual data satisfy the assumptions of the experimental data model calls
for criteria "to determine if the experiment was well run, not to decide
if the [primary hypothesis or model] has merit" (Suppes 1969, 32).
Likewise, a problem with the experimental model is distinct from a
problem with the primary model. Indeed, errors at all but the level of
the primary model reflect weaknesses in the experiment, not in the
primary hypothesis being tested. Separating out the models relating
data and hypotheses goes a long way toward achieving the aim of correctly apportioning blame (as well as praise). Frederick Suppe, in discussing Suppes, puts it this way:

> The actual experimental setup will be a putative realization [model]
> of the theory of experimental design and the *ceteris paribus* conditions;
> the actual correction procedures used to put the raw data in canonical
> form will be putative realizations of the theory of data. And the pre-
> dictions yielded from the [physical] theory will be putative realiza-
> tions of the theory of the experiment. . . . In case of anomalous or
> disconfirming experimental results, the source of the anomaly may
> be the result of the experimental procedures failing to be realizations
> of any of these theories or as a result of the theory's empirical falsity.
> (Suppe 1977, 108–9)

Why do I find it illuminating to express Duhem's problem in terms
of the hierarchy of models whereas I had denied that characterization
of the Bayesian formulation (in chapter 4)? Because this expression
gives a realistic picture of how such problems are effectively grappled
with. To remind us of the Bayesian solution, exemplified by Dorling,
we can utilize example 5.1, on oral contraceptives. The test hypothesis

H_0 was that the birth control pill does not increase the risk of clotting disorders (or does so negligibly). Suppose that the result showed a risk increase RI that was statistically significant among women receiving the pill. Let auxiliary hypothesis A assert that the auxiliary factors are not responsible for the observed increase. To assess whether H_0 or A is most disconfirmed, the Bayesian calculates the posterior probabilities. But these depend on prior degrees of belief. If the Bayesian researcher strongly believes that the pill imposes no increased risk (and thus has a high enough prior degree of belief in H_0), then one has the makings of a situation where the Bayesian analysis disconfirms A—the blame falls on the auxiliaries rather than on the pill.

In our framework, the question of hypothesis A arises at the stage of checking whether the data obey the assumptions of the experiment. Separate work to show that the auxiliaries are responsible is required before disconfirming A. There is no consideration (here or elsewhere, for that matter) of what we called the Bayesian catchall factor. One's belief in the hypothesis that the pill does not increase the risk of clotting disorders simply does not enter into the separate statistical tests to check whether the influence of each auxiliary factor is responsible for the observed effect.

This is not to say that background theories or, if one likes, background beliefs, play no role in the experimental inference. Background knowledge (e.g., concerning the pill's action and the nature of blood-clotting disorders) is what led researchers to look for risk increases and not decreases, and to control for some factors (e.g., diet) and not others. But whereas specifying experiments may well reflect subjective beliefs, pragmatic factors, or something else, what the results *do and do not indicate* is not a matter of subjective beliefs. By reporting the features of the test specifications, researchers enable others to scrutinize a result, to perhaps find experimental assumptions violated or a result misinterpreted or irrelevant (to some primary question that interests them). Eddington believed that the aberrant result (from the second set of plates at Sobral) was due to the sun's distortion of a mirror. But the only way this belief enters, the only way it ought to enter, is in motivating him to make a case for blaming the mirror. And that case should not depend on the correctness of the primary hypothesis that the data are supposed to test. If it does, a different alarm goes off, now at the level of the experimental model to test H— the resulting test of H may fail to be severe.

Methodological Rules Must Be Assessed in Context

Just as data statements need to be scrutinized within the appropriate data model, rules and principles of experimental methodology need to be queried by reference to the proper experimental node. In section 5.1 I listed the key questions that correspond to the different models of inquiry. Different principles of experimental methodology relate to one or more of these questions, and proposed rules should be assessed in relation to these tasks. Some refer to before-trial experimental planning, others to after-trial analysis of the data. The former include rules about how specific errors are likely to be avoided or circumvented; the latter, rules about checking the extent to which given errors are committed or avoided in specific contexts.

If I am right, then we can understand why the task of appraising methodological rules (metamethodology) has run into such trouble as of late. Take a much discussed methodological rule, sometimes called the rule of predictive novelty: prefer hypotheses that predict unknown facts to those that are constructed to account for known facts. The way to assess naturalistically this and other rules, say some (e.g., Laudan 1987, 1990b, 1996), is to see how often a rule correlates with successful hypotheses and theories. Scientific episodes that violate the prediction rule and yet end up with successful hypotheses are taken as evidence against the efficacy of the rule. The present account rejects this kind of appraisal as misleading or uninformative.[5] Aside from the difficulty of inferring the efficacy of a rule by finding means-end correlations, such a strategy of appraisal is vitiated by the "false-negatives" that result. That is, even where a rule is effective for a given end—say, for attaining the aim of reliable hypotheses—we should expect the historical record to provide us with violations (i.e., with cases where the end is achieved even where the rule is violated). Why? Because insofar as there is progress in experimental methodology, it is likely that methods are developed with the express purpose of securing reliable results even where a rule is not or cannot be strictly upheld.

For an example, consider an uncontroversial rule: perfect control in causal inquiries serves to obtain reliable causal hypotheses. The more we learn of how extraneous uncontrolled factors can create obstacles to correctly attributing causes, the better we have gotten at avoiding these obstacles even without literal control. (Such ways include randomizing treatments, double-blinding, and stratification.) As with violating strict controls so it is with violating predictive novelty. With respect to certain test procedures, constructing hypo-

5. For one thing, it is far from clear how to determine when a rule has been applied or followed in reaching a hypothesis and when it has been violated. For another, there is the problem of determining whether a hypothesis resulting from a rule is successful or reliable. Moreover, even if applying a rule is found to be correlated with a successful hypothesis, it would not show that obeying the rule is the cause of that hypothesis's success. Leplin (1990) raises related criticisms.

theses to fit known facts is wholly unproblematic because the overall reliability of the inference is sustained. (I take up this rule in chapter 8.)

Rules cannot be properly appraised apart from their specific tasks
in inquiry. Since the rules are claims about strategies for avoiding mistakes and learning from errors, their appraisal turns on understanding
how methods enable avoidance of specific errors. One has to examine
the methods themselves, their roles, and their functions in experimental inquiry.

5.4 CANONICAL INQUIRIES INTO ERRORS

An adequate epistemology of experiment should shed some light
on the problems that arise at each level of the hierarchy of models. A
good place to start is with the task of breaking down substantive inquiries into manageable pieces. These piecemeal experimental inquiries are guided by canonical inquiries into the four types of errors
I have marked out: errors about real as opposed to accidental effects,
about parameter values, about causes, and about experimental assumptions (of the type addressed in checking data models). Canonical questions about standard types of errors, and the methodological
tools available for addressing them, are heuristic guides for how to
proceed.

An inquiry into a possible error is itself an example of an experimental inquiry. Hence we can set out the canonical models for inquiring about errors by employing the hierarchical framework of models
of inquiry. We can do this by letting the primary question of the inquiry be asking after the specific error. In carrying out a substantive
inquiry, the methodologist's arsenal includes exemplary or paradigm
examples of inquiries that have as their primary question asking about
key errors. These are the canonical models of error or, more correctly,
models of error inquiry.

The first error I have listed is mistaking artifacts or "mere chance"
for real or systematic effects. In distinguishing real from accidental effects, a common model to employ, either directly or by analogy, is the
Binomial or "coin-tossing" model of experiment. In the next section I
will give the bare essentials of the components of a Binomial experiment.

A Binomial Model of Experiment

So-called games of chance like coin tossing and roulette wheels
are exemplary processes used in modeling a variety of questions about

errors. Take coin tossing. At each trial of a standard coin-tossing experiment, there are two possible outcomes, "heads" and "tails." The Binomial model labels such dichotomous outcomes as "success" and "failure," which may be abbreviated as s and f, with the probability of success at each trial equal to some value p.[6] The population here is really the coin-tossing mechanism or process, but because this process is modeled by the Binomial distribution, we often speak of the *population distribution*. Running n trials of the experimental process, in other words, is tantamount to taking n samples from this population distribution. As with all statistical distributions, the Binomial distribution assigns a probability to each possible outcome, here to each possible number of successes out of n trials. Once the probability of success on each trial, p, is given, this probability assignment follows. So the key quantity or *parameter* of the Binomial distribution is p, and assertions about the value of p are statistical hypotheses defined in terms of the Binomial model.

An example of a primary hypothesis framed in terms of this model is the hypothesis that p equals .5 (the standard "fair coin" hypothesis). This is a typical example of a *test* or *null hypothesis*, and can be abbreviated as H_0:

H_0: the value of the parameter of the Binomial model, p, equals .5.

Alternative hypotheses are defined in relation to H_0.

Experimental Model: The Binomial Experiment

To be more explicit, let us consider a tiny experiment consisting of only three independent tosses of a coin, and where H_0 is true, that is, the probability of heads p on each trial is equal to .5. Then we can calculate the probability distribution of the experimental variable or statistic $\overline{X}$—the proportion of heads in the 3 tosses. The proportion of heads in n tosses can also be expressed as the *relative frequency* of heads in n tosses. The experimental distribution of statistic $\overline{X}$ consists of a list of possible values of $\overline{X}$ together with a probability assignment to each value.[7] To arrive at this experimental distribution, first consider each

6. Such trials are called Bernoulli trials, with p the *parameter*.
7. Rather than list the probability for each possible number k of successes out of n trials, a general formula is given. The probability of getting k successes out of n Binomial trials (each having a probability of success equal to p) equals

$$P(\text{relative frequency of success} = k/n) = \frac{n!}{k!(n-k)!} p^k(1-p)^{n-k}.$$

TABLE 5.3 Binomial Experiment

Experimental Outcome	Value of $\overline{X}$	Probability
f,f,f	0	$\dfrac{1}{8}$
f,f,s	$\dfrac{1}{3}$	$\dfrac{1}{8}$
f,s,f	$\dfrac{1}{3}$	$\dfrac{1}{8}$
f,s,s	$\dfrac{2}{3}$	$\dfrac{1}{8}$
s,f,f	$\dfrac{1}{3}$	$\dfrac{1}{8}$
s,f,s	$\dfrac{2}{3}$	$\dfrac{1}{8}$
s,s,f	$\dfrac{2}{3}$	$\dfrac{1}{8}$
s,s,s	1	$\dfrac{1}{8}$

of the 8 possible outcomes—each a sequence of 3 ss or fs. The probability of each is $(\frac{1}{2}) \times (\frac{1}{2}) \times (\frac{1}{2})$ or $\frac{1}{8}$ (table 5.3).

$\overline{X}$ has 4 distinct values. The probability of each value of $\overline{X}$ is calculated by summing the probabilities of all the individual outcomes that lead to that value. This yields the experimental (or sampling) distribution of $\overline{X}$ (table 5.4).

Data Model

A given experimental outcome is only one of the 8 possible sequences of 3 trials (e.g., s,s,f), yielding only one of the 4 possible values for $\overline{X}$. In order for it to be a random sample, the experimental assumptions must hold. That is, each of the 3 trials must be an independent Binomial trial with p equal to $\frac{1}{2}$.

It should be remembered that a *statistic*, such as the proportion of successes $\overline{X}$, is a property of the experimental outcome, whereas a *parameter*, such as p, is a property of the population from which the outcome is a sample. If assumptions are sufficiently met, however, the value of the statistic can be used to learn about the population parameter. A highly important relationship for all of statistics is that between the sample mean $\overline{X}$ and the population mean. It is this: the average or

TABLE 5.4 Experimental Distribution of $\overline{X}$ (with $n = 3$, $p = 0.5$)

Value of Random Variable $\overline{X}$	Probability
0	$\frac{1}{8}$
$\frac{1}{3}$	$\frac{3}{8}$
$\frac{2}{3}$	$\frac{3}{8}$
1	$\frac{1}{8}$

mean value of the average $\overline{X}$ is itself equal to the population mean. This relationship is entirely general. Since, in the Binomial distribution, the population mean is equal to p (the probability of success on each trial) we get:

The mean of $\overline{X}$ = the population mean = p.

Thanks to this, an observed value of the sample mean $\overline{X}$ is used to learn about the mean of $\overline{X}$, and thereby to learn about the population mean p (to which the latter is equal).[8] The standard deviation of the statistic $\overline{X}$ is also related in a known way to the standard deviation of the population, e.g., in the Binomial experiment the standard deviation of $\overline{X}$ is the square root of $\frac{p(1 - p)}{n}$.[9]

In explaining the key features of statistical tests, a central role is played by paradigm examples or exemplary cases where tests supply tools for inquiring about common types of errors. Statistical models not only offer paradigms for asking about errors, they also embody much of our ordinary understanding of these errors. The following is a variant on Fisher's famous example of the "Lady Tasting Tea."

Example 5.2: A Lady Tasting Tea: Mistaking Chance for a Real Effect

To avoid mistaking a merely chance effect for a genuinely systematic one, we need a way of discriminating between the two. Fisher

8. The mean of $\overline{X}$ is also called the *expected valued* of $\overline{X}$, written $E(\overline{X})$. So we have $E(\overline{X}) = p$. The mean is defined as the arithmetic average.
9. The standard deviation of X, the population distribution, equals the square root of $p(1 - p)$.

describes one way by telling his story of the "lady tasting tea."[10] Imagine that a lady maintains that she can tell by tasting a cup of tea with milk whether the milk or the tea was first added to the cup. Suppose that her claim is not that she will always be right, but that she can do better than chance. In other words, the lady is claiming that she can do better than someone who is merely guessing. To say that her claim is in error is to say that in fact she can do no better than someone who is merely guessing, that her pattern of successes is merely "due to chance." To express this in a canonical way, denying her claim asserts that the lady's procedure is one that produces no more successful predictions than the process of guessing.

The standard null hypothesis here formalizes the claim that the effect is "like guessing" or "due to chance" by appealing to a familiar chance process. It asserts that the process that produces the lady's pattern of successful discernments is like the process that produces the pattern of heads and tails in fair tosses of a coin. They are alike in their respective relative frequencies of success. This suggests running an experiment where the Binomial model can be utilized as a canonical model of error.

The Binomial Model as a Canonical Model of Error

The proposed experiment in Fisher's lady tasting tea example involves recording the success or failure of each of her n predictions about the order of tea and milk in the infusion. The hypotheses of interest are claims about the value of the parameter p of the Binomial model. The possible values for p, the parameter space, range from 0 to 1. Let us say that the interest is only in whether p is .5 or greater.[11] The hypothesis that the lady is guessing or, more generally, that the process is of the chance variety is the null hypothesis H_0. H_0 asserts that p, the probability that the lady will give the correct answer, equals .5. Within the statistical model of experiment it may be stated as follows:

H_0: p equals .5 (the lady is merely guessing),

while the lady's claim H' is of the form

H': p is greater than .5 (the lady does better than one who is merely guessing).

10. It is discussed in Fisher 1947. An extended introductory treatment of this example can be found in Bailey 1971.

11. The treatment is analogous if the interest is in p differing from .5 in both directions. In the corresponding test (which is called a "two-tailed" test) the null hypothesis is rejected when the success rate is *either* much larger than .5 or much smaller than .5.

In the canonical Binomial model of experiment, each possible outcome is a sequence of n successes or failures, one for each trial of the experiment. The procedure of running a trial, say trial i, and recording either success or failure, is represented in the model as the result of observing the value of a Binomial (random) variable X_i. Each n-fold sequence of outcomes may be represented as a vector of length n, X_1, X_2, . . . , X_n where each X_i is a Binomial variable. (Each is assumed to be independent of the other, and p is the probability of success at each trial.) There are 2^n such possible samples.

Experimental Statistic

One needs to consider what question can be put to the experiment in order to extract information about the parameter of interest. The function of the data—the statistic—that would be canonical in tests about the parameter p is the *proportion* or *relative frequency* of successes in the n experimental trials $\overline{X}$. The statistic $\overline{X}$ condenses each n-fold outcome into a single quantity. Whether any information about p is lost by this condensing process is an important question that cannot be gone into here. It turns out that $\overline{X}$ is an example of a statistic containing all the information in the sample concerning p. It is a *sufficient* statistic.

In specifying the experimental model we must decide upon the number of samples, n. Suppose we let $n = 100$.[12] The lady will be given 100 cups of tea to test her discriminatory powers. Thus the test (or null) hypothesis, H_0, essentially asserts that the experimental process is like that of 100 tosses with a fair coin. More specifically, H_0 says that the probability of r percent successes in 100 trials with the tea-tasting lady is the same as the probability of r percent successes in 100 trials of tosses of a fair coin (where success is identified with the outcome heads). In other words, the experimental statistic $\overline{X}$ has the same distribution as the proportion of heads in 100 (independent) tosses of a fair coin (i.e., the probability of heads on each toss is .5). This is the *experimental distribution*.

Data Model

The data of the canonical Binomial experiment may be modeled in different ways. That is, the same data may be considered as belonging to a number of different experimental populations, each corresponding to a different way of cutting up or partitioning the possible

12. Fisher used much smaller sample sizes in presenting the lady tasting tea example. The use of sample size 100 is to simplify each of the calculations as well as to ensure the assumption of Normality of the sample means.

experimental outcomes or sample space. When learning about parameter p, the relevant set of experimental outcomes consists of the different possible proportions of successes in 100 trials. (There are 101 such results, one for each possible number of successes out of 100 trials.) On a single realization of the experiment, only one of the members of the sample space is obtained, that is, only one list of 100 outcomes is recorded,

$$x_1, x_2, \ldots, x_{100},$$

where x_i is the result of trial i. Each is condensed into a single value: the observed proportion (or relative frequency) of successes. Say, for example, that 60 percent successes are recorded. Then we have $\bar{X} = .6$.

The probability of any such result could be calculated if one knew the value of the parameter p, the probability of success on each trial. But that is what we are trying to find out. However, for any hypothesized value of p, the probability of each outcome may be calculated, which is all that is required for testing claims such as H_0. Still, this calculation assumes that the actual experiment really is a Binomial experiment—that the realized data are a model of the data. With respect to the actual experiment, this formal assumption corresponds to satisfying the experimental assumptions.

The worry about experimental assumptions occurs at the level of the data model in the hierarchy. However, the work that goes into satisfying the experimental assumptions would be placed at the levels below the data model. In the ideal experimental design, all the cups of tea are exactly alike except in the factor of interest: the order in which the tea and milk are placed in the cup. Such a requirement, as always, is impossible. A good deal of experimental design and data analysis have been developed to construct experiments that, although not literally controlled, allow arguing as if they are. Thanks to these methods, one may secure data that are as accurate and reliable as if the experiment were controlled, and the degrees of accuracy and reliability can be made precise.

Some sources of relevant differences in the tea-tasting experiment might be the amount of milk or tea in each cup, the amount of sugar, if any, the type or color of the cup, and the temperature of the mixture. Once specified, it may be possible to largely avoid the difference and so control for the given factor. But it would typically be impossible to check each of the possible factors that might influence a result in a given experiment. This leads some to object that auxiliary hypotheses can never fully be ruled out in pinpointing the source of an experimental result. Two things must be noted: First, it is a mistake to suppose that the elimination of disturbing factors is necessarily desirable. Analyses based on deliberately ignoring certain factors are often as or more reliable and accurate than those that eliminate them. One of the aims of statistical methods of experimental design is to tell us which factors are worth controlling for and to what extent. Second, it matters only that the ability to learn from the experiment not be impeded by such disturbing factors.

One possible strategy in this example would be to randomly assign 50 of the 100 cups to the group that is to have tea poured in the cup first and let the remaining 50 have tea poured second. In this way, each cup has the same chance of being in the "tea-first" group.[13] These aspects of the theory of the experimental design, we see once again, refer to the pretrial methods to satisfy experimental assumptions. For each of the assumptions of the Binomial experiment there are separate, after-trial statistical tests to check whether they are satisfied. Often more than one test will be carried out to test a single assumption, say, of independence. Naturally, the tests of assumptions will themselves have assumptions, but they may be designed to be extremely easy to satisfy (e.g., as with distribution-free tests). Moreover, since different tests have different assumptions, if none detect that a given assumption is violated, then that will give excellent grounds for taking the assumption as satisfied. Notice once again that we are arguing from error: there is a very high probability that at least one of the tests of assumptions would detect a violation, say, in independence, if one were committed. So if none do, the assertion "independence is satisfied" passes a severe test.

Do we have to take additional samples to carry out separate tests of assumptions? Possibly, but in the canonical experimental models a typical strategy is to use the same data set and model it differently. That is why many different data models may be utilized in the context of a single experimental inquiry.

Experimental Tests

It is time to say more about how data models can be used to answer primary questions by means of tests. I locate this link between data models and a primary question in the experimental model. A *test* will always be defined within a context of an experimental inquiry. A formal statistical test is generally defined with respect to a single data set and a single question. When we talk about the severity of tests of sub-

---

13. In Neyman's discussion of this case, each trial consists of a pair of cups—differing only in the order of the tea and milk infusion. Although this pairwise design seems more satisfactory, I keep my discussion closer to the simpler design used by Fisher.

stantive hypotheses, however, we will generally be referring to the test provided by the experimental inquiry as a whole—where this generally includes several tests. (For example, we will speak of the 1919 eclipse test of Einstein's gravitational theory.) Just now, however, I am focusing on a single statistical test. The notion of *an experimental test statistic* is a neat way to define tests.

An experimental test statistic is a function of the experimental outcome that links the (modeled) data (in this case, the sample mean $\overline{X}$) with the null hypothesis H_0. This test statistic may be seen as a *distance measure*, or as a measure of fit (or, perhaps, misfit).[14] It measures, for each possible experimental outcome, how far the data are from what is expected under H_0 (in the direction of some alternative hypothesis H'). For example, in testing our hypothesis H_0: the lady tasting tea is just guessing, a sensible distance measure is the (positive) difference between the observed mean, $\overline{X}$, and the expected mean, .5.

Distance measure $D(\overline{X}) = \overline{X}$ observed $- \overline{X}$ expected (i.e., .5).

In this example, the larger the value of difference D, the more indication there is that H_0 is not true—that her pattern of successes is *not* like the pattern that would be expected if she were guessing.

The significance level of a difference is a prime example of a formal distance measure, with this inversion: the smaller the significance level the greater the difference. When the experiment is run, some particular value of D is observed. Let us abbreviate the observed difference as D_{obs}. We can then define the statistical significance level associated with D_{obs}:

Statistical significance level of the difference D_{obs} (in testing H_0) = the probability of a difference as large as or larger than D_{obs}, given that H_0 is true.

Recall the discussion of the "significance question" in Galison's neutral current example in chapter 3. That analysis, as well as the birth-control-pill example in this chapter, are spin-offs of the canonical question articulated in the lady tasting tea exemplar. Recall that the significance question asks, What is the probability that, if hypothesis H_0 is true, a result even more distant (from H_0) than the observed result is would occur in trials of the experimental procedure? The experimental distribution provides the answer.

Suppose that the observed proportion of successes in the tea-

14. The idea of a test statistic as a measure of distance or deviation is emphasized in Cramér 1974 and Kempthorne and Folks 1971.

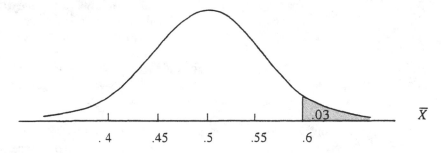

FIGURE 5.2. The experimental distribution of the binomial statistic $\overline{X}$ with $n =$ 100, given that H_0 is true. $P(\overline{X} \geq .6 \mid H_0) = .03$

tasting example equals .6. The difference from what is expected, under H_0 (.5) is .1. Since the needed significance probability is under the assumption that p equals .5 in a Binomial experiment, it is easy to calculate this value. The probability of a result even further (from .5) than .6 is .03. A graph of the experimental distribution with 100 trials is shown in figure 5.2.

Figure 5.2 makes plain why the significance level is often referred to as a "tail-area" probability. It gives the probability of outcomes at the tail end of the distribution.

The test can simply report the observed significance level or set a cutoff point beyond which the null hypothesis H_0 is rejected. A plausible cutoff point might be 60 percent successes (out of 100 trials), corresponding to a significance level of about .03.

Error Probabilities

Suppose that the test rejects H_0 just in case the difference between hypothesized and observed proportions of successes exceeds .6. To reject H_0 when in fact H_0 is true is an error—in particular it is a *type I error*. To reject H_0 only if a significance level of .03 is reached is to have a test with a .03 probability of committing a type I error. The methodology of experimental testing, in the standard (Neyman-Pearson) approach, specifies tests, not only with an eye toward a low type I error. It also considers the probability the test has of detecting a real effect if it exists. If the test makes it so hard for the lady to demonstrate her power, it may have little or no chance of detecting it even if it exists. That is, we also want to avoid what is called a *type II error*, failing to detect an effect (accepting H_0) even when an effect exists. For each possible value of p in excess of .5 (for each particular value contained in alternative H') we can calculate the probability that a given test would fail to detect it, even if it existed. Before the trial, we can specify the test so that it has a sufficiently high probability of correctly detecting effects of interest. It should, in the Neyman-Pearson terminology, have sufficiently high *power*.

I will put off discussing these two formal errors further until chapter 11. By then we will have a clear understanding of the notion of severity, which will enable us to understand the role and rationale of error probabilities in experimental arguments.

Let us summarize. A (null) hypothesis H_0 that hypothesizes that it would be an error to consider her ability better than .5 says, in effect: any better-than-guessing performance is due to chance. There is a distance measure, D, that measures the difference between what is expected, assuming that H_0 is true, and an actual experimental outcome. In simple significance tests only the associated significance level is reported. Alternatively, a test may report "H_0 fails" or "reject H_0" whenever this distance exceeds some specified value. The minimum distance that results in H_0 failing is specified so that H_0 would rarely be rejected erroneously. In statistics, rejecting H_0 erroneously is called a type I error. The test is specified so that the probability of a type I error is low. At the same time the test is chosen to have a suitably high probability of detecting departures from H_0 if they exist. After the trial, these two error probabilities are crucial for interpreting the test result. In order to do so correctly, however, it is necessary to consider a test's severity with regard to a particular hypothesis. We will take this up explicitly later.

Checking Data Models: The Testability of Experimental Assumptions

A statistical significance test may also be used to check that the data satisfy the assumptions of the experimental model. The null hypothesis would now assert that a given assumption is met by the data, and the data would now be the actual, not the condensed, data set. Suppes describes the pattern of (goodness-of-fit test) reasoning as follows:

> For each possible realization Z of the data, we define a statistic $T(Z)$ for each question [about assumptions]. This statistic is a random variable with a probability distribution. . . . We "accept" the null hypothesis if the obtained value of the statistic $T(Z)$ has a probability equal to or greater than some significance level α on the assumption that indeed the null hypothesis is true. (Suppes 1969, 29)

The kind of analysis that Suppes describes is quite typical in testing assumptions. It was, for instance, the mode of testing extraneous factors in the study on the pill and blood clotting. It can perhaps be seen from this brief sketch why I have separated out mistakes of experimental assumptions in my proposed list of key error types (section 1.5, error type *d*). First, the appeals are to very low level, very often distribution-free, methods, and second, when it comes to ruling out these mistakes, it is sufficient to *accept* (or fail to reject) a null hypothesis. In other words, it is generally sufficient here to be able to say that the actual data are not too far from what would be expected if the assumption in question holds. In well-designed experiments, in canonical inquiries, assurances of that order are sufficient for the key goal—obtaining a severe test of a primary hypothesis (or of hypotheses relating to a primary one). Threats that would call for a more substantive inquiry should already have been taken care of by choosing well-understood and well-tested procedures of data generation.

At this point, two contrasts with the subjective Bayesian Way bear noting. First, the statistical significance tests for checking data-model assumptions are part of the battery of standard statistical methods. "For the purposes of this paper," Suppes remarks, "it is not important that some subjectivists like L. J. Savage might be critical of the unfettered use of such classical tests" (Suppes 1969, 30). One wishes he had elaborated. For our purposes it is important, and it highlights the real and substantial difference in the positions of standard error statistics and subjective Bayesianism.

Second, the ability and need to test the experimental-model assumptions are an important part of why standard statistical methods are often referred to as "objective." The "skeptical experimenter," as Suppes calls him (p. 30), can always charge that data may pass all the tests of assumptions and still be guilty of violating them. Suppes points out, however, that the properties of actual procedures and checks have been thoroughly investigated, which "makes such a dodge rather difficult" (ibid.). In the terms I have been developing, it can often be argued that passing a battery of tests is overwhelmingly improbable if assumptions are violated by more than a specified extent. In any event, experimental assumptions are part of the statistical report and can be challenged by others; they are not inextricably bound up with subjective degrees of belief in the final sum-up.

5.5 MODELS OF PROBABILITY, RELATIVE FREQUENCY, AND EXPERIMENTATION

The statistical theory of experiment deals only with certain kinds of experiments insofar as their behavior may be characterized by certain parameters. A characteristic of key interest is the relative frequency with which an outcome occurs, or would occur, in a sequence of applications of the experiment in question.[15] (Note that in the Binomial experiment, the average number of successes in n trials, i.e., the mean $\overline{X}$, is the same as the relative frequency of success.) It is because of the interest in relative frequencies that probabilistic or statistical models arise in learning from error.

The Relevance of Relative Frequencies

There are several reasons why relative frequencies in series of experiments, often hypothetical, are of interest. In chapter 4 I advanced a frequentist construal of claims about the probable success of hypotheses and contrasted it to that of Salmon and Reichenbach. What is of interest and is really obtainable, I said, was information about the relative frequency with which a series of applications of the hypothesis H would yield results in accordance with what H asserts. A specific description of a successful outcome could be given by way of a suitable measure of fit or distance. This kind of relative frequency knowledge describes one thing that experimental data teach us about the physical system at the top of the hierarchy. That is, we learn about a hypothesis by learning about how well and how often it would succeed in a sequence of applications of interest. (This is experimental knowledge.) On the way to obtaining that kind of knowledge, relative frequencies of outcomes also play a central role in experimental inquiry.

Raw experimental data, we know, are finite, and are distorted by errors and "noise" from extraneous factors. A chief role for statistics is to estimate the likely extent to which extraneous factors influence results in the experiment at hand. We can then "subtract them out" to see the effects of the parameters of interest. In the case of the lady tasting tea we are effectively subtracting out the effect of mere guessing. Discriminating backgrounds, signal from noise, real effect from artifact, we said, is the cornerstone of experimental knowledge. Beginning with canonical examples such as the lady tasting tea, one arrives at spin-off strategies for accomplishing these ends in diverse fields.

Recall our discussion of Galison's neutral currents case. The experimental result was the recorded ratio of muonless to muonful events, namely, 54/56. We want to know whether observing so many muonless events (in this one experiment) may be due to something other

15. In some cases repetitions of the same experiment are of interest, in others a sequence of applications of different experiments with the same distribution is of interest.

than the existence of neutral currents (e.g., escaping muons). We would like this one result to reveal its system of origin. We ask, Did you come from a system where there are neutral currents? by asking, Are you a fairly typical product of an experiment where there are *no* neutral currents? This is what the significance question addresses. As we saw, the significance question is, How often, in a series of experiments such as the one done by the HWPF group, would as many muonless events be expected, given no neutral currents? If this is the sort of occurrence that is rather typical even if no neutral currents exist, then it should not be regarded as good grounds for the existence of neutral currents.

But what is typical if no neutral currents exist? This question led to the Monte Carlo simulation to estimate the effect of the escaping muons (the effect of the artifact). The estimate supplied by this simulation plays precisely the role that .5 plays in the lady tasting tea example—it is the expected effect due to an error of interest (guessing, and escaping muons, respectively). The test or null hypothesis asserts that the error is committed in the experiment at hand. We can use our new apparatus to capture the highlights of the neutral current experiment.

The null (or test) hypothesis asserts that no neutral currents exist, and the predicted or "calculated" number of muonless events is arrived at by assuming that the null hypothesis is true. The calculated or expected number (stemming from the simulation) of muonless events, was 24, giving an expected ratio of 24/56. Next there is the distance measure, in this case the observed ratio minus the calculated ratio. If the experimental procedure used in obtaining the observed ratio were to be repeated, different values of this distance measure would result. However, the probability with which different values would occur is fairly well known because the distance statistic has a known distribution. Hence the significance question was answerable. This, in turn, answers the question we initially put to the data: how often (i.e., how frequently), in experiments such as this one, would as impressive a result as you represent be expected (i.e., one with as many muonless events), even if there are no neutral currents (and the real reason for the recorded muonless events is that muons fail to reach the detector)?

With this we have hit upon the central role of the concept of *relative frequencies* of an observed difference *in a hypothetical series of repetitions* of an experiment (whether they are actually performed or not). A single experiment gives just one value of a statistic of interest, here the difference statistic. But because the distance statistic is a function of the data and a hypothesized answer (H) to a primary question (e.g., a distance measure), the experimental distribution of this statistic lets the data speak to how typical or rare its appearance would be, if the answer (*H*) were true. And it allows doing so without already knowing the answer. It does so by telling us, approximately, the relative frequency with which so large a difference would occur in a hypothetical sequence of experiments, were *H* correct.

The reason why the theory of probability has so important a role in the current approach to experimental learning is that probabilistic models afford very effective tools for approximating the relative frequencies in hypothetical series of experiments of the sort that are of interest. I am talking of the entry of probability in experimental learning, not in science altogether. For the task of experimental learning, *the chief reason that probabilistic models are of value is that they are apt tools for asking about key errors that need to be addressed in learning from experiments.*

Both the examples of the lady tasting tea and of neutral currents address questions of an effect's being due to something other than the factor of interest, guessing rather than discrimination, escaping muons rather than neutral currents. Probabilistic models are apt tools for telling us *what it would be like* if the experiment were one in which it would be a mistake to attribute the effect to one factor rather than another. In the lady tasting tea example, the familiar paradigm of coin tossing (with an unbiased coin) tells us, in the experiment at hand, what it would be like if she were merely guessing. In the neutral currents example, the Monte Carlo simulation tells us what it would be like if the muonless events were due only to escaping muons. In both cases the "what it would be like" question is answered in terms of the relative frequency with which certain results would occur in a (hypothetical) series of experiments. It is answered by "displaying" a probability distribution. Having such answers is of value because of their links to the corresponding statistical tools of analysis (tests and estimation methods).

How do we know that probabilistic models are able to inform in these ways about relative frequencies? The notion of probability embodied in these models and its uses in the standard error statistics approach are often called *frequentist probability* and *frequentist statistics,* respectively. The probability of an experimental result is understood as the relative frequency of that result in a hypothetical long run of applications or trials of the experiment. Probabilistic models serve the roles just outlined because it turns out that they are excellent models for certain types of experiments that we can actually perform. One need not look any deeper to justify their use.

Real Random Experiments and the Empirical Law of Large Numbers

Jerzy Neyman paid a lot of attention to the empirical basis for the use of statistical models of experiment in the frequentist approach. Although my interest here is solely with the use of statistical models in experiment, let us look at what Neyman says in general about the source and substance of frequentist ideas.

Neyman (1952) begins by noting that there are real experiments that "even if carried out repeatedly with the utmost care to keep conditions constant, yield varying results" (p. 25). These are real, not hypothetical, experiments, he stresses. He cites examples such as a roulette wheel (electrically regulated), tossing coins with a special machine (that gives a constant initial velocity to the coin), the number of disintegrations per minute in a quantity of radioactive matter, experiments based on random sampling numbers, and the tendency for properties of organisms to vary despite homogeneous breeding. Although we cannot predict the outcome of such experiments, a certain pattern of regularity emerges when they are applied in a long series of trials. ("Long" here does not mean infinitely long or even years, but that they are applied often enough to see a pattern emerge.) The pattern of regularity concerns the relative frequency with which specified results occur. The regularity being referred to is the long-run stability of relative frequencies. The mathematical theory of probability, Neyman says, applies to any data that may be modeled as the result of such random experiments.

In the beginning (and for quite some time) there were only relative frequency distributions, that is, records of different results and the relative frequency with which they were observed to occur. Probability models came later. Studying the observed relative frequency distributions revealed that relative frequencies of specified results tended to be stable in long series of trials (performed as uniformly and as independently of each other as possible). This stability of relative frequencies (e.g., of means) observed in a wide variety of domains just happens to be well modeled by the same models used to capture the canonical random experiments from games of chance.

It is this empirical fact of long-run stability, Neyman explains, that gives the mathematical models of probability and statistics their applicability. "It is a surprising and very important empirical fact that whenever sufficient care is taken to carry out" a large number of trials as uniformly as possible, the observed relative frequency of a specified outcome, call it "a success," is very close to the probability given by the Binomial probability model (Neyman 1952, 26). As Hodges and Lehmann (1970) put it, "Conceptually, the probability p of success [in a Binomial experiment] represents in the model this stable frequency" (p. 209). Indeed, one can determine the number of trials so that, with high probability, the observed relative frequency is within a specified distance ε from p. This is of key relevance in specifying experiments. (I shall come back to this shortly.)

Our warrant for such a conceptual representation is captured by the *law of large numbers* (LLN). While this law applies quite generally (see note 16), it is typically introduced by reference to the Binomial model, and I shall proceed this way as well. Listen to Neyman:

> The justification for speaking of the [mathematical notion of probability] . . . in terms of imaginary random experiments lies in the empirical fact which Bortkiewicz insisted upon calling the "law of large numbers." This law says that, given a purely mathematical definition of a probability set . . . we are able to construct a real experiment, possible to carry out in any laboratory, with a certain range of possible results and such that if it is repeated many times, the relative frequencies of these results and their different combinations in small series approach closely the values of probabilities as calculated from the definition. . . . Examples of such real random experiments are provided by the experience of roulette. (Neyman 1952, 18)

"These examples show," Neyman continues, "that random experiments corresponding in the sense described to mathematically defined probability sets are possible. However, frequently they are technically difficult. . . . Whenever we succeed in arranging the technique of a random experiment, such that the relative frequencies of its different results in long series approach" sufficiently the mathematical probabilities in the sense of the LLN, we can say that the probabilistic model "adequately represents the method of carrying out the experiment" (pp. 18–19). That is, the actual experiment adequately satisfies the probabilistic model of experiment. Hodges and Lehmann (1970) put it thus: "Long-run stability is an empirical fact. The law of large numbers merely asserts that our model for probability is sufficiently realistic to agree with this fact" (p. 210).

Teddy Seidenfeld (1979a) offers a good discussion of these same passages from Neyman. Neyman's appeal to the law of large numbers, Seidenfeld observes, is to "guarantee that knowledge of the empirical frequencies" is available (p. 33). Given the random experiment for generating experimental outcomes, the LLN shows that we can use the probabilities given from the probability model of the experiment to approximate the relative frequencies of outcomes in a long sequence of trials (p. 33).

Consider the canonical example of Binomial trials, say in coin tossing by an appropriate mechanism. The formal statement of the law of large numbers says that if each trial is a Binomial trial with a constant probability of success equal to p, then the relative frequency of success (i.e., sample mean) will, with high probability, be (specifiably) close to p in a long series of trials. In actual coin tossings performed by a mechanism judged adequate, the accordance between observed relative frequencies and p is repeatedly found to hold: the sample mean varies from p about as often as the LLN says it would. The Binomial model, therefore, is an excellent model of this coin-tossing mechanism and can be used to estimate the expected relative frequencies. In general, Neyman concludes, if in repeatedly carrying out a series of random experiments of a given kind we find that they always conform to the empirical law of large numbers, then we can use the calculus of probability to make successful predictions of relative frequencies.

We are allowed to describe or model the results of such real canonical experiments as random samples from the population given by the probability model. If we remember that this is merely a convenient formal abbreviation for a lengthy statement about an experiment's conforming to the law of large numbers, we can see the reason for calling experimental variables used to model such experiments *random variables*. They are called random variables, in effect, *because their results are (or can be modeled as being) determined by the results of (real) random experiments.* The term random variable is not the misnomer it is often thought to be.

From the study of real random experiments—in which the canonical assumptions needed for randomness hold—we have learned to develop tools to check experimental assumptions when they are *not* known to hold. By varying a known Binomial (or other) process so as to violate one of the assumptions *deliberately* (e.g., cause it to violate independence), we can observe the pattern of outcomes that results. Tests can then be developed that would very probably detect such a violation should it occur. All kinds of relevant properties of these tests may be recorded in the "tool kit" of experimental checks. The appropriate test can then be selected and relied on to test a given assumption in experiments where it is not known to hold.

Now for more about how Monte Carlo studies work.[16] They are often based on so-called random number tables, created by a procedure known to give each number from 0 to 9 an equal chance of being selected. Computers, rather than tables, are now more commonly

16. See, for example, Snedecor and Cochran 1980.

used. But the interesting part of their use is this: We can use the random number table to select successive samples of a specified size, say 10, from a population that we know. Say, for instance, that we know the mean value μ of some characteristic in a population. For each sample selected we can calculate the sample mean. We might draw 100 random samples, each of size 10, thereby calculating 100 sample means. We can then consider the relative frequency of the different values of the sampling mean. This yields the *frequency (i.e., relative frequency) distribution* of sample means. This tells us how frequently sample means (in these 100 samples) differ from the known population mean by different amounts. We can use this knowledge in applying the random number table or generator to cases where the population μ is unknown. When the frequency distributions obtainable by such Monte Carlo methods can be approximated using probability models, we can avoid the laborious process of studying the actual frequency distribution associated with a given method of data generation. We can just compute them from the probability models.

This should also clarify the relationship between the law of large numbers construed as an empirical law and as a mathematical law whose proof is given in many textbooks on probability. (Richard von Mises [1957] discusses these two construals at some length.) The mathematical law is proved for experiments where the assumptions hold, where the experimental trials constitute an idealized random sample. But there are certain types of actual experiments, as Neyman emphasizes, that are well approximated by the mathematical law, as shown, for example, in Monte Carlo studies. Moreover, for a given real experiment the question of whether it obeys the law is open to an empirical test along the lines of the tests of experimental assumptions (at the level of data models). Ronald Giere's terminology for relating hypotheses to real systems is apt: The empirical law of large numbers asserts that certain real experiments are similar to probabilistically modeled ones in the respects stipulated in the law. This similarity may be demonstrated by means of statistical tests. And for certain experiments the law is found to hold.

Neyman is talking quite generally about the empirical justification of probabilistic models. The use of such models in our area of interest—standard experimental testing—turns out to be even less onerous to justify. We will see this in the consideration of two famous results, the law of large numbers and the central limit theorem (CLT). While I am striving to keep formal results to an absolute minimum, our discussion would be incomplete without some discussion of these two remarkable results.

Two Remarkable Results

The Binomial Law of Large Numbers. The law of large numbers, credited to James Bernoulli, can be simply stated with respect to the Binomial model. It shows that with a sufficiently large sample size n, the relative frequency of success will, with as high a probability as one likes, be specifiably close to the probability p in the Binomial model. But it does more. It also lets us calculate the sample size n that will work.[17]

Let $\overline{X}_n$ be the relative frequency of success in n independent Binomial trials, with probability p of success at each trial. (This is the same as the mean number of successes out of n trials.) I am inserting the sample size n as a subscript now, rather than just writing $\overline{X}$, because we want to make some claims about the relationship between n and certain error bounds. Consider the difference, in *either* direction, between $\overline{X}_n$ and p: Let $|D|$ abbreviate this difference, that is, $|D|$ = the absolute value of the difference between $\overline{X}_n$ and p. Now $|D|$ is a statistic and may take on different values. Consider the event that this statistic exceeds some value c, as small as one likes.

$|D|$ exceeds c.

The Binomial LLN sets an upper bound to the probability of this result. It can be stated as follows:

$$P(|D| \text{ exceeds } c) \leq \frac{p(1 - p)}{nc^2}.$$

This upper bound reaches its highest value when $p = .5$. Therefore, we can state, *even without knowing p:*[18]

17. There are several good discussions of the LLN. Here I follow the discussion in Parzen 1960. Suppose that $\overline{X}_n$ is the mean value of n independent samples from a population distribution with mean μ. As we have already said, $\overline{X}_n$ itself has a mean (or expected) value equal to μ. (The average value of $\overline{X}_n$ is equal to μ.) The standard deviation of $\overline{X}_n$ is also related in a known way to the standard deviation of the population distribution. In particular, the standard deviation of $\overline{X}_n$ equals the population standard deviation divided by the square root of n (the sample size). From this it follows that the standard deviation of the sample mean $\overline{X}_n$ gets smaller and smaller as the sample size n gets bigger. That is, the larger the sample size, the less the sample mean differs on average from the population mean. From a result called Chebychev's inequality, it follows that by taking a large enough sample size n, the sample mean is approximately equal to the population mean with a probability as close to 1 as desired.

18. That is because for any value of p, the quantity $p(1 - p) \leq \dfrac{1}{4}$. Note that $\dfrac{p(1 - p)}{n}$ is the variance of $\overline{X}_n$. So its maximum value is $\dfrac{1}{4n}$.

$$P(|D| \text{ exceeds } c) \leq \frac{1}{4nc^2}.$$

In words, it says that the probability that the difference between the sample mean $\overline{X}_n$ and parameter p exceeds c is less than or equal to $\frac{1}{4nc^2}$.

The LLN can be used to tell us how to ensure that the relative frequency of success will, with high probability, be within a specified distance from the population mean p. Although much lower bounds can almost always be calculated (using the central limit theorem), it is quite magical that bounds can be set altogether. Here's a typical homework problem[19] for a conservative specification for n:

Example 5.3: How many (independent) trials of a Binomial experiment should be performed to ensure a .95 or better probability that the observed relative frequency of success will differ from the probability of success p (on each trial) by no more than .02?

The formula for the answer is

$$n \geq \frac{1}{4c^2(1 - \alpha),}$$

where $(1 - \alpha)$ is the desired probability that the observed relative frequency will differ from p by no more than a preset distance c. Substituting .95 for $(1 - \alpha)$ and .02 for c, we get $n \geq 12,500$.

From the LLN, we know that $\overline{X}_n$ is well approximated by p, the probability of success at each trial in the sense that $\overline{X}_n$ is expected to be close to p in the long run. While "being expected to be close in the long run" might seem a complicated kind of property, it is precisely what frequentist statistical methods are designed for. We know that $\overline{X}_n$ will be close to its mean, whatever it is, in the long run. Statistical methods use a single value of $\overline{X}_n$ to learn about the value of $\overline{X}_n$ to be expected in the long run. The value of $\overline{X}_n$ expected in the long run equals p, whatever it is. So a single value of $\overline{X}_n$ can be used to learn about p, whatever it is. It should be noted that the more general LLN asserts that with increasing n, a random variable becomes increasingly close to *its mean*. In the case of the Binomial random variable $\overline{X}_n$, the mean is identical to the parameter p.

A much more powerful result for both pretrial specification of sample size and inference to population means is the central limit theorem.

19. See, for example, the text by Parzen (1960, 231).

The (Astonishing) Central Limit Theorem (CLT). Textbooks are not known for using exalted language in their statements of theorems of probability and statistics—with one exception: the central limit theorem. The central limit theorem tells us that the sample mean $\overline{X}_n$ is approximately Normally distributed regardless of the underlying (population) distribution of variable X. The mean of this distribution of $\overline{X}_n$ equals the mean of X itself, and its standard deviation is the standard deviation of X divided by the square root of n. What makes probability and statistics texts refer to this result as "remarkable" and "astonishing" is that it holds regardless of the underlying distribution of the random variable:[20]

> *Central limit theorem* (CLT): Let $\overline{X}_n$ be the arithmetic mean of n independent random variables, each variable X from a distribution with mean μ and finite nonzero variance s^2. Then $\overline{X}_n$ is approximately normally distributed with mean μ and standard deviation $\dfrac{s}{\sqrt{n}}$.

For most cases of interest (where the underlying distribution is not terribly asymmetrical), the Normal distribution gives a good approximation even for small sample sizes—it is generally quite good with samples of around 30. The CLT actually holds for many cases that do not satisfy the above conditions of randomness (i.e., that each X_i is independent and identically distributed).[21] The finite nonzero variance assumption is practically no restriction at all, and even this has been shown to be capable of being relaxed.

Using the Normal approximation yields a much smaller required sample size than the rough bounds given using only the LLN. The corresponding problem in example 5.3 requires a sample size of only $n \geq 2,500$.

The CLT is at the heart of why the distribution of the sample mean (the relative frequency) is so central in the experimental model. It is what links a claim or question about a statistical hypothesis (a population distribution) to claims about what relative frequencies would be expected in a hypothetical sequence of applications of the experiment. Owing to this link, we can answer the significance question with regard to a great many problems about errors. For any observed $\overline{X}_n$ we can ask it to tell us what values of the population parameter it is im-

---

20. The rate at which $\overline{X}$ approaches Normality, however, is influenced by the underlying distribution.

21. See Feller 1971 and Cramér 1974.

probably far from, which it is typical of, and so on. This is the basis of ampliative inference.

The Standardized Difference. Let us state the CLT in its more useful form, which will also allow me to define an important notion: the *standardized distance.*

Let D be the difference between the sample mean $\overline{X}_n$ and the population mean: D equals $\overline{X}_n - \mu$. It is extremely useful to put D in standard units, that is, to express it in units of the standard deviation of $\overline{X}_n$. This yields a standard distance, abbreviated by D^*:

$$\text{Standardized distance } D^*: \frac{\overline{X}_n - \mu}{\text{standard deviation of } \overline{X}_n}.$$

The CLT says that D^* is approximately Normally distributed with 0 mean (so the curve is centered on 0) and standard deviation 1. This is called the *standard Normal distribution,* and because we can always standardize a Normal variable, it suffices to use tables of standard Normal distributions to look up probabilities. Hence, the value of canonical Normal tables.

Specifying the Long-Run Series

Which of the canonical models is appropriate depends upon the specified procedure for generating experimental outcomes and the way those outcomes are to be modeled. This dependency on the data generation procedure is the error statistician's way of addressing the familiar problem of the reference class. The reference class, upon which probabilities of experimental results are based, consists of the possible outcomes of the data generation procedure. The need to specify such a procedure and the fact that probability is always relative to how the procedure is modeled are often a source of criticism of the use of frequentist probabilities. But those criticisms pertain to a use of frequency probability quite different from how it is used in frequentist statistics. The criticism arises, for example, if the goal is to assign a probability to a particular case and *detach* the resulting claim (e.g., the probability that Roger is from Leuven). But the use of probability in frequentist statistics—which refers to inferences based on frequentist probability models—always has them attached to the experimental processes or test procedures they characterize. (It is a testing approach, not an E-R approach.)

As for choosing how to model the experimental procedure, this will depend upon the twin considerations of what one wants to learn (e.g., the type of canonical error being investigated) and the kinds of data it is reasonable to collect. We have already seen how different ways to model data from an experiment arise in different types and stages of inquiry (e.g., in using n coin-tossing trials to test a hypothesized value of p as opposed to testing assumptions of that test). The ability to vary data models, to use several even in a single inquiry, is not a hindrance but precisely what allows error statistical methods to check their own assumptions and to accord so well with the needs of scientific practice. We shall explore this further in the experiments on Brownian motion (chapter 7) and in revisiting statistical tests (chapter 11).

Avoiding Statisticism

In section 5.5 I have attempted to take some of the mystery out of the idea of random experiments and their modeling by way of random variables. This should free us to use these and other notions from probabilistic models of experiments, remembering, as Neyman stresses, that they are really only "picturesque" ways of talking about actual experiments. When I speak of the chance of a result or a chance occurrence I am simply appealing to a statistical model of an experiment, and do not mean to impute any chance agency to the situation although at times that would be apt. We must be wary of what P. D. Finch (1976) refers to as "statisticism." The aim of representing variable quantities (quantities that can take on different values) by means of random variables, he stresses, "is simply to describe one type of variability in terms of another more familiar one" (p. 9)—one whose variability is known. Statistical models of experiments may be used so long as the experiments accord with them in the right ways. It just so happens that certain real random experiments or, more correctly, questions that can be posed about them are well modeled by the same statistical models that fit games of chance. This gets to the heart of the justification for the use of these statistical models and methods in experimental inference. When we're ready to wrap up our story, we will return to the issue of justification (chapters 12 and 13).

Next Step

We are ready to explore in detail the notion of a severe test defined within a hierarchy of models of an experimental inquiry. In so doing we will take up a central problem that has been thought to stand in the way of ever saying a test is truly severe—namely, the possibility of alternative hypotheses that pass the test equally well.

CHAPTER SIX

Severe Tests and Methodological Underdetermination

The basic trouble with the hypothetico-deductive inference is that it always leaves us with an embarrassing superabundance of hypotheses. All of these hypotheses are equally adequate to the available data from the standpoint of the pure hypothetico-deductive framework.

—W. Salmon, *The Foundations of Scientific Inference*, p. 115

A MAJOR PROBLEM that has been thought to stand in the way of an adequate account of hypothesis appraisal may be termed the *alternative hypothesis objection:* that whatever rule is specified for positively appraising H, there will always be rival hypotheses that satisfy the rule equally well. Evidence in accordance with hypothesis H cannot really count in favor of H, it is objected, if it counts equally well for any number of (perhaps infinitely many) other hypotheses that would also accord with H.

This problem is a version of the general problem of underdetermination of hypotheses by data: if data cannot unequivocally pick out hypothesis H over alternatives, then the hypotheses are underdetermined by evidence. Some have considered this problem so intractable as to render hopeless any attempt to erect a methodology of appraisal. No such conclusion is warranted, however. There is no general argument showing that all rules of appraisal are subject to this objection: at most it has been successfully waged against certain specific rules (e.g., the straight rule, simple hypothetico-deductivism, falsificationist accounts). Since chapter 1 I have been hinting that I would propose utilizing a test's severity to answer the underdetermination challenge. It is time to make good on this promise. Doing so demands that we be much clearer and more rigorous about our notion of severity than we have been thus far. Indeed, by exploring how an account of severe testing answers the alternative hypothesis objection, we will at the same time be piecing together the elements needed for understanding the severity notion. In anticipation of some of my theses, I will argue that

1. the existence of hypotheses alternative to H that entail or accord with evidence e (as well as or even better than H) does not
   prevent H from passing a severe test with e;
2. computing a test's severity does not call for assigning probabilities to hypotheses;
3. even allowing that there are always alternative hypotheses that
   entail or fit evidence e, there are not always alternatives equally
   severely tested by e.

As important as many philosophers of science regard the alternative hypothesis challenge, others dismiss it as merely a " philosopher's
problem," not a genuine problem confronting scientists. In the latters'
view, scientists strive to find a single hypothesis that accounts for all
the data on a given problem and are untroubled by the possibility of
alternatives. Granted, there are many examples in which it is generally
agreed that any alternative to a well-tested hypothesis H is either obviously wrong or insignificantly different from H, but this enviable situation arises only after much of the work of ruling out alternatives has
been accomplished. Anyone seeking an account adequate to the task
of *building up* experimental knowledge, as I am, must be prepared to
deal with far more equivocal situations. Moreover, an adequate philosophical account should be able to explain how scientists are warranted, when they are, in affirming one hypothesis over others that
might also fit the data.

Grappling with the alternative hypothesis objection will bear other
fruit. Appealing to a test's severity lets us see our way clear around
common misinterpretations of standard statistical tests. In section 6.5,
for example, the question of how to interpret statistically insignificant
differences is addressed.

6.1 METHODOLOGICAL UNDERDETERMINATION

The "alternative hypothesis objection" that concerns me needs to be
distinguished from some of the more radical variants of underdetermination. Some of these more radical variants are the focus of a paper by
Larry Laudan (1990a), "Demystifying Underdetermination." "[O]n the
strength of one or another variant of the thesis of underdetermination," Laudan remarks, "a motley coalition of philosophers and sociologists has drawn some dire morals for the epistemological enterprise."
Several examples follow.

> Quine has claimed that theories are so radically underdetermined by
> the data that a scientist can, if he wishes, hold on to *any* theory he
> likes, "come what may." Lakatos and Feyerabend have taken the un-

derdetermination of theories to justify the claim that the only difference between empirically successful and empirically unsuccessful
theories lay in the talents and resources of their respective advocates. . . . Hesse and Bloor have claimed that underdetermination
shows the *necessity* for bringing noncognitive, social factors into play
in explaining the theory choices of scientists. (Laudan 1990a, p. 268)

Laudan argues that the Quinean thesis that "any hypothesis can
rationally be held come what may" as well as other strong relativist
positions are committed to what he calls the *egalitarian thesis.* "It insists
that: *every [hypothesis] is as well supported by the evidence as any of its rivals*"
(p. 271). Nevertheless, Laudan maintains that a close look at underdetermination arguments shows that they at most sustain a weaker form
of underdetermination, which he calls *the nonuniqueness thesis.* "It holds
that: *for any [hypothesis] H] and any given body of evidence supporting [H],
there is at least one rival (i.e., contrary) to [H] that is as well supported as [H]*"
(p. 271). Laudan denies that the nonuniqueness thesis has particularly
dire consequences for methodology; his concern is only with the extreme challenge "that the project of developing a methodology of science is a waste of time since, no matter what rules of evidence we
eventually produce, those rules will do nothing to delimit choice" (p.
281). I agree that the nonuniqueness thesis will not sustain the radical
critique of methodology as utterly "toothless," but I am concerned to
show that methodology has a severe bite!

Even if it is granted that empirical evidence serves *some* role in
delimiting hypotheses and theories, the version of underdetermination
that still has to be grappled with is the alternative hypothesis objection
with which I began, that for any hypothesis H and any evidence, there
will always be a rival hypothesis equally successful as H. The objection,
it should be clear, is that criteria of success based on methodology and
evidence alone underdetermine choice. It may be stated more explicitly as the thesis of methodological underdetermination (MUD):

> *Methodological underdetermination:* any evidence taken as a good test of
> (or good support for) hypothesis H would (on that account of testing
> or support) be taken as an equally good test of (or equally good sup-
> port for) some rival to H.

While not alleging that anything goes, it is a mistake to suppose that
the MUD thesis poses no serious threat to the methodological enterprise. The reason formal accounts of testing and confirmation ran into
trouble was not that they failed to delimit choice at all, but that they
could not delimit choice sufficiently well (e.g., Goodman's riddle).
Moreover, if hypothesis appraisal is not determined by methodology

and evidence, then when there is agreement in science, it would seem to be the result of extraevidential factors (as Kuhn and others argue).

The existence of alternative hypotheses equally well tested by evidence need not always be problematic. For example, it is unlikely to be problematic that a hypothesis about a continuous parameter is about as well tested as another hypothesis that differs by only a tiny fraction. In the following discussion of my account of severe testing, I will focus on the seemingly most threatening variants of the MUD challenge.

Clearly, not just any rule of appraisal that selects a unique hypothesis constitutes an adequate answer to the challenge. Not just any sort of rule is going to free us from many of the most troubling implications of MUD. That is why the Bayesian Way does not help with my problem. Its way of differentially supporting two hypotheses that equally well entail (or otherwise fit) the data is by assigning them different prior probabilities.[1] But, as I argued in chapter 3, prior probabilities, except in highly special cases, are matters of personal, subjective choice—threatening to lead to the relativism we are being challenged to avoid (inviting a MUD-slide, one might say).

Summary of the Strategy to Be Developed

How does appealing to the notion of severity help? While there are many different conceptions of severe tests, such accounts, broadly speaking, hold the following general methodological rule:

> Evidence e should be taken as good grounds for H to the extent that H has passed a *severe test* with e.

What I want to argue is that the alternative hypothesis objection loses its sting once the notion of severity is appropriately made out.

It is easy to see that the alternative hypothesis objection instantiated for a method of severe testing T is more difficult to sustain than when it is waged against mere entailment or instantiation accounts of inference. The charge of methodological underdetermination for a given testing method, which I equate with the alternative hypothesis objection, must show that *for any evidence test* T *takes as passing hypothesis* H *severely, there are always rival hypotheses that* T *would take as passing equally severely*. While MUD gets off the ground when hypothesis appraisal is considered as a matter of some formal or logical relationship

---

1. Indeed, if two hypotheses entail the evidence, then the only way they can be differently confirmed by that evidence by Bayes's theorem is if their prior probability assignments differ.

between evidence or evidence statements and hypotheses, this is not so in our experimental testing framework.

The cornerstone of an experiment is to do something to *make* the data say something beyond what they would say if one passively came across them. The goal of this active intervention is to ensure that, with high probability, erroneous attributions of experimental results are avoided. The error of concern in passing H is that one will do so while H is not true. Passing a severe test, in the sense I have been advocating, counts for hypothesis H because it corresponds to having good reasons for ruling out specific versions and degrees of this mistake.

Stated simply, *a passing result is a severe test of hypothesis* H *just to the extent that it is very improbable for such a passing result to occur, were* H *false.* Were H false, then the probability is high that a more discordant result would have occurred. To calculate this probability requires considering the probability a given procedure has for detecting a given type of error. This provides the basis for distinguishing the well-testedness of two hypotheses—despite their both fitting the data equally well. Two hypotheses may accord with data equally well but nevertheless be tested differently by the data. The data may be a better, more severe, test of one than of the other. The reason is that the procedure from which the data arose may have had a good chance of detecting one type of error and not so good a chance of detecting another. What is ostensibly the same piece of evidence is really not the same at all, at least not to the error theorist.

This underscores a key difference between the error statistics approach and the Bayesian approach. Recall that for the Bayesian, if two hypotheses entail evidence e, then in order for the two hypotheses to be differently confirmed there must be a difference in their prior probabilities. In the present approach, two hypotheses may entail evidence e, while one has passed a far more severe test than the other.

6.2 The (Error)-Severity Requirement

The general requirement of, or at least preference for, severe tests is common to a variety of approaches (most commonly testing approaches), with severity taking on different meanings. To distinguish my notion from others, I will sometimes refer to it as *error-severity.*

a. *First requirement:* e *must "fit"* H. Even widely different approaches concur that, minimally, for H to pass a test, H should agree with or in some way fit with what is expected (or predicted) according to H. We can apply and contrast our definition with that of other approaches by allowing "H passes with e" to be construed in many ways (e.g., H is supported, e is more probable on H than on not-H, e is far from the denial of H on some distance measure, etc.[2]). Minimally, H does not fit e if e is improbable under H.

b. *Second requirement:* e's fitting H *must constitute a good test of* H. Those who endorse some version of the severity requirement concur that a genuine test calls for something beyond the minimal requirement that H fits e. A severity requirement stipulates what this "something more" should be.

Following a practice common to testing approaches, I identify "having good evidence (or just having evidence) for H" and "having a good test of H." That is, to ask whether e counts as good evidence for H, in the present account, is to ask whether H has passed a good test with e. This does not rule out quantifying the goodness of tests.[3] It does rule out saying that "e is a poor test for H" and, at the same time, that "e is evidence for H."

c. *The severity criterion (for experimental testing contexts).* To formulate the pivotal requirement of severe tests, it is sufficient to consider the test outputs—"H passes a test T with experimental outcome e" or "H fails a test T with experimental outcome e." I am assuming that the

---

2. This allows us to state the first requirement for H to pass a test with e as

a. H fits e,

with the understanding that a suitable notion of fit, which may vary, needs to be stipulated for the problem at hand. While some accounts of testing construe the fit as logical entailment (with suitable background or initial conditions), except for universal generalizations this is rarely obtained. One way to cover both universal and statistical cases is with a statistical measure of fit, such as e fits H to the extent that $P(e \mid H)$ is high. (The entailment requirement results in $P(e \mid H)$ being 1.) Because $P(e \mid H)$ is often small, even if H is true, passing a test is commonly defined comparatively. Evidence e might be said to fit H if e is more probable under H than under all (or certain specified) alternatives to H.

There is nothing to stop the hypothesis that passes from being composite (disjunctive). For example, in a Binomial experiment H may assert that the probability of success exceeds .6, i.e., $H: p > .6$. The alternative H' asserts that $p \leq .6$. In such cases, e fits H might be construed as e is further from alternative hypothesis H' than it is from any (simple) member of H, where "further" is assessed by a distance measure as introduced in chapter 5.

3. The question of whether H's passing a test with result e provides a good test of H may alternately be asked as the question of whether e provides confirmation or support for H. However, when the question is put this way within a testing approach it should not be taken to mean that the search is for a quantitative measure of degree of support—or else it would be an evidential-relationship and not a testing approach. Rather, the search is for a criterion for determining if a passing result provides good evidence for H—although the goodness of a test may itself be a matter of degree.

underlying assumptions or background conditions for a test—whatever they are—are located in the various data models of an experimental inquiry, as delineated in chapter 5. This frees me to characterize the severity requirement by itself. The severity requirement is this:

> *Severity requirement*: Passing a test T (with e) counts as a good test of or good evidence for H just to the extent that H fits e and T is a *severe test* of H,

and the severity criterion (SC) I suggest is this:

> *Severity criterion* 1*a*: There is a very high probability that test procedure T would *not* yield such a passing result, if H is false.

By "such a passing result" I mean one that accords at least as well with H as e does. Its complement, in other words, would be a result that either fails H or one that still passes H but accords less well with H than e does. It is often useful to express SC in terms of the improbability of the passing result. That is:

> *Severity criterion* 1*b*: There is a very low probability that test procedure T would yield such a passing result, if H is false.

One may prefer to state the SC in terms of the measure of accordance or fit. (1*a*) and (1*b*) become

> *Severity criterion* 2*a*: There is a very high probability that test procedure T would yield a worse fit, if H is false.

> *Severity criterion* 2*b*: There is a very low probability that test procedure T would yield so good a fit, if H is false.

While the *a* versions express severity in terms of the test's high probability to detect the incorrectness of H, the equivalent *b* versions express severity in terms of the low probability of its failing to detect the incorrectness of H.

d. *The Severity Criterion in the Simplest Case (SC*): A Test as a Binomial Statistic.* Standard statistical tests are typically framed in terms of only two possible results: H passes and H fails, although "accept" and "reject" are generally the expressions used rather than "pass" and "fail." This reduction to two results is accomplished by stipulating a cutoff point such that any particular result e that differs from H beyond this cutoff point is classified as failing H; all others pass H. The test, in short, is modeled as a Binomial (or pass-fail) procedure. The severity criterion for this special case is simpler to state than for the general case:

(SC*) The severity criterion for a "pass-fail" test: There is a very high probability that the test procedure T fails H, given that H is false.[4]

Modeling tests in this "Binomial" manner may be sufficient for specifying a test with appropriate error probabilities. However, it is often too coarse grained for interpreting a particular result, which is why its use leads to many criticisms of standard error statistics—a point to be explained in chapter 11. The trick is to be able to calculate the severity achieved by some *specific outcome* among those the test would take as passing H. That is the reason for my more cumbersome definition.[5] Nevertheless, the severity criterion for the pass-fail (or Binomial) test (SC*), because of its simplicity, is the one I recommend keeping in mind even in arguing from a specific passing result. One need only be clear on how it may be used to arrive at the general SC, the calculation we really want. Let us illustrate.

That H passes with a specific outcome e may be regarded as H having passed with a given *score*, the score it gets with outcome e, just like a score on an exam. Suppose we want to calculate the severity associated with that particular passing score e. We can divide the possible scores into two: scores higher than the achieved score e, and those as low as or lower than e. We have now (re)modeled our test so that it has only two results, and we can apply the simple severity calculation for a pass-fail test. We have

> SC*: The probability is high that test T would *not* yield so high a score for H as e, given that H is false.

Alternatively, in terms of the complement (b) we have

> SC*: It is very improbable that H would have passed with so successful a score as e, given that H is false.

We have arrived at the calculation that the more general severity criterion (SC) demands.

How to understand the probabilities referred to in our severity criterion is a question whose answer may be found in the discussion of frequentist probability in the last chapter. A high severity assignment asserts that were we experimenting on a system where H is false, then in a long series of trials of this experiment, it is extremely rare (infrequent) that H would be accorded such a good score; the overwhelming

---

4. Calculating SC* considers the probability that an outcome would reach the cutoff for failing H, even if H is false.

5. This will be clarified further in distinguishing severity from "power" in chapter 11.

preponderance of outcomes would accord H a worse fit or a lower score.

Minimum (0) and Maximum (1) Severity

We can get at the commonsense rationale for desiring high severity and eschewing low severity by considering extreme cases of violating or satisfying severity. Here the probabilities of H not passing when false may be shown to be 0 and 1 (or practically so), respectively. I begin with the first extreme case, that of a *minimally severe* or a *zero-severity* test.

> *Passing a minimally severe (zero-severity) test:* H passes a zero-severity test with e if and only if test T would always yield such a passing result even if H is false.

In the present account, such a test is no test at all. It has no power whatsoever at detecting the falsity of H. If it is virtually impossible for H to receive a score less than e on test T, even if false, then H's receiving score e *provides no reason* for accepting H; it fails utterly to discriminate H being true from H being false.

That a test would always pass H even if H is false does not entail that H's passing is always erroneous or that H is false. H may be true. I may even have a warrant for accepting H, on *other* grounds. It is just that passing with a zero-severity test itself does not warrant such an acceptance. (That is, one can be right, but for the wrong reasons.)

These remarks accord well with familiar intuitions about whether passing marks on an exam warrant merit of some sort. Consider a test to determine whether a student can recite all the state capitals in the United States; say the hypothesis H is that the subject can correctly recite (aloud) all fifty. Suppose that a student passes the test so long as she can correctly assert the capital of any one state. That a person passes this test is not much of a reason to accept H, because it is not a very severe test. Suppose now that a student passes the test so long as she can recite *anything* aloud. Granted, being able to recite all fifty capitals entails being able to speak aloud (H entails e), but this test is even less severe than the first. It is easier (more probable) for a pass to occur, even if the student is *not* able to recite all the state capitals (H is false).[6]

Alternatively, if a student passes a test where passing requires reciting all fifty capitals correctly, certainly that is excellent support for hypothesis H, that the student can correctly recite them all. This identifies the other extreme, that of a maximally severe test:

6. Popper (1979, 354) also uses an exam analogy to make this point.

Passing a maximally severe (100 percent severity) test: H passes a maximally severe test with e if and only if test T would never yield results that accord with H as well as e does, if H is false.

A test is maximally severe if the results that the test takes as passing H cannot occur (in trials of the given experimental test), given that hypothesis H is false. It is a maximally reliable error probe for H. That passing a maximally severe test warrants accepting H may seem too obvious to merit noting. After all, in such a test passing with e entails H! Nevertheless, as will be seen in chapter 8, not all accounts of testing countenance maximally severe tests as good tests.

Let us move from 100 percent severity to merely high severity and see whether the reasoning still holds. Consider two tests, T_1 and T_2.

T_1 is known to have a very high, say a .99, probability of failing a student (giving her an F grade, say) if the student knows less than 90 percent of the material. That is, 99 percent of the time, students ignorant of 10 percent of the material fail test T_1.

Test T_2, let us suppose, is known to have only a 40 percent probability of failing a student who knows less than 90 percent of the material.

T_1 is obviously a more severe test than T_2 in our ordinary use of that term, and likewise in the definition I have given. Passing the more severe test T_1 is good evidence that the student knows more than 90 percent of the material. (For, if she were to know less than 90 percent, test T_1 would, with high probability, .99, have detected this and failed her.) Clearly, all else being equal, better evidence of the extent of a student's knowledge is provided by the report "She passes test T_1," than by the report "She passes test T_2." Passing test T_2 is an altogether common occurrence (probability .6) even if the student knows less than 90 percent of the material.

An Analogy with Diagnostic Tools

Tools for medical diagnoses (e.g., ultrasound probes) offer other useful analogies to extract these intuitions about severity: If a diagnostic tool has little or no chance of detecting a disease, even if it is present (low severity), then a passing result—a clean bill of health—with that instrument fails to provide grounds for thinking the disease is absent. That is because the tool has a very high probability of issuing in a clean bill of health even when the disease is present. It is a highly unreliable error probe. Alternatively, suppose a diagnostic tool has an extremely high chance of detecting the disease, just if present—suppose it to be a highly severe error probe. A clean bill of health with that kind of tool provides strong grounds for thinking the disease is not present. For if the disease were present, our probe would almost certainly have detected it.

It is important to stress that my notion of severity always attaches to a particular hypothesis passed or a particular inference reached. To ask, How severe is this test? is not a fully specified question until it is made to ask, How severe would a test procedure be, if it passed such and such a hypothesis on the basis of such and such data? A procedure may be highly severe for arriving at one type of hypothesis and not another. To illustrate, consider again a diagnostic tool with an extremely high chance of detecting a disease. Finding no disease (a clean bill of health) may be seen as passing hypothesis H_1: no disease is present. If H_1 passes with so sensitive a probe, then H_1 passes a severe test. However, the probe may be so sensitive that it has a high probability of declaring the presence of the disease even if no disease exists. Declaring the presence of the disease may be seen as passing hypothesis H_2: the disease is present. If H_2 passes a test with such a highly sensitive probe, then H_2 has *not* passed a severe test. That is because there is a very low probability of *not* passing H_2 (not declaring the presence of the disease) even when H_2 is false (and the disease is absent). The severity of the test that hypothesis H_2 passes is very low.

Some further points of interpretation are in order.

Severity and Arguing from Error

Experimental learning, I have been saying, may be addressed in a formal or informal mode, although those might not be the best terms. In its formal mode, experimental learning is learning about the probabilities (relative frequencies) of specified outcomes in some actual or hypothetical series of experiments—it is learning about an *experimental distribution*. In its informal mode, experimental learning is learning of the presence or absence of errors. Experimental learning, in its formal mode, is learning from tests that satisfy the severity criterion (SC). In its informal mode, it is learning by means of an argument from error, one variant of which was given in section 3.2. Here are two versions:

> It is learned that an error is absent when (and only to the extent that) a procedure of inquiry (which may include several tests taken together) that has a very high probability of detecting an error if (and only if[7]) it exists nevertheless detects no error.

7. The "only if" clause is actually already accommodated by the first requirement of passing a severe test, namely, that the hypothesis fit the data. If the fit required is entailment, then the probability of passing given the hypothesis is true is maximal.

> It is learned that an error is present when a procedure of inquiry that has a very high probability of not detecting an error if (and only if) none exists nevertheless detects an error.

That a procedure detects an error does not mean it definitely finds the error. It is generally not known whether the procedure gets it right. It means that a result occurs that the procedure takes as passing the hypothesis that an error is present. An analogous reading is intended for detecting no error.

In the canonical arguments from error, the probabilistic severity requirement may capture the argument from error so well that no distinction between so-called formal and informal modes is needed. In general, however, asserting that a hypothesis H passed a highly severe test in this formal sense is but a pale reflection of the actual experimental argument that sustains inferring H. The purpose of the formal characterization is to provide a shorthand for the actual argument from error, which necessarily takes on different forms. The formal severity criterion may be seen to represent a systematic way of scrutinizing the appropriateness of a given experimental analysis of a primary question. Referring to the Suppean hierarchy of models from the last chapter, it is a critique at the level of the experimental testing model.

On the one hand, the informal and often qualitative argument from error takes central stage in applying our severity criterion to actual experiments. On the other hand, there are many features of the formal characterization of severity that offer crucial guidance in doing so. This latter point is as important as it is subtle, and to explain it is not as simple as I would wish. Let me try.

In an informal argument from error one asks, How reliable or severe is the experimental procedure at hand for detecting an error of interest? To answer this question, it is essential to be clear about the (probabilistic) properties of the experimental procedure. Our informal thinking about such things may be anything but clear, and formal canonical models (from standard random experiments) may come to the rescue. For example, at the heart of a number of methodological controversies are questions about whether certain aspects of experimental design are relevant to appraising hypotheses. Does it matter whether a hypothesis was constructed to fit the data? Does it matter when we decide how much data to collect? These are two examples that will be taken up in later chapters.

The formal severity criterion, by reminding us that the test procedure may be modeled as a random variable, comes to our aid. For we know that we cannot determine the distribution of a random variable without being clear on what it is that is being taken to vary from trial

to trial. Is it just the sample mean that varies (e.g., the different proportions of heads in n trials)? Or is the very hypothesis that a test procedure winds up testing also varying? Formally modeled (canonical) experiments demonstrate how error probabilities and, correspondingly,
severity can be altered—sometimes dramatically—by changing what
is allowed to vary. (Doing so is tantamount to changing the question
and thereby changing the ways in which answers can be in error.)
Carrying a few of the formally modeled test procedures in our experimental tool kit provides invaluable methodological service.

   The distinction between the formal model and informal arguments
from error also frees us to talk about a hypothesis being true without
presuming a realist epistemology. Within a canonical experimental
test, the truth of H means that H gives an approximately correct
description or model of some aspect of the procedure generating experimental outcomes. Precisely what this statement of experimental
knowledge indicates about the system of interest will vary and will
have to be decided on a case by case basis. The main thing to note is
that our framework allows numerous interpretations to be given to the
correctness of H, as well as to the success of H. Realists and nonrealists
of various stripes can find a comfortable home in error testing. Aside
from varying positions on realism, a variety of interpretations of "H is
true" (and, correspondingly, "evidence indicates that H is true") are
called for because of the very different kinds of claims that may be
gleaned from experiments. (The Kuhnian normal scientist of chapter
2, for example, may view "H is true" as asserting that H is a satisfactory
solution to a normal problem or puzzle.)

   Despite this room for diversity, there is uniformity in the pattern
of arguments from error. We can get at this uniformity, I propose, by
stating what is learned from experiment in terms of the presence or
absence of some error (which may often be a matter of degree). For
example, a primary hypothesis H might be

   $H:$ the error is absent,

and not-H, that the error is present. (Alternatively, H can be construed
as denying that it would be an error to assert H.) When an outcome is
in accordance with H and (appropriately) far from what is expected
given not-H, then the test passes H. Error now enters in a second way.
The error of concern in passing H is that one will do so while H is not
true, that the error will be declared absent though actually present.

   When a test is sufficiently severe, that is, when an argument from
error can be sustained, the passing result may be said to be a *good indication of* (or good grounds for) H. The resulting knowledge is experimen-

tal knowledge—knowledge of the results to be expected were certain experiments carried out.

We now have to tackle the "other hypothesis" objection. For the existence of alternative hypotheses that accord equally well with test results may be thought to strangle any claim purporting that a test's severity is high.

6.3 IS THE OTHER HYPOTHESIS OBJECTION AN OBJECTION TO SEVERITY?

The thrust of the "other hypothesis" objection is this: the fact that data fit hypothesis *H* fails to count (or to count much) in favor of *H* because the data also fit other, possibly infinitely many, rival hypotheses to *H*. The above characterization of severe tests suggests how this objection is avoidable: mere fitting is not enough! If hypotheses that fit the data equally well were equally well supported (or in some way credited) by the data, then this objection would have considerable weight. But the very raison d'être of the severity demand is to show that this is not so.

Still it might be charged that demanding severity is too demanding. This is Earman's (1992) criticism of me. Examining his criticism allows me to address an anticipated misunderstanding of the severity criterion, namely, the supposition that it requires what I called the Bayesian "catchall" factor (section 4.3).

Earman's Criticism of Error-Severity

In order for hypothesis *H* to pass a severe test, the test must have a low probability of erroneously passing *H*. (This alludes to the *b* forms of SC.) Earman's criticism of my severity requirement seems to be that it requires a low probability to the Bayesian catchall factor. The Bayesian catchall factor (in assessing *H* with evidence *e*), recall, is

$P(e \mid \text{not-}H)$.

However, satisfying SC does not require computing the Bayesian catchall factor.

The catchall, not-*H*, refers to all possible hypotheses other than *H*, including those that may be conceived of in the future. Assessing the probability of *e* on the catchall requires having a prior probability assignment to the catchall. Assigning a low value to the Bayesian factor on the catchall, while all too easy for a personalist—it is sufficient that he or she cannot think of any other plausible explanation for *e*—is too difficult for a tempered subjectivist or frequentist Bayesian, for it requires, recalling Salmon's remark, that we "predict the future course of the history of science" (Salmon 1991, 329).

Earman grants the *desirability* of a low assignment to the Bayesian catchall factor, because, as we said, the lower its value, the more Bayesian confirmation accrues to *H*. The difficulty he sees is in obtaining it. While I agree that this presents an obstacle for the Bayesian approach to support, satisfying the severity criterion SC does not require computing the Bayesian catchall factor. Because of this, alternatives in the catchall that might also fit the evidence are not the obstacle to obtaining high severity that Earman thinks they are.

Consider the example Earman raises in this connection (I substitute *e* for his *E* to be consistent with my notation):

> If we take *H* to be Einstein's general theory of relativity and *e* to be the outcome of the eclipse test, then in 1918 and 1919 physicists were in no position to be confident that the vast and then unexplored space of possible gravitational theories denoted by ¬GTR does not contain alternatives to GTR that yield the same prediction for the bending of light as GTR. (Earman 1992, 117)

In fact, he continues, there is an endless string of such alternative theories. *The presumption is that alternatives to the GTR that also predict light bending would prevent high severity in the case of the eclipse test.*

But alternatives to the GTR did not prevent the eclipse results from being used to test severely the hypotheses for which the eclipse experiments were designed. Those tests, to be taken up in chapter 8, proceeded by asking specific questions: Is there a deflection of light of about the amount expected under Einstein's law of gravitation? Is it due to gravity? Are alternative factors responsible for appreciable amounts of the deflection? Finding the answers to these questions in a reliable manner did not call for ruling out any and all alternatives to the general theory of relativity.

Take the question of the approximate deflection of light. If this is the primary question of a given inquiry, then alternative answers to it are alternative values of the deflection, not alternatives to the general theory of relativity. If alternative theories predict the same results, so far as the deflection effects go, as Earman says they do, then these alternatives are not *rivals* to the particular hypotheses under test. If the endless string of alternative theories would, in every way, give the same answers to the questions posed in the 1919 tests, then they all agree on the aspects of the gravitation law that were tested. They are not members of the space of alternatives relative to the primary question being addressed.

This reply depends on a key feature of my account of testing, namely, that an experimental inquiry is viewed as a series of models, each with different questions, stretching from low-level theories of data and experiment to higher level hypotheses and theories of interest. In relation to the hypotheses about the deflection effect, alternatives to the general theory of relativity are on a higher level. The higher-level alternatives are not even being tested by the test at hand. Most important, higher-level alternatives pose no threat to learning with severity what they needed to learn in the specific 1919 experiments.

For a silly analogy, consider a dialogue about what can be inferred from an exam (we assume cheating is ruled out):

Teacher: Mary scored 100 percent on my geography final—she clearly knows her geography.

Skeptic: How can you be so sure?

Teacher: Well, it *is* possible that she guessed all the correct answers, but I doubt that any more than once in a million years of teaching would a student do as well as Mary by merely guessing.

Skeptic: But there is an endless string of childhood learning theories that would predict so good a score. Perhaps it's the new text you adopted or our attempts to encourage girls to compete or . . .

Teacher: My final exam wasn't testing any of those hypotheses. They might be fun to test some day, but whatever the explanation of her performance, her score on the final shows me she really knows her geography.

The general lesson goes beyond answering Earman. It points up a strategy for dispelling a whole class of equally good fitting alternatives to a hypothesis *H*. The existence of alternatives at a higher level than *H* is no obstacle to finding high severity for *H*. The higher-level questions, just like the question about the correctness of the whole of the GTR, are simply *asking the wrong question*.

Testing versus Learning About

Saying that the eclipse tests were not testing the full-blown theory of general relativity does not mean that nothing was learned about the theory from the tests. What was learned was the extent to which the theory was right about specific hypotheses, for example, about the parameter λ, the deflection of light.

This points up a key distinction between experimental learning in the present approach and in the Bayesian approach, which may explain why Earman thinks that error-severity founders on the alternative hypothesis objection. For a Bayesian, learning about a hypothesis or theory is reflected in an increase in one's posterior probability as-

signment to that hypothesis or theory. For a result to teach something
about the theory, say the GTR, for a Bayesian, that theory must have
received some confirmation or support from that result. But that
means the theory, the GTR, must figure in the Bayesian computation.
That, in turn, requires considering the probability of the result on the
negation of the GTR, that is, the Bayesian catchall factor. That is why
Earman's criticism raises a problem for Bayesians.[8]

For the error theorist, in contrast, an experiment or set of experiments may culminate in accepting some hypothesis, say about the existence of some deflection of light. This can happen, we said, if the
hypothesis passes a sufficiently severe test. That done, we are correct
in saying that we have learned about one facet or one hypothesis of
some more global theory such as the GTR. Such learning does not require us to have tested the theory as a whole.

Our approach to experimental learning recommends proceeding
in the way one ordinarily proceeds with a complex problem: break it
up into smaller pieces, some of which, at least, can be tackled. One is
led to break things down if one wants to learn. For we learn by ruling
out specific errors and making modifications based on errors. By using
simple contexts in which the assumptions may be shown to hold sufficiently, it is possible to ask *one question at a time*. Setting out all possible
answers to this one question becomes manageable, and that is all that
has to be "caught" by our not-*H*.

Apart from testing some underlying theory (which may not even
be in place), scientists may explore whether neutral currents exist,
whether dense bodies are real or merely artifacts of the electron microscope, whether F_4 and F_5 chromosomes play any part in certain types
of Alzheimer's disease, and so on. In setting sail on such explorations,
the immediate aim is to see whether at least one tiny little error can
be ruled out, without having to worry about all the ways in which one
could ever be wrong in theorizing about some domain, which would
only make one feel at sea.

Within an experimental testing model, the falsity of a primary hypothesis *H* takes on a specific meaning. If *H* states that a parameter is
greater than some value *c*, not-*H* states that it is less than *c*; if *H* states
that factor *x* is responsible for at least *p* percent of an effect, not-*H*
states that it is responsible for less than *p* percent; if *H* states that an
effect is caused by factor *f*, for example, neutral currents, not-*H* may

---

8. These remarks do not encompass all the ways that the error-severity calculation differs from calculating the Bayesian catchall factor. They simply address the
point that was at the heart of Earman's criticism.

say that it is caused by some other factor possibly operative in the experimental context (e.g., muons not making it to the detector); if *H* states that the effect is systematic—of the sort brought about more often than by chance—then not-*H* states that it is due to chance. How specific the question is depends upon what is required to ensure a good chance of learning something of interest (much like ensuring satisfaction of the Kuhnian demarcation criterion of chapter 2).

I am not denying the possibility of severe tests of higher-level theoretical hypotheses. When enough is learned from piecemeal studies, severe tests of higher-level theories are possible. Kuhn was right that "severity of test-criteria is simply one side of the coin whose other face is a puzzle-solving tradition." The accumulated results from piecemeal studies allow us at some point to say that several related hypotheses are correct, or that a theory solves a set of experimental problems correctly.

Earman (1992, p. 177) discusses for a different reason the progress that has been made in a program by Thorne and Will (1971) to classify theories of gravity, those already articulated as well as other possible theories satisfying certain minimal requirements.[9] Such a program shows which available experiments can eliminate whole chunks of theories (e.g., so-called nonmetric theories of gravity) and which sets of theories are still not distinguished by known experiments, and it indicates how progress might be made by devising experiments to further discriminate between them (e.g., making cosmological observations). Something like this kind of program of partitioning and eliminating chunks of theories is what the present program would call for at the level of large-scale theories.

Much more work is also needed to show how learning in large-scale inquiries proceeds by piecemeal canonical questions. Later I will focus on specific cases, but the philosopher of experiment's search is not for a uniform analysis of high-level testing—at least not in the way that has ordinarily been understood. Still, there are some general strategies for getting at larger questions by inquiring, piecemeal, into smaller errors: testing parameters, estimating the effects of background factors, distinguishing real effect from artifact, and so on with the other canonical errors. There are also general methodological rules for specifying an experimental test from which one is likely to learn, based on

9. The aim of the program Earman describes is "the exploration of the possibility space, the design of classification schemes for the possible theories, the design and execution of experiments, and the theoretical analysis of what kinds of theories are and are not consistent with what experimental results" (Earman 1992, 177).

background knowledge of the types of errors to be on the lookout for, and strategies for attaining severe tests with limited information.

My concern just now is to get small again, to proceed with some standard tools for severe tests in the experimental models laid out in chapter 5. While they may enable us to take only baby steps, they enable us to take those baby steps severely. Such baby steps are at the heart of the experimenter's focus on what we variously referred to as "topical hypotheses" (Hacking) and "normal puzzles" (Kuhn). Moreover, what these baby steps accomplish will be sufficient for the problem of this chapter: methodological underdetermination. For the reason that arguments about evidence underdetermining hypotheses appear to go through is that we have not bothered to be very clear about what specific evidence is being talked about, what specific hypotheses are being tested, and what specific models of experiment and data are available to constrain inferences.

6.4 CALCULATING SEVERITY

To determine what, if anything, is learned from an experimental result, we must ask, What, if anything, has passed a severe test? Consider our Binomial experiment for the tea-tasting example (example 5.2, section 5.4). We would pass hypothesis H'—that the probability of successfully discriminating the order of tea and milk in an infusion, p, exceeds .5— by failing or rejecting the null hypothesis H_0: $p = .5$ (i.e., the lady is merely guessing). Here, notice that H_0 is the denial of the hypothesis H' that we pass.

The question concerned the population parameter p. The possible answers, the parameter space, consists of all the possible proportions for p from 0 to 1, but the question asked divides it into two spaces, $p = .5$ and $p > .5$. A different inquiry might have tested H_0 against a specific alternative, say $p > .8$. With minor modifications (of test specification[10]), this calls for the same basic test as in our original partition. So that is a good place to start. It is just this kind of rough and simple question that provides a standard for distinguishing between experimental effects and backgrounds.

In this test the tea-tasting lady scored 60 percent successes in 100 trials. That is, the distance (in the positive direction) between the observed proportion or relative frequency of successes and the hypothesized proportion of successes (.5) equals 2, in standard deviation units

---

10. In this case we would need to increase the sample size beyond 100 to take a rejection of the null hypothesis as severely indicating $p > .8$. I return to such considerations in chapter 11.

(one standard deviation being .05). We ask ourselves: Suppose we were to pass H' (assert that she does better than chance) whenever the experiment results in 60 percent or more successes. How severe would that test procedure be? Would it often lead to mistaking chance effects for systematic or "real" ones?

The test procedure can be written in several ways. One is

Test procedure T (in Binomial experiment 5.2): Pass hypothesis H': $p > .5$) (fail H_0) if at least 60 percent out of the (100) trials are successful.

We then ask the above questions more formally in terms of our "significance question": What is the probability of the experiment producing so large a difference from what is expected under the null hypothesis H_0, if in fact the null hypothesis is true? The answer, we said, was .03—quite easily calculated using the Normal distribution.[11] We have

P(test T passes H', given that H' is false [H_0 is true]) = .03.

This is the probability of erroneously passing H': the *b* variant of the severity criterion. The state of affairs "such a passing result would *not* have occurred" refers to all of the (100-fold) experimental trials that result in less than 60 percent successes. The probability of this event is 1 minus the probability of erroneously passing H', namely, .97. So the severity for H' is high (.97). This means that in a series of repetitions of the experiment (each with 100 trials), 97 percent of the outcomes would yield less than 60 percent successes, were we in fact experimenting on a population where the probability of success was only .5. We can picture this as the area under the Normal curve to the left of .6, assuming the null hypothesis H_0 to be true (fig. 6.1). By rejecting the null hypothesis H_0 only when the significance level is low, we automatically ensure that any such rejection constitutes a case where the nonchance hypothesis H' passes a severe test. Such a test procedure T can be described as follows:

Test Procedure T: Pass H' whenever the statistical significance level of the difference (between $\overline{X}$ and H_0) is less than or equal to α (for some small value of α).[12]

11. A standard chart on the Normal distribution tells us that a sample mean exceeds the population mean by as much as 2 standard deviations less than 3 percent of the time. The central limit theorem ensures that the Normal approximation is more than adequate.

12. Because of the adequacy of using the Normal approximation it does not matter if we use "less than" or "less than or equal to α." That is because it is a continuous distribution.

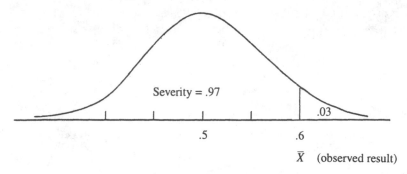

FIGURE 6.1.   The severity for passing H' with $\overline{X} = .6$ equals the probability that test T would yield a result closer to H_0 (i.e., .5) than .6 is, given that H_0 is true.

Calculating severity means calculating 1 minus the probability of such a passing result, when in fact the results are due to chance, that is, when H_0 is true. By definition,

$P(T$ yields a result statistically significant at a level $\le \alpha$, given that H_0 is true) $= \alpha$.

So the severity of the test procedure T for passing H' is $1 - \alpha$.

As might be expected, were the observed success frequency even higher than 60 percent—say she scored 70 percent or 80 percent successes—the severity for H' would be even higher than .97. Here it is enough to see that the severity of passing H' with result .6 (the 2-standard-deviation cutoff) gives a minimum boundary for how severe the test is for H', and that minimum boundary is high. What is indicated in affirming the nonchance hypothesis H' is that the effect is systematic, that the subject's pattern of correct discernments is not the sort typically brought about just by guessing. Granted, to learn this is typically just a first step in some substantive inquiry. Having found a systematic effect, subsequent questions might be: How large is it (perhaps to subtract it out from another effect)? What causes it? and so on. The aim just now was to illustrate how the current framework allows splitting off one question at a time.

Calculating Severity with Infinitely Many Alternatives

In the Binomial experiment above, the hypothesis that passes the test (the nonchance hypothesis H') had only one alternative hypothesis (the "guessing" hypothesis H_0: $p = .5$).[13] In many cases there are

13. That H_0, the null hypothesis, plays the role of the alternative here should cause no confusion despite the fact that it is H' that would generally be called the

several, even infinitely many, alternatives to the primary hypothesis H for which severity is being calculated. In those cases the "not-H" is a disjunction of hypotheses, H_1 or H_2 or H_3 or. . . . How, it will be asked, can severity be made high in such cases? How can we assess the probability that H does not pass, given either H_1 or H_2 or H_3 or . . . ?

The probability of an outcome conditional on a disjunction of alternatives is not generally a legitimate quantity for a frequentist. Its calculation requires a prior probability assignment to each hypothesis H_i. Lest readers declare, "Aha, you are being Bayesian after all!" I had better explain what SC requires in such cases. It requires that the severity be high against *each such alternative*. In other words, the minimum severity against each of these alternative hypotheses needs to be high (the maximum error probability needs to be low), and prior probability assignments are not required to calculate them.

Consider testing the value of a discrete or continuous parameter μ. Specific examples that will arise later are the mean value for Avogadro's number (chapter 7) and the mean deflection of light (chapter 8). The hypothesis H: μ exceeds some value μ', has as its alternative the complex hypothesis made up of all the values less than μ'. That is, "H is false" here means that μ is one of the infinitely many values less than or equal to μ'. Consider the highest of these values, the one that just makes H false, namely, μ'. This corresponds to the simple alternative hypothesis H': μ equals μ'. The probability that the test would *not* pass H, given that this highest valued alternative, H', were true, is calculated in the usual way. A good test (one with a sensible measure of distance from H) yields even higher severity values for each of the alternative μ values less than μ'. In other words, in a good test, if the test has a high chance of detecting that H is just barely false, it has an even higher chance of detecting that H is even more false. This allows us to say that the severity is high against all alternatives. Good error statistical tests provide just such guarantees. Actual experiments can and often do take their lead from these canonical tests. I return to this in chapter 11.

It may be objected that with substantive questions all the possible alternative hypotheses cannot be set out in the manner of alternative values of a parameter. We may not even know what they are. Even where this is so, it does not present an insurmountable obstacle to experimental testing. In such cases what often may be managed is to find a more general or less precise hypothesis such that when *it* is severely

---

"alternative" in formal statistics. When calculating the severity of a test that passes the non-null hypothesis H', the alternative to H' is the null hypothesis H_0.

tested, there are at the same time grounds for rejecting all the alternatives in a manner meeting the severity requirement. The idea is to partition the possible alternatives to learn about the features that any severely tested hypothesis will affirm. What we try to do, in short, is emulate what is possible in canonical experimental tests.

I can rule out the killer's being over six-feet tall without scrutinizing all six-footers. A single test may allow ruling out all six-footers. Using a similar strategy Jean Perrin was able to rule out, as causes of Brownian motion, all factors outside a certain liquid medium. He did so by arguing that if the observed Brownian motion were due to such external factors—*whatever they might be*—the motion of Brownian particles would follow a specified coordinated pattern. His experimental tests, Perrin argued, would almost surely have detected such a pattern of coordination, were it to exist; but only uncoordinated motion was found. In this way, a whole set of extraneous factors was ruled out. This example will be explored in chapter 7.

6.5 USING SEVERITY TO AVOID MISINTERPRETATIONS OF STATISTICAL TESTS: THE CASE OF NEGATIVE RESULTS

I intend the severity criterion to provide a way of scrutinizing what has been learned from applying standard statistical tests. This scrutiny allows us to go beyond merely following the standard conventions of, say, rejecting the null hypothesis on the basis of a statistically significant result. As such, assessing severity is a tool for avoiding common misinterpretations of standard error statistics. Severity considerations, we can say, serve the "metastatistical" function of scrutinizing error statistical results. They can be used to develop standard tools for avoiding canonical mistakes of interpretation (see section 11.6). These mistakes run to type.

A type of mistake particularly appropriate to consider in the present context concerns statistically *insignificant* results—that is, results where the null hypothesis is *not* rejected by the conventional significance test. A classic flaw we need to be on the lookout for in interpreting such "negative" results is the possibility that the test was not sensitive (severe) enough to detect that H_0 was in error in the first place. (The test, it would be said, has too low a *power*.) In that case, just because the test detects no error with hypothesis H_0 is no indication that the error really is absent. I discuss this well-known error in detail elsewhere (e.g., Mayo 1983, 1985a, 1985b, 1989, 1991b), and will address it in discussing statistical tests in chapter 11. Here my concern is to tie our handling of this canonical error with our avoidance of the alternative hypothesis objection. In particular, it will become clear that a hypothesis (e.g., a null hypothesis of no difference) may accord quite well with data (e.g., a negative result) and yet be poorly tested and poorly warranted by that data.

Learning from Failing to Find a Statistically Significant
Difference: The Case of the Pill

A good example is offered by the randomized treatment-control trial on birth-control pills, sketched in example 5.1 (section 5.2). The question of interest concerned parameter Δ, the difference in rates of clotting disorders among a population of women. The question in this study concerned the error of supposing H_0 (no increased risk) when in fact H' is true—there is a positive increase in risk. That is, we tested $H_0: \Delta = 0$ against $H': \Delta > 0$.

The actual difference observed in the Fuertes study was not statistically significant. In fact, the (positive) difference in disease rates that was observed has a statistical significance level of .4. That is to say, 40 percent of the time a difference as large as the one observed would occur even if the null hypothesis is true. (See note 14.)

However, failing to find a statistically significant difference with a given experimental test is not the same as having good grounds for asserting that H_0 is true, that there is a zero risk increase. The reason is that statistically insignificant differences can frequently result even in studying a population with positive risk increases. The argument from error tells us that we may not declare an error absent if a procedure had little chance of finding the error even if it existed. The severity requirement gives the formal analog of that argument.

Of course, the particular risk increase (Δ value) that is considered *substantively* important depends on factors quite outside what the test itself provides. But this is no different, and no more problematic, than the fact that my scale does not tell me what increase in weight I need to worry about. Understanding how to read statistical results, and how to read my scale, informs me of the increases that are or are not indicated by a given result. That is what instruments are supposed to do. Here is where severity considerations supply what a textbook reading of standard tests does not.

Although, by the severity requirement, the statistically insignificant result does not warrant ruling out any and all positive risk increases—which would be needed to affirm $H_0: \Delta = 0$—the severity requirement does direct us to find the smallest positive increase that *can* be ruled out. It directs us to find the value of Δ' that instantiates the following argument:

Arguing from a statistically insignificant increase: Observing a statistically insignificant (positive) difference only indicates that the actual population increase, Δ, is less than Δ' if it is very probable that the test would have resulted in observing a more significant difference, were the actual increase as large as Δ'.

The "if clause" just says that the hypothesis asserting that Δ is less than Δ' must have passed a severe test.

Using what is known about the probability distribution of the statistic here (the difference in means), we can find a Δ value that would satisfy the severity requirement. Abbreviate the value that is found as Δ*. The above argument says that the statistically insignificant result indicates that the risk increase is not as large as Δ*. That is, the hypothesis that severely passes, call it H^*, is

H^*: Δ is less than Δ*.

Let RI be the statistic recording the observed risk increase (the positive difference in disorder rates among treated and untreated women). Then the severity requirement is satisfied by setting Δ* = RI + 2 standard deviations (estimated). For example, suppose that the particular risk increase is some value RI_{obs} and that this result is not statistically significant. Then the hypothesis

H^*: Δ is less than RI_{obs} + 2 standard deviations

would pass a severe test with RI_{obs}.[14] The severity is .97.

Notice that the test result severely *rules out* all increases in excess of Δ* (i.e., all smaller values pass severely). It thereby illustrates the circumstance discussed in the last subsection—how severely ruling out one hypothesis may entail severely ruling out many others as well. (I return to this example and the question of interpreting statistically insignificant results in chapter 11.)

6.6 SEVERITY IN THE SERVICE OF ALTERNATIVE HYPOTHESIS OBJECTIONS

The standard examples of the last two sections have shown both how to obtain and how to argue from high severity. These standard examples, I believe, let us make short work of the variants of the alternative hypothesis objection. For starters, these cases demonstrate the

14. In the Fuertes et al. (1971) experiment, 9 of the approximately 5,000 treated and 8 of the approximately 5,000 untreated women showed a particular blood-clotting disorder at the end of the study. The observed difference is 1/5,000. For a discussion, see Mayo 1985b.

point I made in grappling with Earman's criticism in section 6.3. We can avoid pronouncing as well tested a whole class of hypotheses that, while implying (or in some other way fitting) a given result, are nevertheless not part of the hypothesis space of the primary test. They are simply asking after the *wrong question*, so far as the given test is concerned.

Alternatives That Ask the Wrong Question

Regarding the Binomial experiment on the tea-tasting lady, examples of wrong question hypotheses would be the variety of hypotheses that might be adduced to explain how the lady achieves her systematic effect, such as psychophysical theories about sensory discrimination, or paranormal abilities. That these other hypotheses predict the pattern observed does not redound to their credit the way the results count in favor of H', that she does better than guessing. This shows up in the fact that they would not satisfy the severity criterion.

The procedure designed to test severely whether the effect is easily explained by chance is not automatically a reliable detector of mistakes about the effect's cause. With regard to questions about the cause of a systematic effect, a whole different set of wrong answers needs to be addressed. At the same time, the existence of these alternative (causal) hypotheses do not vitiate the severity assignment regarding hypotheses for which the test is well designed.

This same argument can be made quite generally to deal with alternatives often adduced in raising the alternative hypothesis objection. While these alternatives to a hypothesis H also fit or accord with the evidence, they may be shown to be less well tested than is H. Often these alternatives are at a higher level in the hierarchy than the primary hypothesis under test (e.g., hypotheses about parameter values when the primary question is about a correlation, questions about the direction of a cause when the primary question is about the existence of a real correlation). There are two main points: First, these alternative hypotheses do not threaten a high severity assignment to the primary hypothesis. Second, it can be shown that these alternatives are not equally severely tested. *Because they ask a different question, the ways in which they can err differ, and this corresponds to a difference in severity.* Moreover, if the primary hypothesis is severely tested, then these alternatives are less well tested. It is not that the nonprimary hypotheses themselves cannot be subjected to other severe tests, although there is certainly no guarantee that they can be. It is simply that they are not tested by the primary test at hand. It follows that hypotheses that entail well-tested hypotheses need not themselves be well tested.

A scientific inquiry may involve asking a series of different primary questions, and each will (typically) require its own hierarchy of experimental and data models. One cannot properly scrutinize hypotheses in isolation from the specific framework in which they are tested.

Alternative Primary Hypotheses

It will be objected that I have hardly answered the alternative hypothesis objection when it becomes most serious: the existence of alternative hypotheses to the primary hypothesis. This is so. But we can handle such cases in much the same fashion as the previous ones—via a distinction in severity.

One point that bears repeating is that I am not aiming to show that all alternatives can always be ruled out. Experimental learning is never guaranteed. What I do claim to show, and all that avoiding MUD requires, is that there are not always equally well tested alternatives that count as genuine rivals, and that there are ways to discriminate hypotheses on grounds of well-testedness that get around alternative hypothesis objections.

Maximally Likely Alternatives. A type of alternative often adduced in raising the alternative hypothesis objection is one constructed after the fact to perfectly fit the data in hand. By perfectly fitting the data, by entailing them, the specially constructed hypothesis H makes the data maximally probable (i.e., $P(e \mid H) = 1$). Equivalently, e makes H *maximally likely*. The corresponding underdetermination argument is that for any hypothesis H there is a maximally likely alternative that is as well or better tested than H is.

The "curve-fitting problem" is really an example of this: for any curve connecting sample points, infinitely many other curves connect them as well. (The infamous Grue problem may be seen as one variant.) The problem of maximally likely alternatives was also a central criticism of the account of testing that Ian Hacking championed in Hacking 1965.[15] In this account, evidence e supports hypothesis H_1 more than hypothesis H_2 if e is more probable given H_1 than given H_2.[16] The trouble is, as Barnard (1972) pointed out, "there *always* is such a rival hypothesis, *viz.* that things just had to turn out the way they actually did" (p. 129).

A classically erroneous procedure for constructing maximally likely rivals:

15. It was one of the reasons he came to reject the account. See, for example, Hacking 1972.
16. The probability of e given H_1, $P(e \mid H_1)$, is called the likelihood of H_1. So Hacking's rule of support can also be stated as e supports H_1 more than H_2 if the likelihood of H_1 exceeds the likelihood of H_2.

gellerization. Clearly, we are not impressed with many maximally likely hypotheses adduced to explain given evidence, but the challenge for an account of inference is to provide a general and satisfactory way of marking those intuitively implausible cases. Bayesians naturally appeal to prior probabilities, and for reasons already addressed this is unsatisfactory to us. Moreover, at least from the present point of view, this misdiagnoses the problem. The problem is not with the hypothesis itself, but with the unreliability (lack of severity) of the test procedure as a whole. Infamous examples—both formal and informal—serve as canonical cases of how maximally or highly likely hypotheses can be arrived at in ways that yield tests with low or minimal severity. I call them "gellerized" hypothesis tests.

An informal example is that of the Texas sharpshooter. Having shot several holes into a board, the shooter then draws a target on the board so that he has scored several bull's-eyes. The hypothesis, *H,* that he is a good shot, fits the data, but this procedure would very probably yield so good a fit, even if *H* is false. A formal variant can be made out with reference to coin-tossing trials:

Example 6.1: A gellerized hypothesis test with coin-tossing trials. The experimental result, let us suppose, consists of the outcomes of *n* coin-tossing trials–where each trial yields heads or tails. Call the outcome heads a success and tails a failure. For any sequence of the *n* dichotomous outcomes it is possible to construct a hypothesis after the fact that perfectly fits the data. The primary hypothesis here concerns the value of the parameter *p*—the probability of success on each coin-tossing trial. The standard null hypothesis H_0 is that the coin is "fair"— that *p* is equal to .5 on each coin-tossing trial. Thus any alternative hypothesis about this parameter can be considered an alternative primary hypothesis. In any event, this is what our imaginary alternative-hypothesis challenger alleges.

Let *G(e)* be some such hypothesis that is constructed so as to perfectly fit data *e*. (*G(e)* is constructed so that $P[e \mid G(e)] = 1$.) Suppose that *G(e)* asserts that *p*, the probability of success, is 1 just on those trials that result in heads, and 0 on the trials that result in tails.[17] It

---

17. For example, suppose that *e*, the result of four tosses of a coin, is heads, tails, tails, heads. That is, *e* = *s,f,f,s* where *s,f* are the abbreviations for "success" and "failure," respectively. Then *G(e)* would be: the probability of success equals 1 on trials one and four, 0 on trials two and three. The null hypothesis, in contrast, asserts that the probability of success is .5 on each trial. Another *G* hypothesis that would do the job would assert that the observed pattern of successes and failures will always recur in repeating the *n*-fold experiment. I owe this second example to I. J. Good.

matters not what if any story accompanies this alternative hypothesis. This hypothesis $G(e)$ says that

$G(e)$: p equals 1 on just those trials that were successes, 0 on the others.

The test procedure, let us suppose, is to observe the series of however many trials, find a hypothesis $G(e)$ that makes the result e maximally probable, and then pass that hypothesis. In passing $G(e)$, the test rejects the null hypothesis H_0 that the coin is fair.

Test procedure T (in example 6.1): Fail (null) hypothesis H_0 and pass the maximally likely hypothesis $G(e)$ on the basis of data e.

The particular hypothesis $G(e)$ erected to perfectly fit the data will vary in different trials of our coin-tossing experiment, but for every data set, some such alternative may be found. Therefore, any and all experimental results are taken to fail null hypothesis H_0 and pass the hypothesis $G(e)$ that is constructed to fit data e—even when $G(e)$ is false and H_0 is true (i.e., even when the coin is "fair"). In a long-run series of trials on a fair coin, this test would always fail to correctly declare the coin fair. Hence the probability of passing $G(e)$ erroneously is maximal—the severity of this test procedure is minimal.

To calculate severity in cases where the hypothesis is constructed on the basis of data e, it is important to see that two things may vary: the hypothesis tested as well as the value of e. One must include, as part of the testing procedure, the particular rule that is used to determine which hypothesis to test. When the special nature of this type of testing procedure is taken into account, our severity criterion SC becomes

SC with hypothesis construction: There is a very high probability that test procedure T would *not* pass the hypothesis it tests, given that the hypothesis is false.

To ascertain whether SC is satisfied, one must consider the particular rule for designating the hypothesis to test.

Let the test procedure T be the one just described. The hypothesis that T tests on the basis of outcome e is $G(e)$. There is no probability that test T would *not* pass $G(e)$, even if $G(e)$ were false. Hence the severity is minimal (i.e., 0). In other words, the test procedure T is a maximally unreliable probe when it comes to detecting the relevant error (the error of rejecting H_0 when H_0 is true). This amounts to the defining characteristic of what I call a *gellerized hypothesis*—or, more precisely, a gellerized hypothesis-testing procedure. With a gellerized procedure, the hypothesis selected for testing is one that is constructed to provide an excellent fit for the data, but in such a way that the constructed hypothesis passes a test with minimal (or near minimal) severity.

The manner in which the severity criterion eliminates such gellerized alternatives is important for it hinges on the distinctive feature of error statistical approaches—the centrality of error probabilities. How this contrasts with other approaches will become much clearer later (in chapters 8–11). It should be stressed that gellerized hypotheses are deemed poorly tested by my account not because they are constructed after the fact to fit the data. As I shall argue (in chapter 8), such after-trial constructions ("use-constructed" hypotheses) can pass with maximal severity. They are deemed poorly tested because in gellerized constructions the tactic yields a low or 0 severity test. A primary hypothesis that fits the outcome *less* well than this type of rigged alternative may actually be better, more severely, tested.

The example of gellerization (which comes in several forms), then, is a canonical example of a minimally severe test. As with all canonical examples, it is a basis for criticizing substantive cases that while less obviously fallacious are quite analogous.

Practically Indistinguishable Alternatives. What about alternatives that cannot be distinguished from a primary hypothesis H on the grounds of severity because they differ too minutely from H? This occurs, for example, when H is an assertion about a continuous parameter. My quick answer is this: if there are alternatives to H that are substantive rivals—one differing merely by a thousandth of a decimal is unlikely to create a substantive rival—and yet they cannot be distinguished on the grounds of severity, then that is grounds for criticizing the experimental test specifications (the test was insufficiently sensitive). It is not grounds for methodological underdetermination.

Empirically Equivalent Alternatives. We have yet to take up what some might consider the most serious threat to a methodology of testing: the existence of rival primary hypotheses that are empirically equivalent to H, not just on existing experiments but on all possible experiments. In the case where the alternative H' was said to ask the wrong question, it was possible to argue that the severity of a test of primary hypothesis H is untouched. (If the ways in which H can err have been well probed and ruled out, then H passes a severe test. There is no reason to suppose that such a test is any good at probing the ways in which H' can err.) But the kind of case we are to imagine now is not like that. Here it is supposed that although two hypotheses, H and H', give different answers to the same primary question, both have the same testable consequences. Does it follow that a severity assessment is unable to discriminate between any tests they both pass?

That depends. If it is stipulated that any good test is as likely to pass H, although H' is true, as it is to pass H' although H is true—if it is stipulated that any test must have the same error probabilities for both hypotheses—then it must be granted. In that case no severe test can indicate H *as opposed to* H'. The best example is mathematical, the two hypotheses being Euclidean and non-Euclidean geometry. But apart from certain, not entirely uncontroversial cases in physics, there is no reason to suppose such pairs of rivals often exist in science.[18] Moreover, even if we grant the existence of these anomalous cases, this would fail to sustain MUD, which alleges that the problem exists for *any* hypothesis. There is no reason to suppose that every hypothesis has such a rival.

We can go further. When one looks at attempts to argue *in general* for the existence of such empirically (or testably) equivalent rivals, one finds that severity considerations discriminate among them after all. In fact, one finds that such attempts appeal to tactics remarkably similar to those eschewed in the case of gellerized hypotheses. They too turn out to be "rigged" and, if countenanced, lead to highly unreliable test procedures.

Richard Miller (1987) gives a good example in objecting to alleged empirically equivalent, "just-as-good" alternatives. He asks, "What is the theory, contradicting elementary bacteriology, that is just as well confirmed by current data?" (p. 425) Granted, an alternative that can be constructed is "that bacteria occasionally arise spontaneously but only when unobserved." The severe testing theory dismisses such a general tactic the same way it dismisses an alleged parapsychologist's claim that his powers fail when scientists are watching. Such a tactic (gellerization again!) allows the alternative hypothesis to pass the test, but only at the cost of having no (or a low) chance of failing, even if it is false—at the cost of adopting a minimally severe test. I condemn such tests because one cannot learn from them.

But, the alternative hypothesis objector may persist, doesn't the existence of such an alternative prevent a high-severity assignment to the hypothesis of elementary bacteriology? No. The grounds for assessing how severely this hypothesis passes are a separate matter. It

18. Earman (1993) suggests that the existence even of exotic empirically indistinguishable rivals is enough to make us worry that only a lack of imagination keeps us from recognizing others "all over the map" (p. 31).

passes a severe test to the extent that there are good grounds for arguing that were the bacteriology hypothesis false, then it almost surely would have been found to be false. Such grounds may or may not exist (in the bacteriology case, as Miller notes, it seems that they do). What matters is that no obstacle to such grounds is presented by a rigged alternative, *R*. Hypothesis *R*, in effect, makes the following assertion:

> *Rigged hypothesis R:* a (primary) alternative to *H* that, by definition, would be found to agree with any experimental evidence taken to pass *H*.

Consider the general procedure of allowing, for any hypothesis *H*, that some rigged alternative or other is as warranted as *H*. Even where *H* had repeatedly passed highly severe probes into the ways *H* could err, this general procedure would always sanction the following argument against *H*: all existing experiments were affected in such a way as to systematically mask the falsity of *H*. That argument procedure is highly unreliable. It has a very high (if not a maximal) probability of erroneously failing to discern the correctness of *H*.

Alternatives about Experimental Assumptions

One way of challenging the claim to have severely tested *H* is by challenging the experimental assumptions. Assigning a high severity to a primary hypothesis *H* assumes that the experimental assumptions are approximately met. In fact, the key feature of well-specified experimental tests is that the only nonprimary hypotheses that need to be worried about, for the sake of answering the single question at hand, are challenges to the assumptions of the experimental model. Chapter 5 discusses how to handle these assumptions (they were placed at the stage of checking the data models yet lower down in the hierarchy), so a sketch should suffice.

Again, the procedures and style of argument for handling experimental assumptions in the formal, canonical inquiries are good standards for learning in actual, informal experiments. These experimental procedures fall into two main groups. The first consists of the various techniques of experimental design. Their aim is to satisfy experimental assumptions before the trial is carried out. The second consists of procedures for separately testing experimental assumptions after the trial. Often this is done by means of the "same" data used to pass the primary hypothesis, except that the data are modeled differently. (For instance, the same sequence of trials may be used to answer questions about the assumptions of the Binomial experiment—e.g., is the cause of the effect the color of the cups? The data set is remodeled to ask a different question.) With respect to the statistical assumptions of the two tests studied earlier (the pill and tea-tasting experiments), a whole battery of separate statistical tests is available, often with trivial assumptions. Moreover, we know from the central limit theorem (chapter 5) that with such a large sample size (100), the Normal approximation to the Binomial experiment in the tea-tasting test is easily justified without further checks.

Recall that I initially separated out the error of violating experimental assumptions (the fourth canonical error [error d] in section 1.5) because in general a far less demanding type of argument is needed here. A rough idea of the distribution of the experimental test statistic suffices to say, approximately, how often it is likely to be further from hypothesized values. A host of virtually assumption-free checks often does the job.

In other cases it may be necessary to generate additional data to rule out possible auxiliary factors, such as when the ceteris paribus conditions become suspect. In yet other cases alternative hypotheses may be rejected on the basis of evidence from earlier experiments, now part of the background knowledge, or because they force inconsistent assignments to physical constants. Chapter 7, on Brownian motion, and the discussion of the eclipse experiments in chapter 8 contain illustrations of both types of strategies.

All of this of course requires astute experimental design and/or analysis. By means of the experimental planning, the logically possible explanations of results are effectively rendered actually or practically impossible. The experimentalist whose aim is to get it right does not appeal to hard cores, prior probabilities, or the like; he or she appeals to the various techniques in the experimentalist's tool kit.

To reiterate, I do not hold that relevant alternatives can always successfully be put to rest in these ways. If the threats cannot be ruled out satisfactorily, then the original argument alleging H to be indicated is vitiated. Even so, it does not follow that this alternative hypothesis is itself well tested. To say that the alternative is well tested requires a separate argument showing that *it* has passed a severe test.

6.7 SEVERITY, POPPERIAN STYLE

In appealing to severity to answer the other hypothesis objection, it is clear that the probability in SC does not just fall out from some logical relationships between statements of evidence and hypotheses. We must look at the particular experimental context in which the evidence was garnered and argue that its fitting a hypothesis is very improbable, if that hypothesis is false. This probability refers to the variable behavior of the test rule in (actual or hypothetical) repetitions of the experiment, and the falsity of the hypothesis refers to the presence of some specific error. This relativity to an experimental testing model and the focus on (frequentist) probabilities of test procedures distinguish my account, particularly from others that likewise appeal to probabilities to articulate the criterion for a good or severe test—even from accounts that at first blush look similar, most notably Popper's.

It is important to distinguish Popperian severity from ours because, like the case of the straight rule, Popperian testing has been successfully criticized as open to the alternative hypothesis objection. I explained in chapter 1 why earning a "best tested so far" badge from Popper would not suffice to earn a "well-tested" badge from me. There are, however, several places in which Popper appears to be recommending the same kind of severity requirement as I am. I suspect that Popper's falsification philosophy is congenial to so many scientists because they suppose he is capturing the standard error-testing principles that are at the heart of experimental practice. Less advertised, and far less congenial, is Popper's negativism, that, as he admits, corroboration yields nothing positive, and that it never warrants relying on well-tested hypotheses for future applications. But Popper's most winning slogans are easily construed as catching the error-severity spirit. Here are a few:

> Mere supporting instances are as a rule too cheap to be worth having; they can always be had for the asking; thus they cannot carry any weight; and any support capable of carrying weight can only rest upon ingenious tests, undertaken with the aim of refuting our hypothesis, *if it can be refuted.* (Popper 1983, 130; emphasis added)

> The theoretician will therefore try his best to detect any false theory . . . he will try to "catch" it. That is, he will . . . try to think of cases or situations in which it is likely to fail, if it is false. Thus he will try to construct *severe* tests, and *crucial* test situations. (Popper 1979, 14)

It is not difficult to hear these passages as echoing the goal of severe tests in the sense of SC. Nevertheless, this goal is not accomplished by means of the logical relationships between evidence and hypothesis that Popper calls for. (The particular mathematical formulas Popper offered for measuring the degree of severity are even more problematic and they will not be specifically considered here.) Popper kept to the logical notion of probability, although no satisfactory account of that concept was ever developed. He failed to take what may be called the "error probability turn."

In the next passage, and elsewhere, Popper describes the type of context that he takes as providing grounds for calling a test severe.

> A theory is tested . . . by applying it to very special cases—cases for which it yields results *different from those we should have expected without that theory, or in the light of other theories* . . . those crucial cases in which we should expect the theory to fail if it is not true. (Popper 1962, 112; emphasis added)

Here Popper plainly states that the reason he thinks that hypothesis *H* can be expected to fail if false is that background and alternative hypotheses predict not-*e*—*e* being the result taken to corroborate *H*. That is to say, for Popper, a nonfalsified hypothesis *H* passes a severe test with *e* if all alternatives to *H* that have so far been considered or tested entail not-*e* (or render not-*e* highly probable). A weaker construal requires only that the alternatives say nothing about whether *e* or not-*e* will occur.

Later we will see that the general question of what counts as a severe test is alternately put in terms of the question of what counts as novel evidence for a hypothesis. The answer given by Popper's requirement here is tantamount to requiring that *e* be novel in the sense Alan Musgrave calls *theoretically novel*. The evidence taken to pass *H* is theoretically novel if it is not already derivable from background theories. Lakatos and Musgrave (at times) endorsed both weak and strong construals:

> According to this [theoretical] view, a new theory is independently testable (or predicts a "novel fact") if it predicts something which is not also predicted by its background theory. Hence there are two kinds of independent or novel predictions, tests of which are severe. . . . First, there are predictions which *conflict* with the predictions of the background theory—tests of these will be crucial tests between the new theory and the old. Second, there are predictions concerning phenomena about which the background theory predicts nothing at all—tests of these will also be independent tests and severe ones. (Musgrave 1974, 15–16)

In this view, the requirement for a predicted fact *e* to count as a severe test of *H* may be understood either in the strong form—*e* disagrees with what would be expected given each alternative *H'*—or in the weak form—*H'* says nothing about *e*. I will continue to take the former, stronger version, as the one Popper champions.[19] That is,

---

19. Popper confirmed that this was his view in a private communication. If *H* may be constructed after the data, then so long as *H'* does not entail not-*e* it is easy to see how this condition can be satisfied while the test of *H* nevertheless has low

Popperian severity:
1. H entails e ($P(e \mid H) = 1$) or e is very probable given H
2. Each available H' alternative to H counterpredicts e.

Since it is not clear whether condition 2 requires that the considered alternatives entail not-e or simply that each renders e very improbable, saying that H' counterpredicts e denotes either, except where specifically noted. We might state Popperian severity as follows: e is a severe test of H if H predicts e, while e is anomalous for all other known alternative hypotheses.

There is no demand that a specific testing context be delineated, there are just these two requirements in terms of the logical relationships between statements of evidence and hypotheses. In contrast with the present account, the relevant hypotheses need not be answers to some primary question; they can be anything at all. It is easy to see how the alternative hypothesis objection gets off the ground. Adolf Grünbaum (1978) gets it off quite well.

Grünbaum's "Alternative Hypothesis" Objection to Popperian Severity

Referring to Popper's statement above (Popper 1962, 112), Grünbaum rightly asks how

> qua *pure* deductivist, can Popper possibly maintain without serious inconsistency, as he does, that *successful* results [of tests severe in his sense] should "count" in favor of the theory . . . in the sense that in these "crucial cases . . . we should expect the theory to fail if it is not true"? (Grünbaum 1978, 130)

Passing a severe test in Popper's sense, Grünbaum claims, would leave "the truth of the 'infinitely' weaker disjunction . . . of ALL and only those hypotheses which individually entail [e]" (p. 130).[20] And Popper himself acknowledges the existence of infinitely many alternatives.

In other words, suppose that outcome e is observed. The hypotheses that entail not-e are rejected, and H, which entails e, passes the test. True, H has not yet been refuted—but neither have the infinitely many not-yet-considered other hypotheses that also entail e. Evidence e, it

---

severity. A version of the so-called tacking paradox will serve this function. Simply let $H = H'$ and e. Here H perfectly fits e (i.e., it entails e). But since this can be done for any hypothesis H', such an agreement may be assured whether or not H is false. This criticism appears in Worrall 1978, 330.

20. He asks "of what avail is it to Popper, *qua deductivist*, that by predicting C, H is one of an infinitude of theories . . . incompatible with those *particular* theories with which scientists had been working by way of historical accident?" (Grünbaum 1978, 131).

seems, counts as much for these other alternatives as it does for *H*. As Grünbaum puts it (using *C* for outcome *e*),

> according to Popper's definition . . . the experiment *E* which yielded the riskily predicted *C* does qualify as a "severe test" of *H*. But surely the fact that *H* makes a prediction *C* which is incompatible with the prior theories constituting the so-called "background knowledge" *B* does *not* justify the following contention of Popper's: A *deductivist* is entitled to expect the experiment *E* to yield a result *contrary* to *C*, *unless H is true*. (P. 131)

For even if *H* is false, its falsity is not weeded out. That is because some true (but not yet considered) hypothesis predicts the same outcome that allows *H* to pass. Given the Popperian definition of severity, and given the assumption that there are always infinitely many hypotheses that entail evidence, Grünbaum's worry is well founded. Nor is the situation ameliorated by the additional requirement Popper often advanced, that the hypotheses precede the data (that the data be "temporally novel," to use a term taken up in chapter 8).[21]

The error-severity requirement, in contrast, exists only as part of an experimental account whose central mission is to create situations and specify procedures where we are entitled to expect the experiment to fail *H*, if *H* is false. It is easy to see that satisfying Popper's severity criterion is not sufficient to satisfy ours. One example will suffice.

Cancer Therapies

Each chemotherapeutic agent hypothesized as being the single-bullet cure for cancer has repeatedly failed to live up to its expectations. An alternate, unorthodox treatment, let us imagine, accords with all the available evidence. Let us even imagine that this nonchemotherapeutic hypothesis predicts the chemotherapeutic agents will fail. As such, it may be accorded (one of) Popper's "well-tested" badges. On the error-severity account, the existing data from tests of chemotherapeutic agents provide no test at all of the alternative treatment, because these tests have not probed the ways in which claims made for this alternative treatment can be in error. To count an alternate cancer treatment as well tested simply because it accords with the re-

---

21. For Popper, evidence *e* cannot count as a severe test of *H* if *e* is already explained by other hypotheses. In reality, however, the very newness of a phenomenon may count against the first hypothesis to explain it, because one suspects there may be lots of other untried explanations. Until there is some work on the matter, there are not yet grounds to think that *H* would have failed if it were false. I return to this in chapter 8.

sults of chemotherapeutic trials is to follow a demonstrably unreliable procedure.

Here are the bare bones of this kind of example: A given hypothesis H predicts that there will be a significant correlation between A (e.g., a cancer drug) and B (the remission rate). Alternative hypothesis H' predicts that there will be no significant correlation between factors A and B in the experimental trials. For example, H' may assert that only a new type of laetrile treatment can help. The data, let us suppose, are just what H' predicts—no significant correlation between A and B (e.g., in remission rate). H' passes a Popperian severe test, but for the error-tester it has passed no genuine test at all (or at best a very weak one).

My criticism of Popperian severity is not merely that credit cannot accrue to currently passing hypotheses because there are, invariably, not-yet-considered alternatives that also would pass. It is rather that, as we have just seen, a good test is not constituted by the mere fact that a hypothesis fits the evidence counterpredicted by existing alternatives.[22] Often we can go further and argue that the test is poor because it did not guard against the types of errors that needed to be guarded against.

If H and H' are the only possible alternatives—and H' entails not-e while e occurs—then e is a maximally severe test of H in our sense (and presumably everyone else's). But, in general, Popperian severity is not sufficient for severity in our sense. Neither is Popperian severity *necessary* for error-severity. To consider Popperian severity necessary for a good test would seem to prevent any data already entailed by a known hypothesis to count as severely testing a new hypothesis. (This is argued in Worrall 1978a, 330–31.)

This should finish up the problem with Popperian corroboration first posed in chapter 1. Corroboration, passing a test severe in Popper's sense, says something positive only in the sense that a hypothesis has not been found false—this much Popper concedes. But Popper also suggests that the surviving hypothesis H is the best-tested theory so far. I have argued that Popperian tests do not accomplish this. After all, as Popper himself insists, for a hypothesis to be well tested it must have

---

22. Determining if SC is met by Popper's criterion requires asking, "What is the probability of the conditions for H's passing a Popperian severe test being satisfied (in the case at hand) even if H is false?" SC requires two things of any test rule: first, that we be able to approximately determine the probability that it results in H's passing even if H is false; and second, that this probability is determined to be low. But Popper's severity condition does not provide grounds for assigning a low probability to erroneously passing H.

been put through the wringer. One needs to be able to say that H had little chance of withstanding the inquiry if false. It is a mistake to consider a result counterpredicted by known alternatives to H as automatically putting H through the wringer.

Looking at the problem in terms of the logical relationships between evidence and hypotheses ignores all of the deliberate and active intervention that provides the basis for arguing that if a specific error is committed, it is almost certain to show up in one of the results of a given probe (series of tests). By such active intervention one can substantiate the claim that we should expect (with high probability) hypothesis H to fail a given test, if H is false. And we can do so even if we allow for the possibility of infinitely many alternative hypotheses.

Granted, arguing that a hypothesis is severely tested takes work. In many cases the most one can do is approximate the canonical cases that ground formal statistical arguments. But often it can be argued that a hypothesis is severely tested—even if it means modifying (weakening) the hypothesis. By deliberate and often devious methods, experimenters are able to argue that the test, in context, is severe enough to support a single answer to a single question.

6.8 My Reply to the Alternative Hypothesis Objection

Let us recapitulate how my account of severe testing deals with alternative hypothesis objections that are thought to be the basis for MUD. The MUD charge (for a method of severe testing T) alleges that for any evidence test T takes as passing hypothesis H severely, there is always a substantive rival hypothesis H' that test T would regard as having passed equally severely. We have shown this claim to be false, for each type of candidate rival that might otherwise threaten our ability to say that the evidence genuinely counts in favor of H. Although H' may accord with or fit the evidence as well as H does, the fact that each hypothesis can err in different ways and to different degrees shows up in a difference in the severity of the test that each can be said to have passed. The same evidence effectively rules out H's errors—that is, rules out the circumstances under which it would be an error to affirm H—to a different extent than it rules out the errors to affirming H'.[23]

This solution rests on the chief strategy associated with my experimental testing approach. It instructs one to carry out a complex inquiry

23. This strategy for distinguishing the well-testedness of hypotheses can also be used to resolve the philosophical conundrums known as the Grue paradox and the Ravens paradox.

by breaking it down into pieces, at least some of which will suggest a question that can be answered by one of the canonical models of error. (In some cases one actually carries out the statistical modeling, in others it suffices to do so informally, in ways to be explained.) With regard to the local hypotheses involved in asking questions about experimental mistakes, the task of setting out all possible answers is not daunting. Although it may be impossible to rule out everything at once, we can and do rule out one thing at a time.

Naturally, even if all threats are ruled out and H is accepted with a severe test, H may be false. The high severity requirement, however, ensures that this erroneous acceptance is very improbable, and that in future experiments the error will likely be revealed.

The thrust of experimental design is to deliberately create contexts that enable questions to be asked one at a time in this fashion. In focusing too exclusively on the appraisal of global theories, philosophers have overlooked how positive grounds are provided for local hypotheses, namely, whenever evidence counts as having severely tested them. By attempting to talk about data and hypotheses in some general way, apart from the specific context in which the data and hypothesis are generated, modeled, and analyzed to answer specific questions, philosophers have missed the power of such a piecemeal strategy, and underdetermination arguments have flourished.

Having set out most of the needed machinery—the hierarchy of models, the basic statistical test, and the formal and informal arguments from severe tests—it is time to explore the themes here advanced by delving into an actual scientific inquiry. This is the aim of the next chapter.

CHAPTER SEVEN

The Experimental Basis from Which to Test Hypotheses: Brownian Motion

> My major aim in this was to find facts which would guarantee as much as possible the existence of atoms of definite finite size. In the midst of this I discovered that, according to atomistic theory, there would have to be a movement of suspended microscopic particles open to observation, without knowing that observations concerning the Brownian motion were already long familiar.
>
> —Albert Einstein, *Albert Einstein: Philosopher-Scientist*, p. 47

> I have sought in this direction for crucial experiments that should provide a solid experimental basis from which to attack or defend the Kinetic Theory.
>
> —Jean Perrin, *Atoms*, p. 89

7.1 Brownian Motion: Some Introductory Remarks

A more full-blown example from science can now best elucidate the machinery I have assembled—the hierarchy of models in an experimental inquiry, the simple statistical test, the piecemeal check for errors by splitting off local questions, the use of canonical models of error, and the strategies for arriving at severe tests. An example often discussed by philosophers is the appraisals of hypotheses surrounding the phenomenon of Brownian motion.

Brownian motion, discovered by the botanist Robert Brown in 1827, refers to the irregular motion of small particles suspended in fluid, a motion that keeps them from sinking due to gravitation. Brown thought the particles were alive until he found that the motion occurred in inorganic as well as organic substances.[1] Attempts to explain

---

1. Brown discovered a piece of quartz in which a drop of water had been trapped for millions of years. Observing it under a microscope, he saw numerous particles in ceaseless, irregular motion.

this phenomenon link up with the atomic debates of the late-nineteenth and early-twentieth centuries. The atomic debate being too broad to be taken up here, I restrict my focus to the testing of the Einstein-Smoluchowski (ES) theory of Brownian motion by Jean Perrin. Perrin, who received the 1926 Nobel Prize in physics, provided the long sought after evidence in favor of the molecular-kinetic theory of gases against classical thermodynamics.

Brownian Motion and Paradigm Shifts

This case lends itself not only to explicating the present program of breaking down substantive inquiries into piecemeal canonical questions but also to the exploration of the key role of experimental knowledge in larger scale theory change. The debates about the cause of Brownian motion correlate with a number of disputes between what might be regarded as rival paradigms or disciplinary matrices, as discussed in chapter 2. The disputes were between molecular and phenomenological or energeticist ontologies, mechanical and phenomenological explanation, atomic and continuous metaphysics, statistical and nonstatistical models, realism and instrumentalism, and still others. Correspondingly, the acceptance of the ES theory of Brownian motion led to the changing of the key elements that compose paradigms or disciplinary matrices. It led to a change in beliefs about fundamental entities, the existence of molecules, and the particulate nature of matter. It also led to a change in scientific methodology: a new limit to experimental accuracy due to Brownian fluctuations and "noise" in measuring systems was introduced along with corresponding canonical models of error. The entry of statistical validity into physics also conflicted with the cherished philosophical conception that physics discovers wholly exact laws. (New applications of models from Brownian motion continue up to the present time.[2])

Nevertheless, the dynamics of these changes bore no resemblance to the Kuhnian picture of holistic change, as discussed earlier. At each step of the way experimental methodology and shared criteria of reliability and severity constrained the appraisal. What we see are practitioners from ostensibly rival paradigms learning from and communicating by means of experimental arguments. We cannot make good on these claims, however, until we have properly considered Perrin's

---

2. Statistical models based on Brownian motion are used to understand star clustering, the evolution of ecological systems, and even the fluctuation of retail prices.

much more local work on Brownian motion. It is here that one really finds the locus of the action.

Molecular Reality and Arguments from Coincidence

The Perrin case often has been taken to exemplify variants of the "argument from coincidence" discussed in chapter 3. Gilbert Harman (1965) uses it to introduce inference to the best explanation, Wesley Salmon (1984) to illustrate his argument to a common cause, Nancy Cartwright (1983) to illustrate her inference to the most probable cause. Hacking (1983) ties the Perrin example to his discussion of the argument from coincidence.

They take the Perrin case to illustrate an argument from co-incidence because experiments on many distinct phenomena (e.g., gases, Brownian motion, blue of the sky) gave estimates for Avogadro's number, N (the mean number of molecules per gram molecule), of a similar order of magnitude.[3] Salmon argues:

> If there were no such micro-entities as atoms, molecules, and ions, then these different experiments designed to ascertain Avogadro's number would be genuinely independent experiments, and the strik-ing numerical agreement in their results would constitute an utterly astonishing coincidence. To those who were in doubt about the exis-tence of such micro-entities, the "remarkable agreement" constitutes strong *evidence* that these experiments are not fully independent— that they reveal the existence of such entities. (Salmon 1984, 220)

Cartwright argues along similar lines:

> We have thirteen phenomena from which we can calculate Avo-gadro's number. Any one of these phenomena—if we were sure enough about the details of how the atomic behaviour gives rise to it—would be good enough to convince us that Avogadro is right. Fre-quently we are not sure enough; we want further assurance that we are observing genuine results and not experimental artefacts. This is the case with Perrin. . . . But he can appeal to coincidence. Would it not be a coincidence if each of the observations was an artefact, and yet all agreed so closely about Avogadro's number? (Cartwright 1983, 84)

Salmon and Cartwright view these arguments as supporting realist conclusions. I think their accounts do capture the crux of the argu-ments that are given for the reality of atoms or to "molecular reality," as Mary Jo Nye (1972) puts it. Those arguments, however, were dis-

---

3. Actually, close estimates of N had already been given many years before Perrin's work.

tinct from the experimental arguments Perrin and others grappled with in learning from their experiments. Of course, the arguments for the reality of molecules depended upon the successful inquiries into Brownian motion, but the success of those inquiries did not hinge on the agreement of estimates of Avogadro's number across the thirteen phenomena. For example, the possible error of experimental artifacts was put to rest by Perrin quite apart from the work on the other phenomena (blue of the sky, radiation, etc.). Otherwise he could not have arrived at a reliable estimate of Avogadro's number in the first place. The same was the case for learning the statistical nature of the second law of thermodynamics.

This is not to deny that Perrin utilized arguments from the coincidence of many distinct results—he certainly did. He repeatedly emphasized that one could only put faith in calculations arrived at in several different ways. The several different ways served two functions: to check errors in rather precarious measurements, and to arrive at standard estimates of error (needed for statistical analysis). But first and foremost, Perrin was arguing from coincidence *to obtain experimental knowledge* of the Brownian movement (of microscopic grains). Molecular reality came later. His arguments are really stunning illustrations of the development and use of canonical models of error and of experimental arguments for learning from error.

In the opening passage, Perrin declares himself searching for crucial experiments, but his idea does not fit the mold of a Popperian severe or crucial test. Brownian motion was not only known long before it was used in testing the kinetic against the classical accounts, but it was also accounted for in a number of other ways. Moreover, only after the interesting experimental work had been done could it be seen that the kinetic and classical theories give conflicting predictions. Let us turn to the interesting experimental work.

7.2 SOME BACKGROUND: THE FLURRY OF EXPERIMENTS ON BROWNIAN MOTION

From its initial discovery in 1827, each inquiry into the cause of Brownian motion has been a story of hundreds of experiments. The experiments are of two main classes: experiments that (arrive at and) test hypotheses that attribute Brownian motion either to the nature of the particle studied or to various factors external to the liquid medium in which the Brownian particles are suspended (e.g., temperature differences in the liquid observed, vibrations of the object glass); and experiments that test the quantitative theory of Brownian motion put forward by Einstein and (independently) by M. von Smoluchowski (the ES theory).

Each molecular-kinetic explanation of Brownian motion (first qualitatively proposed by Christian Wiener in 1863) spurred a flurry of experiments by biologists and physicists aiming to refute it. Each nonkinetic hypothesis tried, as by devising a novel way of explaining it by temperature differences, would trigger a new set of experiments to refute the challenge. The enormous variety of organic and inorganic particles studied includes sulphur, cinnabar, coal, urea, India ink, and something called gamboge. Equally numerous were the treatments to which such particles were subjected in the hope of uncovering the cause of Brownian motion—light, dark, country, city, red and blue light, magnetism, electricity, heat and cold, even freshly drawn human milk.

The scientists working on this problem (e.g., Brown, Wiener, Ramsay, Gouy, Perrin, and Smoluchowski) began by carrying out experiments to exclude all exterior causes—checking and rechecking even those suspected factors that already had been fairly well ruled out. (Even after Perrin's work and the general acceptance of molecular theory, experiments using ever-improving methods to observe hundreds of thousands of microscopic grains continued.[4]) By the end of the nineteenth century the most favored explanations in some way attributed Brownian motion to heat (e.g., the theories of Exner, Dancer, Quincke). The need for a molecular explanation began to take hold only at the start of the twentieth century. Ironically, the fact that the same Brownian particles could be used over and over again, sometimes conserved on slides for twenty years, compelled those who had been searching for *nonkinetic* explanations to admit this as strong evidence *for* the kinetic explanation. It indicated that the motion was "eternal and spontaneous," in accordance with the kinetic account. Even so, most researchers required many more quantitative experiments before abandoning nonkinetic explanations.

Only by keeping in mind that a great many causal factors were ruled out experimentally before Perrin's tests (around 1910) can his experiments be properly understood. This will become apparent only when we explicitly consider how Perrin handled the problems of experimental design and control. We can follow the central experimental arguments by means of the hierarchy of models for an experimental inquiry delineated in chapter 5. Except where noted, all references to Perrin are to Perrin [1913] 1990.

---

4. An excellent sourcebook detailing these modern experiments is Wax 1954.

7.3 MODEL OF THE PRIMARY THEORETICAL HYPOTHESIS: THE DISPLACEMENT DISTRIBUTION

The central problem with appraising the kinetic account of Brownian motion was how to formulate testable predictions. The problem was that for many years experimenters were measuring the wrong thing. What they thought had to be checked was whether the molecular effects on the *velocity* of Brownian particles accorded with that hypothesized by the kinetic theory. But this average or mean velocity had been ascertained by trying to follow the path of a Brownian particle, inevitably yielding a measured path much simpler and shorter than the actual path, which changes too fast. An important advantage of the ES theory was that it provided a testable prediction that made no reference to this unmeasurable velocity. At the same time the ES theory explained why the earlier attempts to measure it had failed.

Values obtained for the mean velocity of agitation by attempting to follow the path of a grain as nearly as possible gave the grains a kinetic energy 100,000 times too small. According to Einstein's theory, mean velocity in an interval of time t is inversely proportional to the square root of t; it increases without limit as the time gets smaller. The meaningless results were just what the ES theory says would be expected. As Stephen Brush (1977, 369) remarks, "One can hardly find a better example in the history of science of the complete failure of experiment and observation, unguided (until 1905) by theory, to unearth the simple laws governing a phenomenon."

While Einstein's theory apparently served that guiding role, this does not mean that the tests of the ES theory depended on already accepting Einstein's theory. (I think some people mistakenly suppose that it does.) The reason attempts to measure a particle's velocity were in error was independent of the ES theory. The error turned on a (now) standard statistical point that had been articulated in other contexts but was overlooked. In particular, the point had been made in 1854 by William Thomson (Lord Kelvin) about the problem of laying the Atlantic cable.

Thomson calculated that unless the Atlantic cable is made very thick the transmission of messages between Britain and America would be very slow (Brush 1977, 369). For economic reasons, engineers were unhappy with this recommendation. In addition, attempts to measure the velocity of electricity achieved widely varying results, some appearing to contradict Thomson's prediction. Thomson defended his theoretical prediction: the diverse measurements of the velocity of electricity, he explained, were due to the time spent making the measurements. The time it takes an electric signal to cover a distance is proportional not to the distance itself, but to the square of the distance. (This is Thomson's law of squares.) The greater the length of wire used, the less the apparent velocity of electricity would seem. No wonder the values varied, measured as they were under different conditions. Brush (1977, 369–70) remarks:

> Apparently the scientists who attempted to measure the velocity of particles in Brownian movement later in the nineteenth century had not followed the dispute about Thomson's law of squares in the electric telegraph problem, and they obtained a similar collection of wildly varying results, none of them in agreement with [what would be expected according to the kinetic theory].

Perhaps if a log of canonical experimental errors had been kept, this mistake could have been instructive rather than repeated!

In any event, a key advantage of the ES theory was that it provided a testable prediction that made no reference to this unmeasurable velocity. Instead it was put in terms of the expected (or mean) *displacement* of particles.

Neglecting, therefore, the true velocity, which cannot be measured,

> Einstein and Smoluchowski chose, as the magnitude characteristic of the agitation, the rectilinear segment joining the starting and end points [of a particle]; in the mean, this line will clearly be longer the more active the agitation. (Perrin [1913] 1990, 110)

The displacement of a Brownian particle is the total distance it travels in any direction (say along the x-axis of a graph) as it weaves its zigzagged path. It is a distance that could be measured using the microscopes of the day. (See figure 7.1.) The measurement actually obtained from microscopic observations is the projection of this displacement onto a horizontal plane (e.g., the x-axis). By "observed displacement," I mean this projection onto the x-axis of the given segment.

So the question of interest concerns the quantity (abbreviated as S_t), the displacement (along the x-axis) after t minutes of a Brownian particle from its starting point. If molecular agitation (as described by the kinetic theory of gases) causes Brownian movement, then the displacement of a Brownian particle about its mean (which by symmetry is 0) follows the Gaussian distribution, which is just the familiar Normal distribution. This distribution is given by two parameters, the mean (which is 0) and the variance. The variance is equal to $2Dt$, where D is the *coefficient of diffusion* and t is the time.[5] As with the familiar "bell curve," if the displacement of a Brownian particle is Normally distributed around 0, then displacements near 0 are most probable, while those further from 0 are increasingly improbable.[6] (How probable specific differences are is what the Normal distribution tells us—we need only know D.) As Einstein states it,

> the probable distribution of the resulting displacements in a given time t is therefore the same as that of *fortuitous error,* which was to be expected. (Einstein [1926], 1956, 16; emphasis added)

So the primary theoretical hypothesis is a hypothesized statistical distribution (of displacements of suspended particles):

> The primary hypothesis $\mathcal{H}$: The displacement of a Brownian particle over time t, S_t, follows the Normal (or Gaussian) distribution with $\mu = 0$ and variance $= 2Dt$.

Having provided this hypothesized distribution by which to test the kinetic theory which entails it, Einstein concludes his 1905 paper by remarking, "It is to be hoped that some enquirer may succeed shortly in solving the problem suggested here" (Einstein [1926] 1956, 18). Perrin took up Einstein's challenge:

> It [Einstein's Theory] is well adapted to accurate experimental verification, *provided we are able to prepare spherules of measureable radius.* Consequently, ever since I became . . . acquainted with the theory, it has been my aim to apply to it the test of experiment. (Perrin, 114)

It was Perrin's dogged efforts to prepare grains with measurable and highly uniform radius that made his experimental tests so successful. As we will see, this uniformity of grains was a key assumption of the experimental testing model.

5. For the interested reader, S_t, the displacement after t minutes of a particle in Brownian motion, follows the Normal probability (density) function f:

$$f_{S_t}(x) = \frac{1}{(4\pi DT)^{1/2}} e^{-x^2/4Dt}$$

where D is a constant, the diffusion coefficient. D depends on the absolute temperature and friction coefficient of the surrounding medium. (Strictly speaking, this assumes that t is not too small.)

The mean of S_t, $E(S_t)$, equals 0 and the variance of S_t, $E[S_t^2]$, equals $2Dt$. E here abbreviates the mean or expected value.

6. From our previous discussion we already know more than this. We know, for example, that it differs from its mean by more than 2 standard deviations (in either direction) less than 5 percent of the time.

7.4 Experimental and Data Models

The prediction of the ES theory (for a given type of particle) can be stated as a predicted *standard deviation*[7]—the square root of the variance $2Dt$. Since Avogadro's number, N, is a function of D, once D is estimated, Avogadro's number can be calculated. The calculated value can then be compared to the value hypothesized by the kinetic theory (N^*).[8] So the crux of Perrin's experimental test of the ES theory is evaluating the statistical hypothesis:

> *H:* The experimental displacement distribution is from a population distributed according to Gaussian distribution M with parameter value a function of N^*.

N^* is the (probable) value for N hypothesized by the kinetic theory (approximately 70×10^{22}).

I do not want to be too firm about how to break down an inquiry into different models since it can be done in many ways. The central point is that a series of models of different types (as delineated in chapter 5) is needed to link actual data with primary hypotheses from a theory.[9] Hypothesis H is what the kinetic theory predicts with respect

7. The standard deviation (square root of the variance) is the displacement in the direction of the x-axis that a particle experiences on average (root mean square of displacement). The importance of this statement of variance for the experimental determination of D is that it states that the mean square displacement of a Brownian particle is proportional to the time t. This suggests that a model for Brownian motion is provided by viewing a particle as taking a *random walk*. We can get a rough idea of how this model leads to the Normal distribution as follows. (I follow the derivation in Parzen 1960, 374–76.)

Let X_i be the displacement of a Brownian particle at step i (projected onto a straight line). Consider the sum S_n where

$$S_n = X_1 + X_2 + \ldots + X_n.$$

S_n represents the displacement of a Brownian particle from its starting point. Since it has the same chance of being displaced a given amount in the positive and the negative direction, the average value of X_i equals 0. From the central limit theorem (chapter 5) we have that the sum of these X_i, namely, S_n, is approximately Normally distributed.

8. The connection is this. Estimates of the diffusion coefficient D indicate the approximate rate at which a particle is moving, from which we ascertain the average number of collisions to which these Brownian particles must be subjected to have caused such diffusion. This indicates approximately how many molecules per unit area there must be, that is, N.

9. Each individual task in an experimental inquiry can be seen as calling for a separate primary inquiry, with its own models of experiment and data. Then the full-blown experimental argument will string together different primary experimental arguments. Alternatively, the full-blown argument can be viewed as a single primary experimental argument, but with subarguments needed at different

to a *given* experimental context E. It may be located one step below the primary theoretical model in our hierarchy: it is part of the experimental model.

Note that hypothesis H makes two assertions: it is an assertion about the distribution M and about the values of parameters. The sample data from experiment E can be used to estimate or test values of N only if they can be seen (i.e., modeled) as the results of observing displacements from the hypothesized Gaussian process.

Correspondingly, I suggest we designate two primary questions that needed to be asked, one about the form of the distribution, the other about parameter values. Perrin's own discussion clearly distinguished between these two tasks, which I shall call step 1 and step 2. In a nutshell, step 1 consists of checking, for each experiment E, whether the results of the experiment actually performed follow the given statistical distribution M, and step 2 involves using estimates of D to estimate or test values of N (Avogadro's number).

Philosophers who discuss this case tend to place the task of step 2 at the forefront. Accordingly, they locate the impressive part of Perrin's argument in his estimates of Avogadro's number, which are close to what the kinetic account predicted. In fact, what made Perrin's results so impressive (at step 2) centered on his arguments for step 1: showing that the distribution of displacements was "completely irregular." Moreover, the argument against the nonstatistical version of the second law of thermodynamics hinged on the results at this first step. (I return to this in section 7.7.) What does step 1 look like?

Step 1: Manipulations on Paper

This is a good example of a case where the substantive question is identified with testing a standard type of statistical hypothesis, one that asserts that the distribution of displacements is of the "chance" (or fortuitous or nonsystematic) variety. Since this statistical hypothesis is one piece of the investigation of H, we had better use a different letter (lower case, to indicate it is a portion of H). Take j:

> j: The data from E approximates a random sample from the (hypothesized) Normal process M.

The *denial* of j, denote it as j', roughly asserts that

> j': the sample displacements of data from E are characteristic of systematic (nonchance) effects.

---

nodes of the experimental context. I choose to model the present example employing the latter way of modeling.

It is noteworthy that here the hypothesis of chance, j, is not rejected but rather "passes" several tests. This is done by rejecting hypothesis j'.

So we have split off from the full problem this one question about a low-level statistical hypothesis j. But appraising this low-level statistical hypothesis—far from being a preliminary side show—was the main event and feature attraction of Perrin's work.

Affirming j, and ruling out j', corresponds to affirming the key assumption about Brownian motion. As Perrin shows, Einstein's derivation of the displacement distribution depends on "making the *single* supposition that the Brownian movement is completely irregular" (p. 112). So ruling out hypothesis j' was the centerpiece of Perrin's work. Asking about j' came down to asking whether factors outside the liquid medium might be responsible for the observed motion of Brownian particles. The general argument in ruling out possible external factors—*even without being able to list them all*—was this: if Brownian motion were the effect of such a factor, then neighboring particles would be expected to move in approximately the same direction. In fact, however, a particle's movement was found to be independent of that of its neighbors. To sustain this argument, Perrin called up experimental knowledge gleaned from several canonical cases of ("real") chance phenomena.

Consider just one of Perrin's experimental tests of j against j'. It was based on an experiment E that consisted of observing 500 displacements of grains of gamboge (a microscopic vegetable particle).[10] Perrin considered these particular grains to be among his most uniform grains. To get the displacements, the positions of the grains (observed with a camera lucida) are recorded every 30 seconds on paper with grids of squares.[11] The path of a single grain might look as in figure 7.1.

The actual data consist of 500 scratch marks, each measuring the displacement of gamboge grains in 4 positions. To turn this into data that can answer questions posed in the experimental model regarding hypothesis j, the observations must be condensed and organized. This takes us to the level of *models of the data*. The idea is to do something that will enable the 500 actual outcomes (displacements) to be seen as a single random sample from the population of possible experimental outcomes (the sample space of the experimental model).[12] If we can

10. Perrin does not say if these 500 displacements are from a larger experimental run.

11. Were the positions recorded at much shorter intervals, each single segment in the figure would be as complicated as the entire figure.

12. If the experiment consists of marking off 500 displacements of a given type of gamboge grain (say at 30-second intervals), then the sample space may be seen to refer to the different 500-fold outcomes that could have resulted. Alternatively,

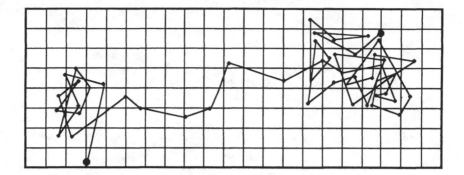

FIGURE 7.1.   Tracing of horizontal projections of lines joining consecutive positions of a single grain every 30 seconds.

perform this feat, then we can ask of this *single* sample of 500 displacements whether it may be seen as a random sample from a population with the hypothesized Normal distribution.

Needed is a characteristic of the data—a statistic—such that this statistic, whose value we *can* observe, will teach us about the parent population that we cannot. Each such statistic refers to a different modeling of the data, and Perrin delineates several such models. That he does so is what makes his work such a treasure for the philosopher of experiment. Here I shall discuss the one data model that Perrin claims gives a "still more striking verification" (p. 117) than the others. It is obtained by looking at the value obtained from shifting each (horizontal) displacement to a common origin, and then counting how many are found at various distances from this common starting point. To sharpen the ability to distinguish between j and j', the data are condensed into 9 pigeonholes, each a different distance from the origin. As Perrin reasons,

> The extremities of the vectors obtained in this way should distribute themselves about that origin as the shots fired at a target distribute themselves about the bull's-eye. [See figure 7.2.]
> Here again we have a quantitative check upon the theory; *the laws of chance enable us to calculate how many points should occur in each successive ring.* (Perrin, 118; emphasis added)

This is a quintessential example of what I mean by "manipulations on paper." Nothing like a bull's-eye is actually observed in experiment

---

if the data are condensed into 9 pigeonholes, each a different size of displacement, then the sample space is the set of different numbers that could be observed in each of the 9 categories or rings.

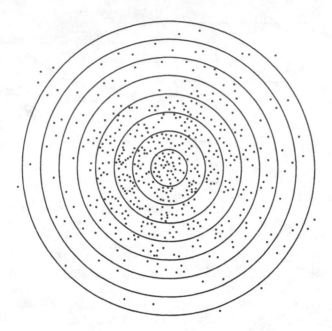

FIGURE 7.2. Manipulations on paper: displacements of 500 Gamboge grains.

E, but rather the displacements at 30-second intervals of several gamboge grains. The bull's-eye picture results not from physical manipulation of the grains, but from manipulations on paper. They serve much the same role in the experimental argument as physical instruments. They allow discerning patterns in the data hidden from an eye looking at 500 scratch marks. The manipulations on paper are warranted not because they represent actual experimental phenomena, but because once they are accomplished (e.g., once the data are manipulated into bull's-eye rings), "the laws of chance enable us to calculate how many points should occur in each successive ring" (Perrin, 118).

  This tactic, so important yet so misunderstood, bears elaboration. The tactic is essentially this: Take a look at the handful of canonical models—in this case they are statistical distributions. In your tool kit are a bunch of random variables that have these distributions. Then think of ways of massaging and rearranging the data until you arrive at a statistic, which is a function of the data and the hypotheses of interest, and has one of the known distributions. (Your tool kit also contains standard ways of massaging and rearranging.) Nothing in front of you needs to actually have the distribution you arrive at, nor need it correspond directly to any actual event. It may simply be the distribution followed by the random variable arrived at through ma-

TABLE 7.1 (adapted from Perrin, p. 119)

Displacement Between	Probability	n calculated	n found
0 & first ring	.063	32	34
1st & 2nd rings	.167	83	78
2nd & 3rd rings	.214	107	106
3rd & 4th rings	.210	105	103
4th & 5th rings	.150	75	75
5th & 6th rings	.100	50	49
6th & 7th rings	.054	27	30
7th & 8th rings	.028	14	17
8th & 9th rings	.014	7	9

nipulations on paper (e.g., averaging, dividing by or adding appropriate numbers, squaring). But that is all you need to assign probabilities to various outcomes on the hypotheses being tested. Once you have this, statistical tests can be run and their error probabilities calculated. And these error probabilities (e.g., severity) *do* refer to the actual experimental test procedure.

Students of theoretical statistics are familiar with this sort of homework problem: Starting with a random variable with an unknown distribution, find a way to alter it (making use of what I call manipulations on paper) so as to arrive at a variable whose values vary in the manner of one of the known distributions.

Perrin's bull's-eye manipulations in testing hypothesis *j* in the Perrin experiment exemplify this tactic. If the statistical hypothesis *j* holds for the actual experiment, then we can deduce the probability that a displacement would fall in each of the 9 rings. We can deduce the number of displacements expected to fall in each ring by multiplying this probability by the number of displacements (500). This number is termed "*n* calculated." That is, as shown in table 7.1 in the column labeled *n* calculated, the expected frequency of displacements falling between the *i*th and the *i* + 1th ring is given by 500 multiplied by the probability of a displacement falling between the *i*th and the *i* + 1th ring.

This provides us with the ingredients for comparing the observed distribution of measurements (*n* found) with a set of hypothesized probable measurements (*n* calculated). The recipe comes from canonical arguments for asking: What should we make of the fit between observed and hypothesized? We want to know, especially when the differences between observed and expected are small, whether they may be merely the fluctuations typical of a sample of that size from a population of the assumed Normal distribution (as asserted in *j*), or whether a serious departure is indicated (as asserted in j'). We tackle this by asking what it would be like if j were a correct description of the experiment. A quantitative argument is given by a statistical significance test much like the ones we have already sketched.

Here the distance measure chosen is a function of the difference between the number found and the number calculated for each of the 9 intervals:

$$\frac{n \text{ found} - n \text{ calculated}^2}{n \text{ calculated}}$$

Summing these up yields a statistic that has a known probability distribution (the chi-square distribution).[13]

What matters is being able to sustain the argument of the significance test (i.e., answer the "significance question"). This we can do because we can determine the probability that a purely random sample of measurements taken from the hypothesized model distribution j would show worse agreement with the model j than is shown by the actual set. If the differences observed are of the sort frequently "caused by" chance (i.e., if they are typical under j), then the sample data are in accord with the hypothesized experimental model. In Perrin's experiment E above, it turned out that the observed differences, or one even larger, are not infrequent but rather are typical, assuming j. That is, the observed difference is not *statistically significant* from what is typical under j. Hypothesis j passes.

Perrin's argument to this effect has weight only because he was able to argue further that if the model was inadequate (if j' was true), we would very often get differences statistically significant from what is typical under j. In other words, he needed to argue that j had passed a severe test. The multiple experiments for which Perrin stresses the need (e.g., Perrin, 96) are deliberately designed so that if one misses an error, another is likely to find it. The experiments are designed, to use David Hull's nifty phrase, to ensure that "errors ramify rapidly." The error of concern at this stage is that some regularity of Brownian motion has been concealed.

As is typical, Perrin's argument for severity was substantiated by reference to other tests. The overall argument goes beyond any single statistical significance test. Here is where the multitude of deliberately varied additional tests plays a particularly important role. In fact, the number of tests, checks, and rechecks Perrin performed amounted to statistical

---

13. The chi-square test, introduced by Karl Pearson in 1900, was actually in use before that. It was alluded to in testing assumptions in chapter 5.

overkill. He was not being overly cautious; he was well aware that others would unearth any weak point upon which to attack his arguments.

To this end, Perrin describes numerous sets of statistical analyses. Some made use of the *same* 500 measured displacements, only modeled in different ways; others involved further recorded displacements on the same gamboge preparation. Still others dealt with totally distinct gamboge experiments, where the key features were deliberately varied. (See table 7.2.) While the same question is being asked (Is *j* adequate?), by phrasing it differently each test is designed to check if there are mistakes in the answers from other analyses.

Viewing Brownian Particles as Taking Random Walks

Familiarity with standard chance mechanisms—quite independent of the ES theory—provided knowledge of experimental phenomena that are correctly described by hypothesis *j*, as well as phenomena more correctly described by *j'*. The experimental knowledge stemmed from the statistical theory of *random walk* phenomena (in one dimension). From this statistical theory about a very general type of fluctuation phenomenon, strategies for the experimental and data models in testing *j* emerged. We see here just the sort of appeal to "real" random experiments that Neyman talked about (chapter 5).

The question in step 1, whether the displacements of Brownian particles could be characterized as the hypothesized Gaussian process *M* (i.e., whether *j* is adequate) is tantamount to asking, Can the displacements be modeled as a problem concerning a *simple random walk?* Such random walk problems were understood at that time. Einstein ([1926] 1956) had presented his derivation of the displacement distribution in $\mathcal{H}$ by means of a model of a random walk in one dimension[14] (see notes 7 and 22).

The displacement of a particle may be seen as the result of *k* steps where at each step the particle has an equal chance of being displaced by a given amount in either a positive or negative direction. (This is called a *simple* random walk). Since it has the same chance of being displaced a given amount in the positive and the negative direction, on average, after *k* steps, the displacement would be 0. Occasionally, more steps will be in one direction than the other, yielding a nonzero total displacement. That the variance is proportional to the time (the number of steps) corresponds to the fact that the more steps taken, the larger the value this nonzero total displacement can have.

14. For other clear derivations see Chandrasekhar 1954, 7 and Parzen 1960, 374–76.

A second type of canonical random experiment was used in deriving the statistical distribution in the ES theory—this one from gambling. Here one capitalizes on the fact that the displacement of Brownian particles is distributed like the winnings of a gambler who stands to win or lose a fixed amount x with equal probability on each game. The more games played, the larger the gambler's loss can be.

There were yet other statistical derivations of the Normal distribution in the ES theory, all fascinating in their own right. I shall resist going into them here. My present purpose is to illuminate the strategy by which the hypothesized distribution in $\mathcal{H}$ is linked with experimental tests of H by way of tests of hypothesis j. For each of several different experiments, abbreviated as $E_1, E_2, \ldots, E_n$, the predicted displacement distribution (in the experimental model) is given by hypothesis j. That is, for a given experiment E_i, we have

> j: the distribution of the n displacements in E_i is from a population distributed according to the Gaussian model M.

In other words, if displacements of Brownian particles follow the random walk model hypothesized in $\mathcal{H}$, then each experiment E_i is, in the relevant respects, just like experimenting on known chance mechanisms. (This assumes of course that experimental assumptions are met, but we will deal with this separately in section 7.6.)

This probabilistic linkage between $\mathcal{H}$ and j has two main functions. First, it allows deriving, for each experiment E_i, the experimental displacement distribution. This, recall, was the basis of the statistical significance test of j. Second, it is the basis for using the different experiments to cross-check and strengthen each test of j.

These linkages are at the heart of Perrin's argument that j had passed a reliable (severe) test. The argument goes like this: if experiment E was not correctly described by hypothesis j (i.e., if j' were true), there would *not* be an equal chance of being displaced by an amount in either direction for each particle: there would be some *dependencies*. But we know what it is like to interact with a mechanism with such dependencies. We know what would be expected in those sorts of experiments—we can even "display" it. We can actually generate the (frequency) distribution of the outcomes from experiments (on ball rollings or other chance apparatus)—where, by design, the probability of being deflected to the right is *not* equal to that of the left. (This would be an example of Neyman's "real" random experiments.) Or we can simulate these results by Monte Carlo methods or (now) by computer simulations. If we are observing such dependencies, then, we calculate, it should be fairly easy (frequent) to generate statistically significant differences in Perrin's various gamboge experiments from j'.

So we can argue that were j', and not j, the case it is extremely improbable that none or even very few of the experiments E_1, E_2, . . . , E_n would have indicated this. It is very probable that a few would have shown differences statistically significant from what is expected under j. That is to say, the test was severe for j—where "the test" includes the results from several individual experiments. The pattern of arguing from error is clear. The experiments conducted by Perrin and his researchers had a very high probability of detecting a statistically significant difference from j', were there dependencies in the motion, yet such differences were not detected. On the basis of such considerations, j' was ruled out (j was affirmed) by Perrin and others. After describing several different experiments (see table 7.2), all of which j passed, Perrin remarks:

> Further verifications of the same kind might still be quoted, but to do so would serve no useful purpose. In short, the irregular nature of the movement is quantitatively rigorous. Incidentally we have in this one of the most striking applications of the laws of chance. (Perrin, 119)

7.5 Step 2: Estimating and Testing Avogadro's N

Having found that hypothesis j passed the tests, Perrin then asks, "But does it lead to these values for the molecular magnitudes that we look for?" (p. 103). Because of what was accomplished in step 1, Perrin is again in a position to split off a manageable piece from the full-blown problem in carrying out step 2. The error of concern at step 2 is that the ES theory is mistaken about the values of the magnitudes (e.g., N). Having chosen each experimental displacement distribution E to be a function of the parameter of interest, Avogadro's number N, step 2 becomes a standard problem of estimating the parameter governing the displacement distribution. More precisely, step 2 involves using data models to test a hypothesis about the statistical parameter governing the probability distribution affirmed in step 1—the variance.[15]

The variance of the distribution, recall, is $2Dt$. Estimates of D are experimentally obtained by calculating the mean-square displace-

15. To say that one or more parameters "govern" a statistical distribution means that if the values of the parameter(s) are given, then the probability of each possible outcome (i.e., the probability distribution) can be calculated.

ments of the numerous particles recorded.[16] Plotting the mean-square displacement against time on a graph yields a nearly straight line, whose slope is an experimental estimate of D. The remaining task, then, is to test a hypothesis about D which is a function of N, Avogadro's number. (Since D varies for different grains, it is preferable to work with N.) Perrin remarks:

> To verify Einstein's diffusion equation, it only remains to see whether the number [obtained for N by substituting the estimate of D into the equation $D = \dfrac{RT}{N}\left(\dfrac{1}{6\pi a\xi}\right)$] is near 70×10^{22}. (Perrin, 132)[17]

The value 70×10^{22} is the value for N that is hypothesized by the ES theory, abbreviated N^*. So the single task of interest has been boiled down to testing a simple statistical hypothesis h expressed in the experimental model

$h: N = N^* \ (70 \times 10^{22})$.

The falsity of h, i.e., "not-h," asserts that Avogadro's number is not near N^*. Given that Perrin wants to affirm h so long as the data indicate that N "is near," and not necessarily exactly equal to, N^*, it might be thought that h should be an interval around N^*. For mathematical reasons it is easier to express h as a "simple" or "point" hypothesis, and then take the "nearness" into account in the testing rule. That is what the standard statistical test does.

So step 2 may be broken down into an application of one or more canonical models for detecting errors in a parameter value. Although much of the modern apparatus for solving a standard estimation problem was not yet developed, Perrin's use of those notions that were available (e.g., probable errors) shows his argument to be based on standard error probability ideas. Perrin uses the data from each of the experiments E to derive estimates of N with known error properties (e.g., a known probable error[18]). (There is certainly no attempt to state prior probabilities and multiply them by likelihoods to yield posteriors.)

Interestingly, the estimates at step 2 come from observations on the same grains that were prepared and used in performing step 1.

16. The displacements of all the particles recorded in an experiment at some time t are squared and the arithmetic average or mean is calculated. This is the mean square displacement at time t (or the variance). The *root* mean square displacement is the square root of this (or the standard deviation).

17. ξ is the viscosity of the fluid; T, its absolute temperature; R, the gas constant; a, the radius of the particles.

18. The probable error is around .7 of the standard deviation.

Only now a different question is asked of these grains and, correspondingly, different data models are formed. Here one needs to count the relative number of grains observed at different levels of the emulsion. The canonical statistical model appealed to is the exponential distribution. A complete investigation of this step would elaborate further on the variety of statistical models Perrin uses here and the shrewd manner by which he combines them to arrive at a reliable argument at step 2. Perhaps for our purposes enough has been said.

"As a matter of fact," Perrin reports, the number he obtains for N "is equal to 69×10^{22} to within ± 3 per cent" (p. 132). This is a good fit to N^*. Referring to the good fit of estimates from a number of experiments on Brownian motion (of emulsions) Perrin concludes:

> This remarkable agreement proves the rigorous accuracy of Einstein's formula and in a striking manner confirms the molecular theory. (P. 123)

In each case the "confirmation" of Einstein's hypothesis is based on a standard statistical argument. The difference between the estimated and hypothesized values of N is the distance measure. Because "the discrepancy is well below the possible error introduced by the somewhat loose approximations . . . in making the calculations" (p. 127), hypothesis h passes the test.

In describing the several sets of experiments from both steps 1 and 2, Perrin emphasizes again and again the deliberate attempt to vary several aspects of the experiment:

> I have carried out personally, or directed in others, several series of measurements, varying the experimental conditions as much as I was able, particularly the viscosity and the size of the grains. (P. 122)

He summarizes them in the following (abridged) table (table 7.2).

Nearly all the estimates of N from these experiments were (statistically) insignificantly far from that predicted by the kinetic theory, N^* (i.e., 70×10^{22}). Perrin declares:

> It cannot be supposed that, out of the enormous number of values *a priori* possible, values so near to the predicted number have been obtained by chance for every emulsion and under the most varied experimental conditions. (P. 105)

It should be supposed, instead, that the reason values so close to the predicted value N^* can be repeatedly generated is that N^* is approximately correct. In terms of experimental knowledge, this means that in a (hypothetical) population or series of experiments, the mean value

TABLE 7.2 (from Perrin, p. 123)

100ξ	Nature of the Emulsion	Radius of the grains μ	Mass $m \times 10^{15}$	Displacements Recorded	$\dfrac{N}{10^{22}}$
1	I. Gamboge grains	.50	600	100	80
1	II. Gamboge grains	.212	48	900	69.5
4 to 5	III. The same grains in sugar solution	.212	48	400	55
1	IV. Mastic grains	.52	650	1,000	72.5
1.2	V. Very large grains (mastic) in urea solution	5.50	750,000	100	78
125	VI. Gamboge grains in glycerine	.385	290	100	64
1	VII. Gamboge grains of very uniform equality (two series)	.367	246	1,500	68.8
				120	64

ξ = the mean value of the viscosity

for Avogadro's number would be $N*$,[19] and in actual experiments, the deviations from $N*$ would be of the pattern ascribable to chance.

Let us summarize the argument in step 2. Step 1 affirmed that the experimental distribution is the Gaussian one hypothesized in j. Given step 1, we know how to design experiments that make it very difficult to generate results close to $N*$ unless we really are sampling from a population where N is approximately $N*$, that is, unless hypothesis h is true. This allows us to design an experimental test of h such that if h passes, then h has passed a severe test.

The denial of hypothesis h asserts that there is genuine discord between the hypothesized value, $N*$, and the "true" value, where the true value refers to the mean or expected value for N in a population (or long series) of experiments. If there is such discord (the extent of which can be made rigorous), then our experiment has given it a good chance (often) to manifest itself. It would manifest itself by producing an observed discrepancy statistically significantly beyond the possible experimental error introduced in estimating N. So each experimental test itself is a reasonably severe test of h. The several experiments taken together are further checks and thereby strengthen the overall severity.

The argument follows the pattern of arguing from error. If we are wrong in any single experiment that results in passing h, then it should

19. That is because, on the average, our estimate of N equals the population mean $N*$. That is, the average (i.e., mean) value of our estimate equals the population mean. This was discussed in chapter 5.

be very hard to reproduce results "close" to N^* in experiments especially designed to display discordances (by revealing statistically significant differences). But Perrin shows we can generate at will (very frequently) estimates of N near the hypothesized value N^*. Therefore, he can argue for the overall reliability of the argument. In terms of a probability calculation, he can say that

P (such good accordance in experiments E_1, E_2, . . . , E_n | h is false) = very low,

meaning that h passes a severe test.

The multiple experiments listed in table 7.2 rule out mistakes or "other hypotheses" in both steps 1 and 2, but the mistakes differ. In step 1 they were used to rule out systematic effects. In step 2, they improved the reliability of the estimates of N and in addition helped rule out mistakes having to do with generalizability. By being deliberately varied in the experiments, any influences of the liquid, the temperature, the nature and density of the grains, and so on would affect Perrin's estimates in all directions and so cancel each other out (in the mean), leaving an extraneous effect comparable to experimental or chance error. This is *statistical control*. These multiple experiments also check errors regarding the generalizability of results. Perhaps the ES theory applies only to particular grains or experimental circumstances. Appropriate variations let us rule this out.

Now for the questions and problems that arise in checking experimental assumptions. They correspond to problems and models that would be placed "below" the data models, and our analysis would be seriously incomplete if we did not address them.

7.6 CHECKING EXPERIMENTAL ASSUMPTIONS: EXPERIMENTAL DESIGN AND DATA GENERATION

A key problem I have placed at the level of data models in an experimental testing context is that of checking that the various experimental assumptions are satisfied. Their violation may introduce alternative explanations for the results and may thereby vitiate experimental arguments. Checking on experimental assumptions sends us to considerations "below" the data models, to those of experimental design or data analysis. We have to look at how the experimental objects—grains in various solutions—were generated and measured to produce raw data.

In attending to the actual data-generation procedures, it becomes plain that to begin the analysis with the estimates of N (Avogadro's number) as the data is too simple. Estimating N calls for a full-blown inference in its own right. Each estimate of N depends on being able to obtain estimates of a number of other quantities, along with their associated errors. Most notably, the experiments turn on being able to accomplish an experimental tour de force—obtaining Brownian particles each with a fairly uniform radius.

What makes Perrin's discussion so valuable for us is his careful explication of the labors required to justify experimental assumptions. He constantly stresses the need to search for and rule out errors by multiply-connected checks and tests. Here one finds ample illustration of before-trial procedures of experimental design (e.g., to ensure that the gamboge preparations are likely to be useful) and after-the-trial checks of whether assumptions are approximately met. It is impossible to appreciate the full force of Perrin's tests without delving into these details. Here are some highlights.

Measuring Microscopic Grains of Gamboge

After unsuccessful attempts to use the substances usually studied, Perrin hit upon gamboge:

> Gamboge (which is prepared from a dried vegetable latex) when rubbed with the hand under water (as if it were a piece of soap) slowly dissolves giving a splendid yellow emulsion, which the microscope resolves into a swarm of spherical grains of various sizes. (P. 94)

The force of Perrin's results, at bottom, hinged on his uncanny ability to ensure that the particles were of approximately the same size, that they could be counted and weighed, and that a host of extraneous factors could be controlled or "subtracted out"—even Einstein expressed surprise. Most impressive of all, perhaps, was the preparation of grains of uniform size. The key was the special technique Perrin developed for "fractional centrifuging."

> The emulsion having been obtained, it is subjected to an energetic centrifuging (as in the separation of the red corpuscles and serum from blood). (P. 94)

A top layer of sediment is formed and poured off, and the grains are again suspended in (distilled) water. The centrifuging and pouring off processes are repeated again and again until the emulsion is practically pure water.

> But the purified emulsion contains grains of very various sizes, whereas a *uniform* emulsion (containing grains equal in size) is required. (Pp. 94–95)

By further centrifuging, it is possible to separate out the grains according to size, the first layers of sediment containing the largest grains. Getting an adequate separation in this manner was an extremely lengthy process. In his most careful measurements, Perrin tells us, he labored over his gamboge for several *months* of treatments:

> I treated 1 kilogramme of gamboge and obtained after several months a fraction containing a few decigrammes of grains having diameters approximately equal to the diameter I wished to obtain. (P. 95)

Reliably Measuring the Density, Volume, and Weight of the Grains

Numerous interconnected checks and rechecks were used in scrutinizing this and other assumptions. The key to ruling out errors was a deliberate variability. Ascertaining the volume of the gamboge grains exhibits the standard pattern:

> Here again, as with the density, it is possible, on account of the smallness of the grains, to place confidence only in results obtained by several different methods. (P. 96)

The different methods desired are those that allow arguing that any error present is very likely to be detected by at least one method. When, through the several methods, Perrin obtained concordant results, he could rule out experimental artifacts. Here he is clearly arguing from coincidence along the lines sketched in Hacking's example of dense bodies (chapter 3); at this stage there is no need for a formal assessment of the degree of severity.

One of the methods used in measuring volume involved measuring the radius of the grains in the camera lucida.

> Considerable error is involved in the measurement of isolated grains (owing to the magnification by diffraction . . .). This source of error is very considerably minimised if it is possible to measure the length of a known number of grains in a row. (P. 96)

If Perrin could find a way to get his grains to line up in a row, he could appeal to a canonical technique for counting objects that had nothing to do with Brownian particles. He discovered that if he let a drop of the emulsion nearly evaporate, capillary forces made the grains run together

> and . . . collect together into groups a single grain in depth and more or less in rows, in the same way that the shot are arranged in a horizontal section through a pile of shot. (P. 97)

It was then possible to count the number of grains lying in a row.

This exemplifies another thread woven through Perrin's work. The counting procedure did double duty: it was also used to check a measurement somewhere else. Perrin continues:

> At the same time a general check upon the uniformity of the grains sorted out by the operation of centrifuging is obtained. The method gives numbers that are perhaps a little too high (the rows not being quite perfect); but owing to its being so direct it cannot be affected by large errors. (P. 97)

For Perrin to learn what he is after, he needs to count his carefully prepared grains. If he can make them arrange themselves like a pile of shot, he can not only count but check on his centrifuging results. Perrin may be said to have invented his techniques for preparing and working with his grains, but the models for analyzing the errors were standard and did not belong to any one domain.

In some cases mistakes were made and later detected. For instance, one estimate of N, while close to the predicted value, was invalidated because it was noticed "during the course of some measurements on some preparations . . . that the proximity of a boundary checked the Brownian movement. (Einstein's theory presupposes an unlimited fluid.) . . . These measurements will be repeated" (p. 124). In other experiments, Perrin deliberately exploited the phenomenon of grains sticking to the walls!

I hope this suffices to get the flavor of how tactics of observation and measurement may be pieced together into an experimental argument that allows learning about primary hypotheses. The complexity of the hierarchy of models in an experimental inquiry can be grouped into two sets of arguments. One links data models with primary hypotheses (via severe tests), a second substantiates the assumptions of the data models.

In addition to the checks of measurement errors, Perrin sought ways to check himself. To avoid a type of experimenter's bias when selecting which grain to follow at steps 1 and 2, Perrin explains that

> in order not to be tempted to choose grains which happened to be slightly more visible than the rest . . . , which would raise the value of N a little, I followed the first grain that showed itself in the centre of the field of vision. (P. 124)

He was trying to obtain a randomly selected grain.

In this sketch of Perrin's experimental arguments two themes I have been tracing surface. First, knowledge of the ways in which one can go wrong leads to multiple procedures that allow errors to be circumvented or dealt with. At the level of data generation and measurement, in this example, the procedures refer more to actual physical manipulations (e.g., spinning the grains, lining them up, forcing some to get stuck on a barrier) than to the "manipulations on paper," which I said was the hallmark of data modeling. Second, validating experimental assumptions is much less a matter of ensuring that errors are not made than it is of knowing how much error is likely to be introduced by the various data generation procedures. Perrin's genius as an experimenter was largely a matter of his skill at catching himself and his students making mistakes, as well as knowing when they do not matter much. These two themes become even more pronounced in considering the ceteris paribus conditions.[20]

7.7 The Most Risky Decision of All: Ceteris Paribus Conditions

Now for the ceteris paribus conditions, at the bottom of the Suppean hierarchy, those factors not included in the systematic checks of experimental assumptions discussed above. The manifold factors, known and unknown, that are part of this soup are often thought to be the locus of a set of alternative hypotheses that cast an ever-present shadow on any primary inference. Given the way I have broken down experimental inquiries, this set of alternative hypotheses consists of threats to the experimental assumptions (or "initial conditions") of some primary inquiry. Because some such alternative ceteris paribus factors are assumed always to exist as threats to experimental assumptions, and because the ability to pinpoint what is learned hinges on these assumptions, ceteris paribus factors are thought to threaten the correct attribution of blame or credit to primary hypotheses. As Lakatos put it:

> The plight [of the methodological falsificationist] is most dramatic when he has to make a decision about *ceteris paribus* clauses, when he has to promote one of the hundreds of "anomalous phenomena" into a "crucial experiment," and decide that in such a case the experiment was "controlled." (Lakatos 1978, 27)

20. Although in delineating the framework of models in chapter 5 I combined checking experimental assumptions with checking ceteris paribus conditions, it is often useful to make a distinction between the two within this level, as I am doing here. The former refer to those experimental assumptions that are amenable to formal statistical testing—as in the case of affirming the Normal distribution M. The latter (ceteris paribus factors) refer to the variety of influences that are either known to be controlled, are subtracted out, or are tested by more informal, domain-specific means.

Affirming the ES hypothesis about the distribution of Brownian particles did deny a nonstatistical version of the second law of thermodynamics—something we will discuss in the next section. So the present example involves the very thing Lakatos worried most about: turning a mere anomaly into a severe test.

Lakatos, recall, gives up on *justifying* control; at best we decide—by appeal to convention—that the experiment is controlled. While I have no desire to revive "methodological falsification," I reject Lakatos and others' apprehension about experimental control. Happily, the image of experimental testing that gives these philosophers cold feet bears little resemblance to actual experimental learning. Literal control is not needed to correctly attribute experimental results (whether to affirm or deny a hypothesis). Enough experimental knowledge will do. Nor need it be assured that the various factors in the experimental context have no influence on the result in question—far from it. A more typical strategy is to learn enough about the type and extent of their influences and then estimate their likely effects in the given experiment.

How was this problem dealt with in Perrin's experiments? First remember that a host of experiments on factors suspected of influencing Brownian motion had already been conducted before Perrin's tests (around 1910). (See section 7.2.) Much was already known about the influences of light, heat, magnetism, electricity, shaking, noises of various sorts, and so on. Michael Faraday (in an 1829 lecture) and others recognized Brown's experiments as having ruled out all the causes of the motion suggested up until that time (e.g., unequal temperatures in the water, evaporation, air currents, heat flow, capillarity, motions of the observer's hands).[21] It is worth noting that these early experiments on the possible cause of Brownian motion were not testing any full-fledged theories. Indeed, it was not yet known whether Brownian motion would turn out to be a problem in chemistry, biology, physics, or something else. Nevertheless, a lot of information was turned up and put to good use by those later researchers who studied their Brownian motion experimental kits.

As I am imagining it, in one's bag of experimental tricks, along with the experimental tools and past experimental mistakes, would be a log of the extant experimental knowledge of the phenomena in question. Astutely using the kind of log I have in mind, Perrin dispelled numerous threats to experimental control. An imaginary (but not far from actual) dialogue quoting Perrin might have gone like this (the

21. See, for example, Jones 1870, 403.

names in parentheses are those of the researchers who worked extensively on the question of interest, and Perrin's lines are direct quotes):

Questioner: How do you know your results are not due to variations in temperature throughout the experimental emulsion? I believe there must be some temperature variations.

Perrin: It makes no difference whether great care is taken to ensure uniformity of temperature throughout the drop; all that is gained is the suppression of the general convection currents, which are quite easy to recognise and which have no connection whatever with the irregular agitation under observation (Wiener, Gouy). (P. 84)

Questioner: Might it not be something in the composition of your grains?

Perrin: The nature of the grains appears to exert little influence, if any at all. In the same fluid two grains are agitated to the same degree if they are of the same size, whatever the substance of which they are composed and whatever their density (Jevons, Ramsay, Gouy). (P. 85)

Questioner: What about vibrations of the glass containing the emulsion? Heavy vehicles passing by have made your table shake.

Perrin: The Brownian movement, again, is produced on a firmly fixed support, at night and in the country, just as clearly as in the daytime, in town and on a table constantly shaken by the passage of heavy vehicles (Gouy). (P. 84)

Numerous other factors are likewise shown to be either irrelevant or accounted for.

Perrin also uses his Brownian motion log to explain why, at one time, certain factors were erroneously thought to have influenced Brownian motion. For example, it had been thought that adding impurities such as acids to the emulsion influenced the motion. The error arose from the fact that the impurities caused the particles to stick to the glass vessel when they touched the sides. In actuality, "the addition of impurities . . . has no influence *whatever* on the phenomenon (Gouy, Svedberg)" (Perrin, 85, n. 1).

Each of these separate experimental inquiries had its own set of primary questions, experimental and data models, and so on. Had they not been available—as is often the case—then the separate tests would have had to be conducted as part of Perrin's experimental testing context. As it happened, by the time he performed the tests we have been discussing it was sufficient just to cite the familiar studies.

But Perrin did not do that; he repeated nearly all the tests anyway! Why? The primary reason, I suggest, is one that emerged during our discussion of "normal science": Deliberately repeating and getting good at generating *anticipated* results teaches a great deal about interacting with one's experimental objects. It is this kind of deliberate practice, not some mysterious knack, that gives one "a feeling for" one's experimental objects.

How far we have come from sighing over infinitely many alternative hypotheses that fit the data equally well. How far away, too, from the Bayesian requirement to consider all the alternatives in the catch-all, plus our degrees of belief in them, to get a single inquiry going. Clearly, ruling out alternatives is not always possible, and even when it is it typically takes a lot of work and aggressive criticism. The good experimenter notices when the criticism may be lodged in terms of specific experimental questions. These questions find their place at some level and at some node of the hierarchy of models. With respect to such specific experimental questions, the infinitely many alternatives really fall into just a few categories. Experimental methods (for answering new questions) coupled with experimental knowledge (for using techniques and information already learned) enable local questions to be split off and answered. These answers in turn may be used to show that experimental assumptions are sufficiently well met for testing a primary hypothesis severely.

Table 7.3 displays the main aspects of the series of models of inquiry that we have discussed. Here I chose to place the data and experimental models side by side.

7.8 WHAT IS LEARNED ABOUT THE SOURCES OF EXPERIMENTAL EFFECTS

Our experiments teach us or indicate the correctness of any hypothesis that may be deemed to have passed a severe test. Whereas in some cases the assessment of severity comes directly from formal statistical calculations (from a test's error probabilities), in others the argument for severity is based on analogies with known canonical (statistical or other) models. In Perrin's experiments, calculations (based on significance tests) show that the two primary hypotheses in the theoretical model pass highly severe tests. Step 1 teaches Perrin that the experimental results on Brownian motion approximate a random sample from a specified Normal distribution M. Step 2 indicates the values of the parameters of this distribution law.

This tells us, for starters, that the Brownian motion of a variety of types of particles is satisfactorily modeled as the realization of a particular statistical process identified (in model M) as Normal with certain approximate parameter values—as asserted in the theoretical distribution $\mathcal{H}$. Perrin's experiments also indicate how to generate manifestations of that process. We can paraphrase that favorite passage

TABLE 7.3 Selection of Entries From Models of Inquiry: Brownian Motion

PRIMARY MODEL:
Hypotheses: $\mathscr{H}$: the displacement of a Brownian particle over time t, S_t, follows the Normal distribution with $\mu = 0$ and variance $= 2Dt$.

MODELS OF EXPERIMENT:	MODELS OF DATA:
Hypotheses: H: the distribution of n displacements in experiment E_i follows Normal model M with parameters a function of N	Hypotheses: Data set is a random sample from experimental model
Break down into steps: Step 1, test of hypothesis j, and Step 2, test or estimation of parameters D (or N) (using data models) *Problems:* Specify the number of displacements to record, choice of experimental (test) statistics, specify adequate error probabilities	*Data Models:* n displacements, observed distributions of grains, measurements of grains *Problems:* How to condense data from one or more experiments to (a) arrive at suitable data models and (b) check if assumptions of experiment hold in each actual experiment E_i

CANONICAL MODELS: Normal distribution, random walks, bull's-eye model, gambling models, random selection, piles of shot

EXPERIMENTAL DESIGN AND DATA GENERATION
Fractional Centrifuging (may take several months), prepare emulsions, microscopic techniques for following displacement distributions of grains, count and weigh the grains, check for uniform radius.

Ceteris paribus conditions
Miscellaneous factors in experiment: uniformity of temperature, color and intensity of light, vibrations, impurities, the nature of the grains

Problems:
(a) How to manipulate grains and emulsions to arrive at data that satisfies experimental assumptions.
(b) How to ensure control of relevant factors, subtract out their influences, or determine that they need not be controlled. Utilize log of previous experiments on influences of background factors.

from Fisher (section 4.3): From these experiments we know how to bring it about that estimates for D (or for N) will very rarely be significantly far from certain values. This is experimental knowledge. The currently accepted value for N is still close to Perrin's values (within 19 percent).

But there is more to be said about what we learn from passing the theoretical distribution hypothesis $\mathcal{H}$. We do not learn about the distribution function of Brownian motion without also learning something about what produces it—something about molecular motions. The molecules about which the experiment teaches, however, agreeing with Nancy Cartwright, need not be the molecules of some substantive theory. The experiment teaches us about a cluster of causal or experimental properties of molecular motion. Minimally, it teaches that molecular motion reliably manifests itself as a Normal (Gaussian) process in Brownian motion experiments. Molecules in motion possess those properties that enable Perrin's experimental effects (e.g., estimates of N) to be reliably produced and reproduced.

Often, knowing this much (together with background knowledge) lets us go further. It may let us arrive at an understanding, if only approximate or partial, of specific properties of the underlying processes triggered in experiments. Perrin can also argue with severity, although without a precise severity assignment, that the experimental effects indicate that we are interacting with a process with certain characteristics, if only in the aggregate. We are familiar with the characteristic types of statistical processes that, when triggered in the manner of Perrin's experiments, produce data distributions of the sort he finds he can generate. More important, we know that the result of altering the underlying processes in specific ways (e.g., creating slight dependencies) would have been manifested in experiments like Perrin's.

Perrin gives the following analogy:

> Direct perception of the molecules in agitation is not possible, for the same reason that the motion of the waves is not noticed by an observer at too great a distance from them. But if a ship comes in sight, he will be able to see that it is rocking, which will enable him to infer the existence of a possibly unsuspected motion of the sea's surface. Now may we not hope, in the case of microscopic particles suspended in a fluid, that the particles may, though large enough to be followed under the microscope, nevertheless be small enough to be noticeably agitated by the molecular impacts? (Perrin, 83)

Just as the rocking of a ship indicates the motion of the sea's surface, the Brownian motion of microscopic particles indicates the motion in the liquid medium. Perrin puts it this way:

> The objective reality of the molecules . . . becomes hard to deny. At the same time, molecular movement has not been made visible. The Brownian movement is a faithful reflection of it, or, better, it is a molecular movement in itself. . . . From the point of view of agitation,

there is no distinction between nitrogen molecules and the visible molecules realised in the grains of an emulsion. (P. 105)

The microscopic grains are small enough to be noticeably agitated by the molecular collisions yet large enough to be observed under a microscope. Of course, molecular collisions are still occurring on the macroscopic level, but they do not displace a suspended body (indeed, this encouraged the initial skepticism toward the kinetic theory). The reason is that the breadth of surface area on average counterbalances the many collisions in different directions (a consequence of the law of large numbers). With microscopic particles, in contrast, the impulses from the collisions do not generally counterbalance each other; the particles are tossed about irregularly.

From Perrin's experiments, and with the knowledge of fluctuation phenomena, we can delimit at least major aspects of the kinds of things that can produce all of this. It must be something in the liquid medium—a discrete-hit type of process approximating a random walk.[22] This is what we can give a severe argument for, at least limiting ourselves to Perrin's experiments.

In this connection it is important to note that the knowledge of the existence of Brownian motion led to a change in scientific methodology. A new limit to experimental accuracy due to Brownian fluctuations and "noise" in measuring systems was introduced. Methods of testing were revised accordingly, and updated experimental tool kits needed to reflect this. Indeed, Brownian motion was and is one of the most important sources of canonical models of types of errors and fluctuations.

A number of mistakes and ways of overcoming them, all gleaned from Perrin's inquiries, also go into our experimental kit. One not yet mentioned deserves special note. Understanding Brownian motion unearthed a general type of statistical error that many people had overlooked. The error was the basis for an important objection to the kinetic account first raised by Karl Nägeli. The objection was based on the common assumption that in order for the molecules to cause

---

22. The major approximation in the modeling of Brownian processes stems from the fact that it can only be seen as a random walk when the interval of time *t* is not too small. As Einstein notes:

The movements of one and the same particle after different intervals of time must be considered as mutually independent processes, so long as we think of these intervals of time as being chosen not too small. (Einstein [1926] 1956, 12–13)

See also Chandrasekhar 1954, 89.

Brownian motion, they would have to move in a coordinated fashion. Random motion, it was objected, could not explain Brownian motion.

Gouy was the first to come close to answering this objection by citing the law of large numbers (chapter 5). Smoluchowski gave a more rigorous explanation based on a statistical argument of which Nägeli and others had been unaware. The argument is based on the canonical model of random walk phenomena discussed earlier. To use the gambler analogy, what the argument shows is how a gambler can lose a great deal of money, even with an even chance of winning or losing a fixed amount on each game, provided that he plays long enough. Analogously, unlike what Nägeli supposed, the jerks of Brownian particles do not need coordinated motion to explain them; with enough hits, the average displacement can be large, even when each hit has an equal chance of moving the particle a given amount to the right or left.

Models of fluctuation phenomena added considerably to the methods and strategies of experimental tool kits.

7.9 ACCEPTING A STATISTICAL VERSION OF THE SECOND LAW: A BIG SHAKE-UP TURNS ON A "SMALL" RESULT

Perrin's results on Brownian motion are sometimes considered to have provided a crucial test between the kinetic theory and classical thermodynamics taken as a whole. But his experiments themselves are not a severe test of the full kinetic account. There are many ways the full kinetic theory can be in error that Perrin's experiments did not direct themselves to uncovering.

Nevertheless, they do provide a severe and crucial test of one small though key piece of the kinetic account against one piece of the classical one. The piece turned on the severe test that took place at step 1: the test of the hypothesized Normal distribution in j against j'. And what was learned from it, based only on local statistical tests, took on enormous importance in the debates. It provided a severe test in favor of a statistical version of the second law of thermodynamics (also referred to as Carnot's principle). Friedrich Ostwald, Ernst Mach (at times), and the mathematician Ernst Zermelo based at least part of their opposition to the molecular-kinetic theory on the fact that it would allow exceptions to the absoluteness of the second law of thermodynamics. Mach and Ostwald held that a phenomenological description, such as thermodynamics, contains sufficient information while escaping the various problems that plagued atomic theory (e.g., the use of entities deemed hypothetical).

Einstein ([1926] 1956) begins by stressing that the two theories give conflicting predictions about Brownian motion:

> If the movement discussed here can actually be observed (together with the laws relating to it that one would expect to find), then classical thermodynamics can no longer be looked upon as applicable with precision to bodies even of dimensions distinguishable in a microscope: an exact determination of actual atomic dimensions is then possible. On the other hand, had the prediction of this movement proved to be incorrect, a weighty argument would be provided against the molecular-kinetic conception of heat. (Pp. 1–2)

Perrin, as expressed in the second epigraph that opens this chapter, regarded his experiments as providing such a crucial test.

The kinetic theory, in contrast with the classical theory, views dissolved molecules as differing from suspended particles only in their size; their motion would be the same. If Brownian motion could be explained as caused by something outside the liquid medium or something within the particles themselves, it would not be in conflict with the classical theory. If, on the other hand, it could be shown that Brownian motion was caused by a molecular motion in the liquid medium, as given in the kinetic theory, it *would* be in conflict. Moreover, it would show that a statistical process was responsible and that the second law (or Carnot's principle) requires a statistical rendering.

That Brownian motion, if indeed spontaneous, would be an exception to the (nonstatistical version of the) second law was recognized even before the ES theory was formulated. For if it is true that without temperature differences in the system, a Brownian particle (denser than water) rises spontaneously, then it constitutes a case in which part of the heat of the medium is transformed into work. This recognition is explicitly discussed by Gouy around 1890. Jules-Henri Poincaré, persuaded by Gouy's arguments, declares:

> But we see under our eyes now motion transformed into heat by friction, now heat changed inversely into motion, and that without loss since the movement lasts forever. This is the contrary of the principle of Carnot. (Poincaré 1905, 610)

In step 1, Perrin showed that one can generate at will an observable process due to an agitation *not* attributable to the particles or external energy sources. Thus in carrying out step 1, Perrin demonstrates the existence of violations of the nonstatistical version of the second law. Perrin even describes his experiments as methods for generating such violations:

Briefly, we are going to show that sufficiently careful observation reveals that at every instant, in a mass of fluid, there is an irregular
spontaneous agitation which cannot be reconciled with Carnot's principle *except just on the condition of admitting that his principle has the probabilistic character suggested to us by molecular hypotheses.* (Perrin 1950, 57;
emphasis added)

This is precisely what is established by the severe test of the distribution
in hypothesis *j*.

The statistical version involves a standard frequentist interpretation
of probability: it means that it will be violated extraordinarily rarely—
so rarely that it can practically be discounted. Perrin calculates, for example, that to have a better than even chance of seeing a one-kilogram
brick suspended by a rope rise to a level by virtue of its Brownian
motion, one would have to wait more than 10^{10} billion years. "Common sense tells us, of course, that it would be foolish to rely upon the
Brownian movement to raise the bricks necessary to build a house" (p.
87). So, practically speaking, the second law is unaffected. Perrin suggests that we can best understand the law by stating it as follows:

> On the scale of magnitudes that are of practical interest to us, perpetual motion
> of the second kind is in general so insignificant that it would be foolish to take
> it into consideration.(P. 87)

Perhaps enough has been said for our purposes, which were to
illustrate the hierarchy of models in a single experimental testing context, the breakdown of larger inquiries into small pieces, and strategies
for arriving at severe tests. There is ample work by others on how Perrin's results bear on theories higher in the hierarchy as well as on more
global disputes arising from the atomic debates.[23] I limit myself to a
few brief remarks.

Going Higher in the Hierarchy

At yet a higher level in the hierarchy of models one could place
more general questions about the molecular-kinetic theory as a whole.
The more global molecular-kinetic theory refers not only to Brownian
motion but also to theories about gases, radiation, diffusion of light,
and others. Here is where the discussion of the thirteen phenomena
enters. In experiments upon each of these widely different phenomena, estimates of Avogadro's number *N* were obtained, and good
agreement was found. Although Perrin takes several chapters to discuss the tests on these other phenomena, they are distinct from his

23. Examples are Brush 1977; Clark 1976; Gardner 1979; and Salmon 1984.

tests, involving primary hypotheses different from those he considered. While passing these further molecular-kinetic hypotheses adds weight to the Brownian motion tests, they largely come into play only when going beyond the Brownian motion tests that are my focus.

The good agreement among the thirteen phenomena on the molecular magnitudes effectively ruled out the worry that extrapolations from one phenomenon to another would not hold up. It was also at the heart of arguments for the reality of atoms, as Salmon and others have maintained; which is why those arguing for realism began with the argument from the thirteen phenomena. Even ardent antiatomists (probably excepting Mach) construed Perrin's experiments as telling. On the basis of such experimental evidence, even Ostwald reversed himself on the atomic-kinetic theory in 1909:

> I have convinced myself that we have recently come into possession of experimental proof of the discrete or grainy nature of matter, for which the atomic hypothesis had vainly sought for centuries. . . . This evidence now justifies even the most cautious scientist in speaking of the *experimental* proof of the atomistic nature of space-filling matter.[24]

As I have said, I am confining myself to what is given by experimental knowledge, and it is not clear that this does not take one as far as one would like. What did Perrin think? In some passages one hears him arguing for molecular reality (see Achinstein 1994, Nye 1972). But this is not the chief concern of his experimental work. Even on the role of the thirteen phenomena, Perrin has this to say near the end of *Atoms:*

> Yet, however strongly we may feel impelled to accept the existence of molecules and atoms, we ought always to be able to express visible reality without appealing to elements that are still invisible. And indeed it is not very difficult to do so. We have but to eliminate the constant N between the 13 equations that have been used to determine it to obtain 12 equations in which only realities directly perceptible occur. (P. 216)

As an example, Perrin explains that by eliminating the molecular parameter between the equations from black body radiation and Brownian motion, we arrive at an equation that lets us predict the rate of diffusion of Brownian particles in water by measuring the intensity of the light in the radiation issuing from a furnace of molten iron:

24. This translated quotation is from Brush 1977, 381. The source is Wilhelm Ostwald 1909. The quotation is from the "Vorbericht."

Consequently the physicist who carries out observations on furnace
temperatures will be in a position to check an error in the observation
of the microscopic dots in emulsions! And this without the necessity
of referring to molecules. (P. 216)

The thirteen equations make fundamental connections among very
different phenomena, therefore providing an effective way of using
one such phenomenon to check errors regarding vastly different
phenomena. This is a powerful source of progress in experimental
knowledge.

CHAPTER EIGHT

Severe Tests and Novel Evidence

I think that people emphasize prediction in validating scientific theories because the classic attitude of commentators on science is not to trust the theorist. The fear is that the theorist adjusts his or her theory to fit whatever experimental facts are already known, so that for the theory to fit these facts is not a reliable test of the theory.

In the case of a true prediction, like Einstein's prediction of the bending of light by the sun, it is true that the theorist does not know the experimental result when she develops the theory, but on the other hand the experimentalist does know about the theoretical result when he does the experiment!

—Steven Weinberg, *Dreams of a Final Theory*, pp. 96–97

WITHIN TWENTY-FOUR HOURS of the bomb explosion at the World Trade Center in New York City in February 1993 there were nineteen telephone calls from individuals or organizations claiming responsibility. The fact that the calls came after the first news reports of the explosion, it was generally agreed, greatly weakened the credibility of the claims. Our intuition here reflects the common principle that evidence predicted by a hypothesis counts more in its support than evidence that accords with a hypothesis constructed after the fact. Many may say that merely explaining known evidence provides little or no support for a hypothesis altogether. Nevertheless, it is equally clear that less than one week after the bombing the deliberate use of known pieces of evidence, in particular, "three feet of mangled, soot-encrusted steel" (*Newsweek*, 15 March 1993, 28), warranted the investigators to finger the man who rented the van that carried the explosive device.[1]

---

1. Under the grime of this piece of the truck chassis was an identification number. This led authorities to New Jersey and to the Islamic fundamentalist who rented the van.

251

The seeming conflict in these two intuitions is at the heart of a long-standing dispute in philosophy of science. The dispute may be seen to center on the following methodological principle or rule:

Rule of novelty (RN): for evidence to warrant a hypothesis *H*, *H* should not only agree with the evidence, but the evidence should be *novel* (in some sense).

On some accounts the novelty of the evidence is required, whereas for others the claim is comparative: novel evidence accords the hypothesis greater weight than if it were nonnovel. Laudan refers to this rule as "one of the most contested principles in recent philosophy of science" (Laudan 1990b, 57). In fact, the controversy over some version of RN has a very long history, marked by vehement debates between several eminent philosophers of science: between Mill and Peirce, Mill and Whewell, and Keynes and Popper.[2] The current dispute about the relevance of novelty is commingled with a family quarrel among those who endorse RN over the very definition of "novel test" or "novel fact." The disputants here tend to take an all or nothing approach. Proponents of novelty find the novelty of the evidence always relevant in assessing its bearing on hypotheses, holding that evidence is accorded extra weight (of some sort) simply by dint of being novel. Opponents deny that novelty ever matters. There are further disagreements even among proponents who share the basic definition of novelty over what counts as violating novelty and, most importantly, over *why* novelty should matter to the import of evidence.[3]

I will not attempt to survey the literature to which this quarrel about novelty has continued to give rise but will argue for a change of focus in the entire battle—on both of the fronts on which it has been fought. Novelty, I claim, was not the real issue in the first place. What lay behind the intuition that novelty mattered is that severe tests matter. What underlies the basic intuition that if the data are not novel, then they fail to test or support a hypothesis are the various impediments to severity that correlate with violating novelty of one sort or

2. Early proponents often cited are Descartes and Leibnitz. An indication in Descartes is his claim in *Principles of Philosophy* that "we shall know that we have correctly determined these causes when we observe that we can explain, by their means, not only those phenomena which we have considered up to now, but also everything else about which we have not previously thought" (Descartes [1644] 1984, pt. 3, p. 104).

3. For some associated readings on the novelty debate beyond those cited in this chapter, see Campbell and Vinci 1983; Howson 1984; Gardner 1982; Musgrave 1978, 1989; and Redhead 1986.

another. But this correlation is imperfect. Novelty and severity do not always go hand in hand: there are novel tests that are not severe and severe tests that are not novel.

As such, criteria for good tests that are couched in terms of novelty wind up being either too weak or too strong, countenancing poor tests and condemning excellent ones. I believe that our notion of severe tests captures pronovelty intuitions *just where those intuitions are correct.*

Understanding and resolving the dispute about RN is important for several reasons. For one thing, an important task for an adequate epistemology of experiment is to unearth the principles underlying familiar methodological rules. The controversy in this case has had particularly serious consequences. Finding fault with those who argue that novelty in some sense always matters has been taken by some as supporting the opposite view, that it never matters. This discounts the important kernel of rightness underlying those who think novelty matters: aspects of the hypotheses and data generation procedures need to be taken into account in assessing the goodness of tests. They may be relevant to the error probabilities and so to the severity of the overall experimental test.

In addition, the controversy has had disturbing consequences for historically minded philosophers of science. Discovering that several scientific episodes appear to fail so apparently plausible a rule has been taken as grounds for questioning the rationality and objectivity of the episode or, alternatively, as grounds for questioning the viability of the methodological enterprise.

There is a further reason that resolving the novelty controversy is of particular importance for our program. Any philosophy of experimental testing adequate to real experiments must come to grips with the fact that the relationship between theory and experiment is not direct but is mediated along the lines of the hierarchy of models and theories as sketched in chapter 5. At various stages of filling in the links, it is standard to utilize the same data to arrive at as well as warrant hypotheses. It is commonly the case, for example, that raw data are used to construct as well as to test a hypothesis about experimental data, and experimental data are used to construct as well as to support an experimental hypothesis—the basis for a comparison with a theoretical prediction. As a matter of course, then, the inferences involved violate even the best construals of the novelty requirement. This is commonly so for the central task of experimental inference—estimating the effects of backgrounds. Indeed, if one goes down the list in chapter 2 of the standard problems of "normal" science, one finds again and again that they are tasks where hypothesized solutions are routinely affirmed by nonnovel evidence. So RN must be abandoned or qualified.

A realistic picture of the relationship between evidence and hypotheses reveals not only that nonnovel results often figure in altogether reliable inferences, but also that there is as much opportunity for unreliability to arise in reporting or interpreting (novel) results given knowledge of theoretical predictions as there is for it to arise in arriving at hypotheses given knowledge of (nonnovel) results. This is Weinberg's point in the epigraph that opens this chapter. The historian Stephen Brush finds cases where scientists are as concerned about the former as the latter, leading him to suggest that "the preference for forecasting implies a double standard for theorists and observers, based on a discredited empiricist conception of science" (Brush 1989, 1127). Why not be as suspicious of novel results claimed to be in accord with a known theoretical prediction? Lest we be driven to suspect all experimental inference, some distinctions clearly need to be made. These distinctions, I propose, appeal to reliability considerations that scientists standardly employ in devising and interpreting their experiments.

8.1 LOGICAL AND EVIDENTIAL-RELATIONSHIP VIEWS VERSUS HISTORICAL AND TESTING ACCOUNTS

If I am correct that the goal of novelty is severity, then the dispute between those who do and those who do not accept some version of the novelty principle emerges as a dispute about whether severity—or, more generally, error characteristics of a testing process—matters in evaluating the import of evidence. By and large, thinking novelty matters goes hand in hand with thinking severity matters. (Where there are exceptions, there is some question about the consistency of the view. I return to this in chapter 10.) This correlation is borne out in the historical disputes as well as in the current debate between Bayesian and non-Bayesian philosophies of hypothesis appraisal.

Alan Musgrave on Logical versus Historical Theories of Confirmation

Several philosophers have enlightened us on the history of the novelty dispute (e.g., Giere 1983; Lakatos 1978; Musgrave 1974). Musgrave puts his finger on what is common to historical opponents of novelty—they hold what he calls a logical theory of confirmation.

> According to modern logical empiricist orthodoxy, in deciding whether hypothesis *h* is confirmed by evidence *e*, and how well it is

confirmed, we must consider only the statements *h* and *e*, and the logical relations between them. It is quite irrelevant whether *e* was known first and *h* proposed to explain it, or whether *e* resulted from testing predictions drawn from *h*. (Musgrave 1974, 2)

One finds this stated plainly in Hempel 1965.

Some variant of the logical (or logicist) approach was implicitly held by historical opponents to RN. "We find it in Mill, who was amazed at Whewell's view" that successfully predicting novel facts gives a hypothesis special weight (Musgrave 1974, 2). Whereas Whewell held that

> men cannot help believing that the laws laid down by discoverers must be in a great measure identical with the real laws of nature, when the discoverers thus determine effects beforehand, (Whewell [1847] 1967, vol. 2, p. 64)

in Mill's view, "such predictions and their fulfillment are . . . well calculated to impress the uninformed. . . . But it is strange that any considerable stress should be laid upon such a coincidence by persons of scientific attainments" (Mill 1888, bk. 3, chap. 14, sec. 6, p. 356). Keynes, another logicist, similarly held that the "question as to whether a particular hypothesis happens to be propounded before or after examination of [its instances] is quite irrelevant" (Keynes [1921] 1952, 305).[4] Clearly, if confirmation is strictly a logical function between evidence (or statements of evidence) and hypotheses, when or how hypotheses are constructed *will* be irrelevant.

The logical approaches to confirmation ran into problems, however, precisely because they insisted on purely formal or syntactical criteria of confirmation that, like deductive logic, "should contain no reference to the specific subject matter of the hypothesis or of the evidence in question" (Hempel 1965, 10). Enter what Musgrave calls the "historical (or logico-historical) approach" to confirmation.

In Musgrave's neat analysis of the situation, the contemporary accounts of novelty in the Popper-Lakatos school arose out of attempts to avoid the paradoxes of the traditional logical approaches to confirmation by requiring various background considerations in the form of novelty requirements. Musgrave calls such accounts historical because, he believes, "it will presumably be a historical task to determine" what the background knowledge is. In particular, "all variants of the historical approach will make the confirmation of a scientific theory some-

4. This is as quoted in Musgrave 1974, 2.

how depend upon the historical setting in which that theory was proposed" (Musgrave 1974, 7).

E-R versus Testing Accounts of Inductive Inference

The situation twenty years after Musgrave's article is rather similar—with one key difference. Few philosophers of science still harbor
the view that a theory of confirmation is merely a formal relationship
between evidence and hypothesis. What distinguishes their accounts
is the kind of background thought to be needed. Bayesians let the
background enter via prior probabilities. In the Bayesian analogy between induction and deduction from chapter 3 we hear echoes of the
position staked out by logical approaches to confirmation.

In contrast, most who follow what I termed "testing accounts" see
the background as coming in via methodological principles such as
those based on novelty.[5] Members of the Popper-Lakatos school (e.g.,
Watkins, Musgrave, Zahar, Worrall) proposed notions of novelty
where the background knowledge was intended to reflect Popperian
goals. Musgrave tells us that "Watkins begins with Popper's thesis that
genuine confirmations can result only from the failure of genuine attempts to refute a hypothesis, from severe tests" (Musgrave 1974, 5).
The central aim in each case was to use background knowledge to distinguish genuine or severe tests from spurious ones.

Novelty in Testing Accounts of Inference. Three main types of novelty
emerge. A novel fact for a hypothesis *H* may be (1) one not already
known, (2) one not already used in arriving at or constructing *H*, or
(3) one not already predicted (or one counterpredicted) by available
hypotheses. The first corresponds to *temporal* novelty, the second, to
heuristic or *use* novelty, the third to *theoretical* novelty. Each has itself
been construed in a variety of ways, multiplying the number of different definitions. The third view, theoretical novelty, was the Popperian
severity requirement discussed in chapter 6. The debate I will now consider revolves around temporal and heuristic or use novelty.[6]

5. The more fundamental difference between these two approaches to the
background is that testing accounts give a central role to what in statistical language
are the error characteristics of a procedure of testing. This will be explored further.
6. Calling in to take credit for the Trade Center bombing violates both types of
novelty. Not only had the blast already occurred, it had already been reported on
the news before the first phone call came in. It is a violation of use-novelty (or
heuristic novelty) because that information was used to construct the particular
hypothesis by filling in the blank: for example,___is responsible. So eschewing violations of either form of novelty explains our giving little credence to the calls.
However, the mangled piece of steel (and a list of other clues) were also known
and used to arrive at the identity of the driver of the suspected van.

A second school of testers—the "error statisticians"—has also up-held novelty principles in the form of rules of predesignation or rules against double counting of data. An important variant of the argument from error statistics warrants separate attention and will be considered in chapter 9. The argument most familiar to philosophers, represented by Ronald Giere, can be considered now with arguments from histori-cal schools.

Recall that for a testing theorist the task of characterizing a good test is not distinct from saying when a hypothesis is warranted or well supported by evidence. To avoid confusion I have couched my account in terms of tests. However, Worrall and Giere often couch their re-marks in terms of when data *support* hypotheses. They must not be mistaken as attempting to provide an evidential-relation measure. For them, to say when data support *H* is to say when data provide a good test of *H*—although degrees of goodness are still possible. The task for novelty criteria is to tell us what more beyond entailing or fitting evi-dence is required for a test to be genuine. The debate centers on how best to accomplish this.

The central problems are how background knowledge should be brought in so that: the account of testing is not turned into a subjective or relativistic affair, the resulting account accords with important cases of actual scientific appraisal, and there is a clear epistemological ratio-nale for doing so. The ongoing debate has not made much progress in satisfying these three desiderata.

Temporal Novelty

On the view of temporal novelty, "novel fact" meant what it said: a novel fact was a newly discovered fact—one not known before its use in testing.[7] Known by whom? For some, it counts as known only if the general scientific community knows it, for others it is enough that a given investigator putting forth the hypothesis knows it (e.g., Gardner 1982).

The temporal view of novelty has been criticized on all three desid-erata for judging novelty criteria. First, there is the problem of how the temporal novelty of the data can be characterized nonsubjectively. How, it is asked, can temporal novelty be determined without having to look into the psyches of individual scientists to determine what they knew and when they knew it? Second, the temporal novelty require-ment denies special evidential significance to tests that intuitively seem

7. Although a temporally novel fact is sometimes equated with a *predicted* fact, the term "predicted" in science generally does not require this temporal element—a point Stephen Brush (1989) makes.

to possess it. Take our example from the last chapter. Brownian motion was known long before the Einstein-Smoluchowski theory was proposed, yet was considered to have provided significant evidence for it. Likewise for the orbit of Mercury and Einstein's theory. Third, there is the question of its epistemological rationale. Why *should* time matter?

> If the time-order of theory and evidence *was* in itself significant for scientists then we should, I think, be reduced merely to recording this as a brute fact. For why on earth *should* it matter whether some evidence was discovered before or after the articulation of some theory? (Worrall 1989, 148)

In response to such objections to temporal novelty, novel accounts have been proposed in which novelty turns instead on the *heuristic* role of facts: on whether the theory it helped construct was in a certain sense ad hoc. Zahar (1973) suggested that a fact is *"novel with respect to a given hypothesis if it did not belong to the problem-situation which governed the construction of the hypothesis"* (p. 103). Old facts (i.e., facts not temporally novel) could be novel facts in this new sense so long as the theory was not devised to explain them. Musgrave and others criticized this view as being too subjective and psychologistic—even more so than temporal novelty. It seemed to make the answer to the question of whether a test was good relative to the specific aims of the designer of the theory. "To assess the evidential support of a theory 'one has to take into account the way [it] is built and the problems it was designed to solve'" (Musgrave 1974, 12). Furthermore, it seems that in Zahar's view the same evidence might accord a given theory as proposed by one scientist a different amount of support than as proposed by another, according to the heuristic route each scientist takes (p. 14).

Worrall reformulates Zahar's heuristic view: the question is not whether a theory was "devised to explain" a fact but whether the fact was "used to construct" the theory. With this new formulation Worrall intends to signal that although support is to be heuristic-relative, it is *not* to be person-relative (e.g., Worrall 1978b, 51). That is, while granting that the heuristic view allows the same theory, because differently arrived at, to be differently supported by the same evidence, Worrall believes that "the heuristic considerations which led to the construction of a theory can be objectively specified" (p. 51).

The Zahar-Worrall view of heuristic novelty may be called *use-novelty*. It requires that for evidence *e* to support hypothesis *H* (or for *e* to be a good test of *H*), in addition to *H* entailing *e*, *e* itself *must not have been used* in *H*'s construction. Worrall states the position as follows:

> The relation [of empirical support] holds if and only if the factual statement is implied by the theory but is not a member of the set of

factual statements used in the construction of the theory. (Worrall 1978b, 50)

Since strict entailment is generally too strong,[8] I will allow that the use-novelty requirement, UN, for a good test is satisfied when

    i. *H* entails or is a good fit with *e*

and

    ii. *Use-novelty* UN: *e* is not used in the construction of *H*.

(Worrall also holds UN sufficient for a good test, but I leave that to one side. See Mayo 1991a.)

Use-novelty, or something very much like it, is endorsed—at least as a necessary condition—by other use-novelists as well as by temporal novelists. Its violation is commonly termed double-use or double-counting of data. If evidence is used in arriving at *H*, then it cannot be used again in *H*'s support. As a shorthand, let us call a hypothesis constructed to fit evidence *e* (however the construction is done) a *use-constructed hypothesis*. The use-novelty requirement for tests is this:

> *UN requirement:* Data *e* that was used to arrive at a use-constructed hypothesis *H* cannot also count as a good test of *H*.

The UN requirement does seem to reflect our ordinary intuitions in cases such as the after-trial claims of responsibility for the bombing of the World Trade Center and astrological retrodictions. This fact should also make it reasonable to suppose that the rationale for these intuitions, where correct, applies uniformly to day-to-day and scientific hypotheses.[9] The account I recommend does just this. Violating UN is correctly eschewed—whether in science or day-to-day reasoning—only if it results in violating reliability or severity.

Against this, proponents of the UN requirement recoil from ever crediting a use-constructed hypothesis for passing a test it was constructed to pass. Their basic intuition is this:

> If a hypothesis *H* has been arrived at to accord with data *e*, then that same data cannot also provide a good test of (or good support for) hypothesis *H*, *since H could not have failed this test*.

It is not that the constructed hypothesis is considered unbelievable or false but rather that the UN proponent denies that finding the accor-

---

8. Worrall's own discussion bears this out. In any case, my arguments here will not turn on whether (i) requires strict entailment or allows a statistical type of fit.

9. This counts against proponents of novelty who, when faced with counterexamples to UN, maintain that science is different.

dance with data should be credited to *H* because the accordance was assured—*no matter what*. This reasoning, while sounding plausible, is wrong.

Finding It Wrong for the Wrong Reasons

In finding the intuition underlying the UN requirement wrong, I am apparently in an odd sort of agreement with the Bayesian Way. Indeed, recent criticisms of the UN requirement, with few exceptions, have been leveled by those wearing Bayesian glasses. What has scarcely been noted, however, is that the Bayesian critiques are deeply flawed. From our discussion of key differences in aims or even from Musgrave's logical versus historical lesson, it is easy to guess at the flaw.

For Bayesian philosophers of science, just as with the earlier "logicist" approaches, there is no slot in which to take into account the novelty of the data. Thus when the rationale for the UN requirement is judged from this Bayesian standpoint, it is, unsurprisingly, found wanting. Finding UN unnecessary for earning high marks *according to the Bayesian account* of hypothesis appraisal, the Bayesian declares the arguments in favor of UN wrong (e.g., Howson and Urbach 1989, 549; Howson 1990). The problem is that even if the rule of novelty (RN) were good for its intended aims, running it through the Bayesian machinery has no chance of "passing" the rule. On Bayesian principles, if two hypotheses that entail evidence *e* are to receive different amounts of support from *e*, then the difference must lie in the prior probabilities. I differ from the Bayesian and concur with the UN proponent in holding that when a difference in appraisal is warranted, the fault lies in the testing process and not in our priors.

Adding to the confusion, there have been Bayesian attempts to support the non-Bayesian arguments for UN. Shooting holes in these Bayesian defenses have wrongly been taken to vitiate the non-Bayesian arguments. All of this warrants more careful scrutiny; I shall return to this discussion in chapter 10.

8.2 Characterizing UN Violations (Nonsubjectively)

Worrall (1985) concedes that "allowing that heuristics play a role does indeed threaten to make confirmation a dangerously unclear and subjectivist notion" (p. 309). The viability of his position, he grants, rests on being able to find out if a theory is use-constructed by careful historical study, for example, by combing historical documents, notes, and letters without having to explore the psyches of individual scientists.

Three types of UN violations emerge in the accounts of Worrall and Giere:

Parameter fixing. There is a hypothesis with a free parameter x (a quantity or constant in some mathematical equation or other blank not yet fixed), for example, the number of molecules per unit is x. We can write the hypothesis with the blank as $H(x)$. The data e are used to work out the value of this or these parameters so that the resulting hypothesis, $H(e)$, yields, entails, accommodates, renders expected, or otherwise fits data e. This need not be a quantitative parameter. It includes any case where data are used to pin down a hypothesis sufficiently so as to account for or accommodate that same data. An example Worrall gives is the use-construction of a Newtonian explanation of Mercury's perihelion (Worrall 1978b, 48).

Exception barring or incorporation. Exception incorporation may arise when H fails to accord with result e, so e is anomalous for H. The constructed hypothesis, H', is the result of revising or qualifying hypothesis H so that H' accords with e. That is, H' is H plus some qualification or modification. An example would be where H is an alleged psychic's claim to be able to "see" via ESP a drawing in a sealed envelope. When he fails, he qualifies H so that it excludes cases where skeptical scientists are watching him. Thus revised, result e now allows H to pass.

Evidence as constraint. Giere alludes to another way in which UN may be violated, although in a sense it subsumes all the preceding ones. Here the violation occurs whenever the evidence acts as a "constraint" on any hypothesis that is going to be considered. An agent who follows this procedure will only put forward hypotheses that can accommodate a known result or phenomena e. Such a procedure, assuming it ends, must somehow accommodate e.

In each case e is being used to construct a hypothesis H to satisfy the condition that H "fits" e. Alternatively, e is used in constructing H to assure that H passes the test. Whenever the evidence e is taken at the same time as supporting or providing a good test of the use-constructed H, we have a UN violation.

Using Historical Data to Test Novelty Accounts

It is far from clear that these attempts to characterize violations of use-novelty ameliorate the difficulty of determining objectively whether a case is use-novel. In the one historical case that both Giere and Worrall look at with express interest in the question of novelty—

the case of Fresnel's (wave) theory of diffraction—they arrive at opposite pronouncements. Following the traditional reading of this episode, Giere finds that the ability of Fresnel's theory to account for the diffraction effect known to occur with straightedges was given less weight (by a prize committee at the time) than its ability to predict the temporally novel (and unexpected) "white spot" effect. In contrast, Worrall argues that accounting for the known straightedge result was given *more* weight (Worrall 1989, 142). But their disagreement goes further. Whereas Worrall holds up the straightedge effect as a case where UN is *satisfied*,[10] Giere holds it up as a case of a UN *violation* (of the evidence as constraint type).

This disagreement raises the problem of when historical data should be regarded as genuinely testing a methodological claim such as the UN requirement. Worrall takes this episode as evidence in support of use-novelty. Yet, even granting his reading of the case, namely, that temporal novelty did not seem to matter, the episode seems at most to be *consistent* with his position about UN (it is not a very severe test). Going by Worrall's discussion of the case, it is far from clear that the appraisal turned on novelty altogether—whatever the type. Worrall remarks:

> The report recorded that Fresnel had made a series of 125 experimental measurements of the external fringes outside the shadow of a straightedge, and that in this whole series the difference between observation and the value provided by Fresnel's integral was only *once* as much as 5/100 mm, only *three* times 3/100 mm and *six* times 2/100 mm. In all the other 115 cases disagreement between theory and observation did not exceed 1/100 mm. (Worrall 1989, 144)

Members of the prize committee were impressed, it seems, by how *often* Fresnel's predictions came very close to the observed results with straightedges. Their argument clearly assumed, as it required such for it to have weight, something like the following: If Fresnel's hypothesis were wrong, we would expect larger differences more often than were observed. That is, such good accordance would be very improbable in a series of 125 experiments if Fresnel's account was not approximately correct in regard to diffraction. It is this argument—an argument from error—that mattered, and not how Fresnel's account was constructed.

We can agree with Worrall (1989) that to assess support, "we need know nothing about Fresnel's psyche and need attend only to the development of his theory of diffraction as set out in great detail and

10. Worrall's purpose in citing the Fresnel example is to argue that scientific judgments reflect a concern with use-novelty and not with temporal novelty.

clarity in his prize memoir" (p. 154). Psyches have nothing to do with it, and scientific reports (if adequate) are enough. But in my view what we really need to learn from these reports is not the route by which the theory was developed, but *the reliability of its experimental test.* I propose that the whole matter turns on how well the evidence—used or not—*genuinely indicates* that hypothesis H is correct. What matters is how well the data, together with background knowledge, rule out ways in which H can be in error. While this calls for attending to characteristics of the entire testing process, which may be influenced by aspects of the generation of test hypotheses and data, it does not call for reconstructing how a scientist came to develop a hypothesis. In planning, reporting, and evaluating tests, it is to the relevant characteristics of the testing process that scientists need to attend.

But I am ahead of myself. My point now is that it is difficult to argue for what historical cases show about some methodological rule without understanding how the rule functions in experimental learning. Worrall does not stop with testing UN via historical data but goes on, in several papers, to wrestle with its epistemological rationale, to which we will now turn.

8.3 THE (INTENDED) RATIONALE FOR USE-NOVELTY IS SEVERITY

Despite the fairly widespread endorsement of something like the UN requirement (especially among testing accounts), enormous confusion about what might be its epistemological rationale persists. In this chapter I will focus on two variants of a single argument that has received considerable attention by philosophers, as found in discussions by Worrall and Giere. I shall argue that the (implicit or explicit) *intended* rationale for use-novelty is severity.

Violating Use-Novelty and the Ease of Erroneous Passing

An important advantage, Worrall (1989) claims, that use-novelty has over temporal novelty "is that it comes equipped with a rationale" (p. 148). Nevertheless, Worrall fails to come right out and say what that rationale is. The most telling hint is that he intends his use-novelty criterion UN to capture Popper's requirement for a genuine or severe test: "Many of Popper's most perspicacious remarks are . . . based on an intuitive notion of testability" (Worrall 1985, p. 313) embodied in the Zahar-Worrall use-novelty account, which, Worrall says, "Popper has never, I think, fully and clearly realized" (ibid.). Paying attention to the manner of theory construction can fully capture the spirit of Popper's intuition about tests, whereas Popper's purely logical account cannot. Popper's intuition, noted in chapter 6, is that we are to "try to think of cases or situations in which [a hypothesis] is likely to fail, if it is false" (Popper 1979, 14).

At the heart of the matter is insisting on *riskiness* of some sort. Popper's favorite example is the test of Einstein's theory by checking the predicted light deflection during the eclipse of 1919:

> Now the impressive thing about this case is the *risk* involved in a prediction of this kind. . . . The theory is *incompatible with certain possible results of observation*—in fact with results which everybody before Einstein would have expected. (Popper 1962, 36)

Popper contrasts this with the way popular psychological theories of his day seemed able to accommodate any evidence and with how unwilling or unable the latter were at putting their hypotheses to genuine test. Yet Popper's logical requirement for a genuine test, I concur with Worrall, does not capture his own informal remarks about what a good test requires (as seen in section 6.7).

Heuristic novelty, Worrall proposes, does a better job than both temporal and theoretical (Popperian) novelty at capturing the needed risk requirement. Worrall reasons that

> if some particular feature of T was in fact tied down on the basis of e . . . then checking e clearly constitutes no real test of T. . . . In such a case even though e follows from T and hence not-e is, in Popper's terminology, a potential falsifier of T—it wasn't *really* a potential falsifier of T, since T was, because of its method of construction, never at any risk from the facts described by e. (Worrall 1989, 148–49)

Ronald Giere (1984a) makes a parallel assertion. Whether known (nontemporally novel) facts may provide good evidence for a hypothesis, Giere claims,

> depends on whether the known facts were used in constructing the model and were thus built into the resulting hypothesis. If so, then the fit between these facts and the hypothesis provides no evidence that the hypothesis is true. These facts had no chance of refuting the hypothesis even if it were wildly mistaken. (P. 161)

The final sentences of Worrall and Giere's passages can and have been misinterpreted (see chapter 10). They should not be taken to mean that some particular data e could not have but compared favorably with H. For that would be so whenever a hypothesis fits or accords favorably with data—even in the best of tests. After all, if H is in accordance with e, then there is no chance that it is not in accordance with e. What Worrall and Giere must intend to be pointing out about use-constructed cases is that evidence *e*—whatever it is—is guaranteed to accord with hypothesis *H* whenever *H* is deliberately constructed to be in accordance with *e*. Any facts resulting from such a process, Giere is saying, had no chance of refuting the hypothesis (constructed to fit them) "even if it were wildly mistaken." That is, Giere is saying, the test fails to be severe in the sense of error-severity.

But why do they suppose violating UN leads to violating severity? I will consider their arguments in turn, for each represents a well-entrenched position.

Worrall and a False Dilemma

Worrall's (1989) position goes like this: Consider the kind of reasoning we seem to use when we *do* take a theory's empirical success as showing it to be true, empirically adequate, or in some way reflecting "the blueprint of the Universe"—"whether or not it can be given some further rationale" (p. 155).

> The reasoning appears to be that it is unlikely that the theory would have got this phenomenon precisely right just "by chance." . . . The choice between the "chance" explanation and the "reflecting the blueprint" explanation of the theory's success is, however, exhaustive only if a third possibility has been ruled out—namely that the theory was engineered or [use-constructed]. (Worrall 1989, 155)

For, in the use-constructed case, Worrall says,

> the "success" of the theory clearly tells us nothing about the theory's likely fit with Nature, but only about its adaptability *and* the ingenuity of its proponents. (Ibid.)

The presumption seems to be that in use-constructed cases the proponent's ingenuity and/or the theory's adaptability suffice to explain the success, and that such success is likely even if the theory does not fit well with Nature. This is tantamount to asserting that use-novelty is necessary for severity. Let us mark this premise (which UN proponents state in various ways) as (*). Here Worrall states it as follows:

> (*) If *H* is use-constructed, then it cannot be argued that its success-fully fitting the evidence is unlikely if *H* is incorrect.

His argument seems to be that since the success of a use-constructed hypothesis can be explained by its having been deliberately constructed to accord with the evidence, there is no need or no grounds for seeking its explanation in the correctness of *H*. We have "used up" the import of the evidence, so to speak.

   Is there not some confusion here between two senses of explaining a success? Consider an imaginary trial of a suspect in the World Trade Center bombing:

Prosecutor: There is no other way to explain how well this evidence fits with X as the culprit (the twisted metal matching the rented van, the matching fingerprints, etc.) save that X was (at least part of the group) responsible.
Defense: Yes there is. The investigators built up their hypothesis about the guilty party so as to account for all of the evidence collected.

   It matters not, for the sake of making out the equivocation, that the hypothesis here is different from most scientific ones.

   In other words the problem is to find a way of accounting for some evidence or some observed effect. Let us say that the problem is declared solved only when a hypothesis is reached that satisfactorily accounts for it. Now suppose the problem is solved, but that particular features of the effect to be accounted for have been used in reaching the solution. One can ask: Why does the hypothesized solution accord so successfully with the evidence? In one way of reading this question (Worrall's), a perfectly appropriate answer is that *any* solution put forward would accord with the evidence: the solution was use-constructed. Quite a different reading of this question—the one to which the prosecutor is answering in the affirmative—has it asking whether the accordance with the evidence indicates the correctness of the hypothesis. The question, on this second reading, is whether the successful accordance with the evidence satisfactorily rules out the ways in which it would be an error to declare *H*. That *H* was constructed to account for the evidence does not force a "no" answer to this second question. *H* might be use-constructed, but use-constructed reliably.

   Conflating the two renders mysterious ordinary scientific discernments. Finding a correlation between a certain gene and the onset of Alzheimer's disease led Dr. Allen Roses (from Duke University) to hypothesize a genetic cause for certain types of Alzheimer's. What accounts for his hypothesis successfully explaining this correlation? The answer, interpreting the question one way, might be that Dr. Roses found this correlation and used it, as well as known properties of the gene, to develop a hypothesis to account for it. A separate question is why his genetic explanation fits the facts so well. One major Alzheimer's researcher declared (despite Roses's hypothesis going against his own work) that after only ten minutes he could see that the data pointed to Roses's hypothesis. Yes, Roses used the data to construct his hypothesis. The particular way in which he did so, however, showed that it provided at least a reasonably severe test in favor of his hypothesis (that the gene ApoE has a genuine connection with Alzheimer's). Had Roses's evidence been less good—as was the case a few years earlier—the scientists would have (and did) largely dismiss the agreement.[11]

To summarize this subsection, we have good grounds for the correctness of *H* to the extent that the circumstances by which *H*'s assertion would be in error have been well ruled out. Evidence may be used to construct *H* and still do a good job of ruling out *H*'s errors. To suppose that these are mutually exclusive is a false dilemma.

Proponents of Zahar and Worrall's argument for use-novelty might agree with all this. Nevertheless they may insist that except for when the use-construction method is clearly reliable, they are right to require, or at least to prefer, use-novel to use-constructed hypotheses. Let us grant them this and declare that there is no disagreement between us here. But let us push them a little further. What is the worry in those cases where the use-construction method is *not* clearly reliable? The worry is that it is one of those dreadful methods of cooking up hypotheses—the kind of method that is always everywhere available. And what is wrong with the kind that is always everywhere available? It is available whether or not the hypothesis reached is true! (Remember gellerization.)

That is precisely my point. The reason for eschewing use-construction methods is the condemnation of unreliable procedures of data accommodation. The underlying rationale for requiring or preferring use-novelty is the desire to avoid unreliable use-constructing procedures. The best spin I can glean from the Zahar-Worrall rationale for requiring UN shows it to be the one I claim. If there is a different rationale, perhaps its proponents will tell us what it is.

An Argument from Giere

Giere does not beat around the bush but plainly declares (*) from the start. Where Worrall emphasizes the cleverness of proponents and

11. When in 1991 Roses first reported having pinpointed the approximate location of a gene in families with late-onset Alzheimer's, it was given little credence by neuroscientists. It not only went against the generally accepted thinking of the time, but Roses clearly had not yet ruled out the numerous ways in which he might have been mistaken to infer a causal connection from such a correlation. Later, a biochemist at Duke's lab sought out natural substances that chemically bind to amyloid. Perhaps some substance was sticking to amyloid, causing the buildup of plaques in the brain. What he thought was an experimental contaminant was ApoE. The biochemist was able to take advantage of the fact that studies of heart disease had already located and isolated the gene for this cholesterol-carrying sub-

the adaptability of hypotheses in order to explain why it is no wonder
that a success accrued, Giere notes how the tester himself may simply
refuse to consider any theory or model that does not successfully fit
the data.

Although Fresnel's wave model accounted for the known diffraction pattern of straightedges, Giere says this did not count as a good
test of the model because the straight-edged pattern violated use-
novelty in the sense that it "acted as a constraint on his theorizing":

> He [Fresnel] was unwilling to consider any model that did not yield
> the right pattern for straight edges. Thus we know that the probability
> of *any* model he put forward yielding the correct pattern for straight
> edges was near unity, independently of the general correctness of that
> model. (Giere 1983, 282)

What emerges once again, though much more directly put, is a version
of premise (*), that use-novelty is necessary for severity. Because Giere
gives us a separate account of testing, it is possible to extricate his full
argument for the UN requirement.

Giere, at least in 1983, endorsed an error-statistical account of testing (along the lines of Neyman-Pearson).[12] He characterized "an *appropriate test* as a procedure that has *both* an appropriately high probability
of leading us to accept true hypotheses as true and to reject false
hypotheses as false" (Giere 1983, 278). A test should, in short, have
appropriately low error probabilities. We have:

1. A successful fit does not count as a good test of a hypothesis if
   such a success is highly probable even if the hypothesis is incorrect. (That is, a test of *H* is poor if its severity is low.)

---

stance. It turned out that the gene for ApoE was located in the very place Roses
had found the suspect gene in families with Alzheimer's.

12. There are important differences between the error-statistical account I favor and Giere's most current decision-theoretic account of testing (Giere 1988).
First, I reject the idea of modeling scientific inference as deciding to choose one
model over another. (See chapter 11.) Second, Giere's decision strategy is to choose
a model *M* when evidence is very probable were *M* correct while being very improbable were some alternative model correct. Yet evidence may be very improbable
under a rival to model *M* and not count as passing *M* severely. (See, for example,
chapter 6.) Finally, the models in Giere's "model-based" probabilities are allowed
to be full-blown scientific models, such as Dirac's and Schrödinger's models (Giere
1988, chap. 7). The assessments of these probabilities rely more or less on the intuitive judgments of scientists, and as Giere's own discussions show, are subject to
serious shifts (even for a given scientist). In my account, the probability assessments
must be closely tied to experimental models about which the statistics are at least
approximately known.

Violating use-novelty, Giere suggests, precludes (or at least gets in the way of) the requirement that there be a low probability that H is accepted if false. This gives an even stronger version of premise (*) than that found in Worrall:

> (*) If a hypothesis H is use-constructed, then its success is high ("near unity") even if it is false.

From premise (1) and (*) we get the conclusion that use-construction procedures fail to count as good tests (of the hypotheses they reach). I agree with premise (1)—it is premise (*) that I deny.

The Gierean Argument for (*) *the Necessity of UN.* Giere's argument for the necessity of UN seems to be that in a use-constructed case, a successful fit is obviously not unlikely—it is assured no matter what. That is, the basis for (*) is an additional premise (2):

> 2. *Basis for* (*): If H is use-constructed, then a successful fit is assured, no matter what.

Ah, but here is where we must be careful. This "no matter what" can be interpreted in two ways. It can mean

> *a.* no matter what the data are

or it can mean

> *b.* no matter if H is true or false.

Although the assertion in (2) is correct with the replacement in (*a*), thus construed it provides no basis for (*), that UN is necessary for severity. For (2) to provide a basis for (*), the replacement would have to be as in (*b*). However, (2) is false when replaced with the phrase in (*b*). Once this flaw in the pivotal intuition is uncovered it will be seen that UN fails to be a necessary condition for a severe test, and that (*) is false.

To clarify, consider two different probabilities in which one might be interested in appraising the test from which a passing result arises:

> A. The probability that test T passes the hypothesis it tests.
> B. The probability that test T passes the hypothesis it tests, *given that the hypothesis is false.*

Note that here two things may vary: the hypothesis tested as well as the value of e. Now consider a test procedure that violates UN in any of the ways this can come about. To abbreviate, let $H(e)$ be a use-constructed hypothesis—one engineered or constrained to fit evidence *e* (it may be read "*H* fixed to fit *e*"). Then the following describes what may be called "a use-constructed test procedure":

> *A use-constructed test procedure T:* Use *e* to construct *H*(*e*), and let *H*(*e*) be the hypothesis *T* tests. Pass *H*(*e*) with *e*.

Since *H*(*e*), by definition, fits *e*, there is no chance of *H*(*e*) *not* passing a use-constructed test *T*. The relative frequency of passing in a series of applications of a use-constructed test equals 1. That is,

A. The probability that (use-constructed) test *T* passes the hypothesis it tests

equals 1.

But that is different from asserting that the test *T* is guaranteed to pass the hypothesis it leads to testing, *even if that hypothesis is false.* That is, asserting that (A) equals 1 is different from asserting that

B. The probability that (use-constructed) test *T* passes the hypothesis it tests, *even if it is false,*

equals 1.

Yes, the use-constructing procedure always leads to passing one hypothesis or another—provided it ends—but this is not incompatible with being able to say that it never, or almost never, leads to passing a false hypothesis.

Imagine that each experimental test rings a bell if the hypothesis it tests passes, and sounds a buzzer if it fails. (A) is a statement about how often the bell would ring in a series of experimental tests of a certain kind. A use-constructed test procedure would always culminate in a ring—if it ended at all.[13] So (A) equals 1 in the case of use-constructed tests. The probability in (B), on the other hand, asks about the incidence of erroneous bell ringing, where by erroneous bell ringing I mean that the test rings the bell when a buzzer should have been sounded. This need not equal 1, even in use-constructed tests. It can even be 0. Those who hold UN necessary do so because violating UN leads to a test that has to result in sounding the bell and never in sounding the buzzer. The assumption is that severity requires some of the test results to lead to buzzing, but this is a mistake.

The illustration with bells and buzzers is just to bring out, once again, the idea of an experimental (or sampling) distribution. A particular experimental test is viewed as a sample from a population of such experimental tests. The probabilities refer to relative frequencies with

---

13. In many cases procedures can be guaranteed to end.

which a "sample test" has some characteristic, in this (actual or hypo-thetical) population of tests. The characteristic in (A) is "passing one hypothesis or another." The characteristic in (B) might be described as "giving an erroneous pronouncement on the hypothesis passed." How to calculate these probabilities is not always clear-cut, but for my point it is enough to see how, in cases where they *can* be calculated, the two are not identical.

This is a beautiful example of how the informal side of arguing from error can and should lead the way in disentangling the confusions into which one can easily wade in trying to consider the formal proba-bilistic arguments. To understand why (A) differs from (B), and, corre-spondingly, why UN is not necessary for severity, we need only think of this informal side. In particular, we need only think of how a proce-dure could be sure to arrive at some answer and yet be a procedure where that answer is rarely or never wrong. This is the basis for the counterexamples I will now consider.

8.4 Use-Novelty Is Not Necessary for Severity: Some Counterexamples

To give a counterexample to the thesis that UN is necessary for severity, I have to describe a case that violates use-novelty yet provides a severe test of the use-constructed hypothesis.

Example 8.1: SAT Scores

For a trivial but instructive example consider a hypothesis H about the average SAT score of the students who have enrolled in my logic class:

$H(x)$: the average SAT score (of students in this class) = x,

where x, being unspecified, is its free parameter. Fixing x by summing up the scores of all n students and dividing by n qualifies as a case of parameter-fixing yielding a use-constructed hypothesis $H(e)$. Suppose that the result is a mean score of 1121. The use-constructed hypothe-sis is

$H(e)$: the average SAT score $= 1121$.

Surely the data on my students are excellent grounds for my hypothe-sis about their average SAT scores. It would be absurd to suppose that further tests would give better support. For hypothesis H follows de-ductively from e. Since there is no way such a result can lead to passing H erroneously, H passes a maximally severe test with e.

In much the same vein, Glymour, Scheines, Spirtes, and Kelly (1987) allude to the procedure of counting 12 people in a room and constructing the hypothesis that there are 12 people in the room. One may balk at these examples. Few interesting scientific hypotheses are entailed by experimental evidence. But allowing that such cases provide maximally severe tests while violating UN suffices to show that criterion UN is not necessary for severity. It may be asked: Is not UN required in all cases *other* than such maximally severe ones? The answer is no. Tests may be highly severe and still violate UN. The extreme represented by my SAT example was just intended to set the mood for generating counterexamples. Let us go to a less extreme and very common example based upon standard error statistical methods of estimation.

Example 8.2: Reliable Estimation Procedures

We are very familiar with the results of polls in this country. A random sample of the U.S. population is polled (the sample size being specified along the lines discussed in chapter 5), and the proportion who approve of the President is recorded. This is the evidence e. Say that 45 percent of the sample approve of the President. The poll report would go on to state its margin of error, say, of 3 percentage points. (The margin of error is generally around 2 standard deviations.) The resulting report says: Estimate that 45 percent of the U.S. population plus or minus 3 percentage points approve of the President. The report is a *hypothesis* about the full population (that p, the population proportion, is in the interval [.42, .48]). The estimate is constructed to accord with the proportion in the sample polled. The procedure may be characterized as follows: Hypothesize that p, the proportion who approve of the President, is an interval around the data e, the observed proportion who approve. The interval is given by the margin of error. Call it ∂. So the procedure is to infer or "pass" hypothesis $H(e)$ where

$H(e)$ asserts: p is equal to $e \pm \partial$.

This procedure, (confidence) interval estimation, will be discussed more fully in chapter 10. The margin of error corresponds to giving the overall reliability of the procedure. A 95 percent estimation procedure has a .95 probability of yielding correct estimates. What is known is that any particular estimate (hypothesis) this procedure yields came about by a method with a high reliability. Depending on the poll, the uncertainty attached might be .05 or .01 (reliability .95 or .99). The hypothesis thereby reached may be false, that is, the true value of p may be outside the estimated range, but, we argue, if p were outside the interval, it is very unlikely that we would not have detected this in the poll. With high probability we would have.

Now consider this question: Is it possible for the data e—the observed proportion who approve of the president in the sample—to fail to pass the hypothesis $H(e)$ that will be arrived at by this estimation procedure? Given that $H(e)$ is use-constructed (to be within ∂ units from e), the answer must be no. For the procedure, by definition, passes the hypothesis: p is in the interval $e \pm \partial$. Even before we have a specific outcome we know there is no chance that the result of this data generation process will fail whatever (use-constructed) hypothesis $H(e)$ the procedure winds up testing.

Compare this with the test's severity in passing $H(e)$. The fit with the resulting hypothesis (the interval estimate $H(e)$) is given by the specified margin of error ∂, say 2 standard deviation units. It is rare for so good a fit with $H(e)$ to occur unless the interval estimate $H(e)$ is true (i.e., unless the population proportion really is within 2 standard deviation units of e). So the severity in passing $H(e)$ is high. (The reasoning is this: To say that the interval estimate $H(e)$ is not true means that the true population proportion p is not in the hypothesized interval. But the p values excluded from this interval are those that differ from the observed proportion who approve, e, by more than 2 standard deviation units (in either direction), and the probability of e differing from p by more than 2 standard deviation units is small [.05].)[14]

Contrast this estimation procedure with one that, regardless of the observed outcome, winds up estimating that at least 50 percent of the population approve of the president. This "wishful thinking" estimation procedure does suffer from a lack of reliability or severity: it always infers a 50 percent or better approval rating in the full population even if the true (population) proportion is less than that. Regardless of whether use-constructing is involved or not, it is for this reason that we condemn it.

An Anticipated Objection. Against my counterexamples one might hear the following objection. The data in my examples provide *evidence* for the hypotheses reached, but they are not tests. I am confusing tests with evidence.

In my account, which is a testing account, there is no distinction.

14. One must be careful to avoid misinterpreting this probability. The .05 does not refer to the probability that the particular estimate arrived at is true, i.e., includes the true value of parameter p. That probability is either 0 or 1. It is the procedure that formed the estimate of p that has a .05 probability of yielding false estimates.

To insist that any example I might give where evidence is used to construct a hypothesis cannot count as a test is to beg the question at issue. The question is whether there ever exists a use-constructed example where the evidence is, nevertheless, good support for or a good test of the hypothesis. Those who deem UN necessary—at least Giere and Worrall—do not say that non–use-novel data count as great evidence but no test. They say that such data fail to count as good evidence or good support for the hypotheses constructed, and they say this *because* the data fail to test these hypotheses. What is more, the method of standard confidence interval estimation is mathematically interchangeable with a corresponding statistical test. In a nutshell, the parameter values within the interval constructed consist of values that would pass a statistical test (with corresponding reliability). So I am in good company in regarding such procedures as tests.

A Curious Bayesian Aside. Curiously, confidence intervals have also been appealed to (e.g., by Howson) in Bayesian arguments against the necessity of use-novelty. Swept up in the task of showing the intuitive plausibility of use-constructed estimations, Howson allows himself the admission that

> there is no question but that confidence interval estimates of physical parameters, derived via some background theory involving assumptions about the form of the error distribution, are the empirical bedrock upon which practically all quantitative science is built. (Howson 1990, 232)

He is quick to add that the Bayesian can show how to assign a high degree of belief to the correctness of the estimate, as if to say that Bayesians can endorse the estimate as well. But the force of the intuition to which Howson is appealing is plainly the reliance science puts on the standard, Neyman-Pearson, estimates. (Hark back to our analogy of Leonardo da Vinci in chapter 3.) The irony of this Bayesian reliance on the intuitive plausibility of non-Bayesian procedures will not be fully appreciated until chapter 10.[15]

8.5 SUMMARY AND SOME CONSEQUENCES FOR CALCULATING SEVERITY

Tests of use-constructed hypotheses are eschewed because a passing result is assured. But what matters is not whether passing is assured

15. Other examples occur in Howson 1984 and Howson and Urbach 1989. Nickles (1987) makes a point similar to Howson's, informally. Like the use-novelist

but whether erroneous passing is. There is no problem with a test having a high or even a maximal probability of passing the hypothesis it tests; there is only a problem if it has a high probability of passing hypotheses erroneously. Hypotheses might be constructed to accord with evidence *e* in such a way that although a passing result is assured, the probability of an erroneous passing result is low; equivalently, the test's severity is kept high. The common intuition to eschew using the same data both to construct and to test hypotheses (to require UN), I claim, derives from the fact that a test that violates UN is guaranteed to pass the hypothesis it tests—no matter what the evidence. But this does not entail that it is guaranteed to pass some hypothesis *whether or not that hypothesis is false*. Indeed, a use-constructed test may have a low or even no probability of passing hypotheses erroneously. Granted, *if* a test is guaranteed to pass the hypothesis it tests, even if that hypothesis is false (i.e., (B) equals 1), then the test is guaranteed to pass the hypothesis it tests (i.e., (A) equals 1); but the converse does not hold.[16]

To hold UN as necessary is to overlook deliberate rules for use-constructing hypotheses with high or even maximal severity. Consider first a rule for using *e* to construct *H(e)* so as to ensure maximal severity of the test:

> *Rule R-1 for constructing maximally severe tests:* Construct *H(e)* such that a worse fit with *e* would have resulted (from the experiment) unless *H(e)* were true (or approximately true).

(That is, if *H(e)* were false, a worse fit would have occurred.) Such a rule is obviously available only in very special cases. But the point is that to calculate the probability in (B), the probability of erroneously passing the (use-constructed) hypothesis tested, requires taking into account the construction rule employed—in this case rule R-1. That is why the severity criterion is modified for cases of hypothesis construction in section 6.6.

Let us abbreviate a use-construction test that arrives at its hypothesis via rule R-1 as test *T(R-1)*. Test *T(R-1)*, by the stipulated definition, is guaranteed to pass *any* hypothesis fixed to fit *e*. (So (A) equals 1.) Nevertheless, the probability of (B)—the probability needed for calculating severity (by taking 1 minus it)—is this:

---

and unlike me, however, Nickles denies that data used to fix the parameter can count as giving what he calls generative as well as consequential support to a hypothesis so fixed.

16. One might be led to the error of thinking that the converse does hold if one erroneously takes the "if" clause in (B) as a material conditional instead of as the appropriate conditional probability.

(B) in test $T(\text{R-1})$: the probability that test $T(\text{R-1})$ passes the hypothesis it tests, given that hypothesis is false.

This equals 0. As such, the severity of a test of any hypothesis constructed by rule R-1 is 1—the severity is maximal.

The rule used in fixing the mean SAT score of students in my class is an example of rule R-1. While there is always some rule by which to arrive at a use-constructed H (the so-called "tacking" method of use-construction will do), the ability to apply the very special rule R-1 is hardly guaranteed. But if one does manage to apply R-1, the constructed hypothesis to which it leads cannot be false. Although one can rarely attain the security of rule R-1, the experimenter's tool kit contains several use-constructing rules that afford a high degree of reliability or severity, call it π. Such a use-construction rule may be written as rule R-π:

Rule R-π for constructing highly severe tests (e.g., to degree π): Construct $H(e)$ such that the probability is very small $(1-\pi)$ that a result from the experiment would accord as well with $H(e)$ as does e, unless $H(e)$ were true (or approximately true).

Examples of high severity construction rules are found in rules for the design and interpretation of experiments. They are the basis of standard estimation theory, as example 8.2 showed.

Let test $T(\text{R-}\pi)$ be the use-construction test based on a construction rule $(\text{R-}\pi)$. What is the severity of test $T(\text{R-}\pi)$?

The answer, of course, is π.

The Informal Calculation of Severity in Use-Constructed Tests

One need not be able to formally calculate the severity π. The identical rationale underlies informal rules for using evidence. In qualitative experimental tests one may only be able to say that the severity is very high or very low, yet that is generally enough to assess the inference. I might mention a (real) example that first convinced me that UN is not necessary for a good test. Here evidence was used to construct as well as to test a hypothesis of the form

$H(x)$: x dented my 1976 Camaro.

The procedure was to hunt for a car whose tail fin perfectly matched the dent in my Camaro's fender to construct a hypothesis about the likely make of the car that dented it. It yielded a hypothesis that passed a high severity test. I was able to argue that it is practically impossible for the dent to have the features it has unless it was created by a specific type of car tail fin. Likewise for the rule the investigators followed in pinpointing the driver of the van carrying the explosive in the World Trade Center bombing. Such rules violate use-novelty; but they correctly indicate attributes of the cause of the explosion and the dent in my Camaro (and ultimately the identity of the drivers) because they are severe in the sense of rule R-π.

One may object, But science is not like that, there are too many possible alternative hypotheses. Having discussed the problem of alternative hypotheses in chapter 6, I am assuming those points here. If the objector persists that there is no way to obtain reliable knowledge or have a severe test of such and such a scientific theory, then, assuming he or she is correct, it must be agreed that no reliable use-construction procedure can accomplish this either. But this is irrelevant to my thesis that there are reliable experimental arguments that violate use-novelty.

We might connect the point of this chapter to our recasting of Kuhn in chapter 2. What is objectionable is not that practitioners are determined to find a way of accommodating data (to solve a given problem); what is objectionable is an accommodation that is not severely constrained (e.g., that it involves changing the problem), which results in unsolved problems often being declared solved when they are not. Alternatively, in a reliable use-construction one can argue that if $H(e)$ were incorrect, then with high probability the test would not have led to constructing and passing $H(e)$.

In one passage—although it is almost only in passing—Popper seems to capture what I have in mind about warranting severity. He says (in replying to his critics) that

> supporting evidence consists solely of attempted refutations which were unsuccessful, *or of the "knowledge" (it does not matter here how it was obtained) that an attempted refutation would be unsuccessful.* (Popper 1974, 992; emphasis added)

In a reliable use-constructed case one can sustain this "would be" argument. This is just what is wanted to affirm that evidence indicates the "probable success" of the hypothesis (in the sense of chapters 4 and 5).

Proponents of use-novelty often view Whewell and Peirce as forerunners of their view. While to an extent they are right, both Whewell and Peirce also discuss the kinds of cases where use-constructions are allowable. (I shall save Peirce's remarks for our later discussion of him.) Whewell also considered "the nature of the artifices which may be used for the construction of formulae" when data of various types are in hand (Whewell [1847] 1967, 392). The artifices he has in mind correspond to cases where the construction may be seen to ensure high severity. The "special methods of obtaining laws from observations" (p. 395) that Whewell cites include the method of curves, the method of means, the method of least squares, and the method of residues (essentially the statistical method of regression).

Such rules are typical, as would be expected, where violations of UN cannot be helped: where hypotheses are arrived at and affirmed by data, and it is impossible or impractical to obtain additional evidence (e.g., theories about dinosaurs, evolutionary theory, epidemiology, anthropology). As the next section shows, however, violations of UN are required even in cases lauded as models of severe and crucial tests, such as the 1919 eclipse tests of Einstein's gravitational hypothesis. Indeed, once the piecemeal aspect of testing is uncovered, such use-construction rules are indispensable. The same data may be used both to construct and ground hypotheses—so long as it is improbable that reaching so good an agreement is erroneous.

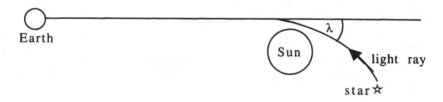

FIGURE 8.1.   Deflection of starlight by the sun.

8.6 THE 1919 ECLIPSE TESTS OF EINSTEIN'S LAW OF GRAVITATION

According to Einstein's theory of gravitation, to an observer on earth, light passing near the sun is deflected by an angle, λ, reaching its maximum of 1.75" for light just grazing the sun. Terrestrial tests of Einstein's gravitation law could not be severe, since any light deflection would be undetectable with the instruments available in 1919. Although the light deflection of stars near the sun (approximately 1 second of arc) *would* be detectable, the sun's glare renders such stars invisible, save during a total eclipse. "But," as Arthur Eddington ([1920] 1987, 113) noted, "by strange good fortune an eclipse did happen on May 29, 1919," when the sun was in the midst of an exceptionally bright patch of stars, providing a highly severe test such as would not recur for many years. Two expeditions were organized: one to Sobral in northern Brazil, another (including Cottingham and Eddington) to the island of Príncipe in the Gulf of Guinea, West Africa.

Eddington, Davidson, and Dyson, the astronomer royal (henceforth Dyson et al. 1920), outline three hypotheses "which it was especially desired to discriminate between" (p. 291). Each is a statement about a parameter, the deflection of light at the limb of the sun, λ (in arc seconds):

1. Gravitation affects starlight according to Einstein's law of gravitation: the deflection at the limb of the sun $\lambda = 1.75''$.
2. Gravitation affects light according to the Newtonian law of gravitation: the deflection of a star at the limb of the sun $\lambda = 0.87''$.
3. Gravitation does not affect light, $\lambda = 0$.

The "Newtonian"-predicted deflection, (2), which stems from assuming that light has a certain mass and follows Newton's law of gravity, is exactly half that predicted by Einstein's law. Before setting out for Príncipe, Eddington suggests that

> apart from surprises, there seem to be three possible results: (1) A deflection amounting to 1.75" . . . which would confirm Einstein's theory; (2) a deflection of 0.87" . . . which would overthrow Einstein's theory, but establish that light was subject to gravity; (3) no deflection, which would show that light, though possessing mass, has no weight, and hence that Newton's law . . . has broken down in another unexpected direction. (Eddington 1918, 36)

A little over one year later, the results are in, and the conclusions given:

> The results of the expeditions to Sobral and Príncipe can leave little doubt that a deflection of light takes place in the neighbourhood of the sun and that it is of the amount demanded by Einstein's generalised theory of relativity, as attributable to the sun's gravitational field. (Dyson et al. 1920, 332)

This capsulizes the two key inferences from the eclipse inquiry: first, that there is a deflection effect of the amount predicted by Einstein as against Newton (i.e., the "Einstein effect"), and second, that the effect was "attributable to the sun's gravitational field" as described in Einstein's hypothesis.

The appraisal of the results by numerous scientists consisted of two corresponding parts or stages, which I label i and ii. Stage i involved inferences about the value of λ and critical discussions of these inferences, stage ii, inferences about the cause of λ and the associated (heated) discussions about these inferences. Each stage involved testing more local hypotheses, first to discriminate between the values of parameter λ, and second to discriminate between causes of the observed λ. Eddington, an adept data analyst, provides lavish and fascinating discussions of the nitty-gritty details of the data extraction and modeling. This, together with the intrinsic importance of the case, makes it an excellent subject for applying the full-blown hierarchy of models framework. Aspects of the data gathering were touched on in chapter 5. Lest I try my readers' patience, however, I will limit my discussion to aspects of the case most relevant to the present issue.

Stage i: Estimating the Eclipse Deflection at the Limb of the Sun

The "observed" deflection (on May 19), as with most experimental "results," is actually a hypothesis or estimate. Due to two major sources of error, arriving at the result is a matter of statistical inference: First, one does not observe a deflection, but at best observes (photographs of) the positions of certain stars at the time of the eclipse. To "see" the deflection, if any, requires learning what the positions of these same stars would have been were the sun's effect absent—a "control" as it were. Eddington remarks:

> The bugbear of possible systematic error affects all investigations of this kind. How do you know that there is not something in your apparatus responsible for this apparent deflection? . . . To meet this criticism, a different field of stars was photographed . . . at the same altitude as the eclipse field. If the deflection were really instrumental, stars on these plates should show relative displacements of a similar kind to those on the eclipse plates. But on measuring these check-plates no appreciable displacements were found. That seems to be satisfactory evidence that the displacement observed during the eclipse is really due to the sun being in the region, and is not due to differences in instrumental conditions. (Eddington [1920] 1987, 116)

Where the check plates could serve as this kind of a control, the researchers were able to estimate the deflection by comparing the position of each star photographed at the eclipse (the eclipse plate) with its normal position photographed at night (months before or after the eclipse), when the effect of the sun is absent (the night plate). Placing the eclipse and night plates together allows the tiny distances to be measured in the x and y directions, yielding ∂x and ∂y (see figure 8.2). These values, however, depend on many factors: the way in which the two plates are accidentally clamped together, possible changes in the scale—due mainly to the differences in the focus setting that occur between the exposure of the eclipse and the night plates—on a set of

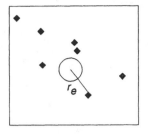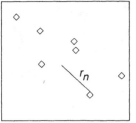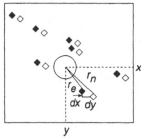

FIGURE 8.2. Comparing the "eclipse plate" and the "night plate" (adapted from von Klüber, 1960, 52). (a) "eclipse plate" with sun and surrounding stars. (b) corresponding "night plate" taken of the same star field when visible at night. (c) both plates combined as they appear in the measuring machine. (From Mayo, 1991a)

other plate parameters, and finally, on the light deflection, λ itself.[17] By what is quite literally a "subtraction" method, it was possible to estimate λ.

A second important source of error stems from the fact that the predicted deflection of 1.75" refers to the deflection of light just grazing the sun; but the researchers only observed stars whose distance from the sun is at least two times the solar radius. Here the predicted deflection is only about 1" of arc. To compare the evidence with the theoretical prediction it is necessary to estimate what the deflection would have been for starlight near the sun.

Thus, despite the novelty of the theoretical prediction of 1.75", to reach the hypothesis about the estimated deflection, the eclipse data themselves must be used, both to fix each of the experimental parameters and to arrive at the extrapolation to the limb of the sun. Furthermore, checking the validity of these inferences requires using, once again, the eclipse data. So the UN requirement is apparently violated. But great pains were taken to ensure that reliability or severity was not. They used only those results for which there were measurements on enough stars (at least equal to the number of unknown parameters in the equations—6) to apply a reliable method of fixing: the statistical method of least squares (regression), a technique well known to astronomers from determining stellar parallax, "for which much greater

17. A detailed discussion of this and several other eclipse tests of Einstein's deflection is provided by H. von Klüber (1960). See also D. Moyer 1979.

accuracy is required" (Eddington [1920] 1987, 115–16) than that for the eclipse test.

(Note also that it was impossible to adhere to the classic requirement to prespecify the sample size. Before obtaining and analyzing the data one did not know how many of the photographed stars would be usable.[18] I will return to the issue of prespecification in chapter 9.)

The Results

Subtracting out the variety of factors algebraically, one arrives at estimates of λ from the different sites, along with their probable errors (or, the measure now used, their standard errors).[19] The "observed results," in short, are actually hypotheses about the expected deflection (at the limb of the sun), λ. The two eclipse results, one from Sobral, one from Príncipe, taken as crucial support for Einstein were, with their standard errors,[20]

Sobral: the eclipse deflection = 1.98″ ± 0.18″.

Príncipe: the eclipse deflection = 1.61″ ± 0.45″.

Using either standardized measure of error allows assigning probabilities to experimental results under different hypotheses about λ. This permits severity to be calculated. Eddington reasons:

> It is usual to allow a margin of safety of about twice the probable error on either side of the mean. The evidence of the Principe plates is thus just about sufficient to rule out the possibility of the "half-deflection," and the Sobral plates exclude it with practical certainty. (Eddington [1920] 1987, 118)

The severity criterion (SC), the formal analog to our argument from error, explains the weight accorded to each result. The pattern of reasoning is one with which we are by now very familiar. An observed difference from a value predicted by a hypothesis H_0 genuinely indicates that H_0 is in error, if so large a difference is very improbable (just) if the error is absent. The appraisal at stage i had several parts. In the portion of the appraisal alluded to in the above passage, H_0, the hy-

18. This was Barnard's point in discussing the eclipse results in Barnard 1971.

19. One probable error equals .68 standard errors. A standard error is the estimate of the standard deviation. The reason for the choice of the probable error as a standard is that a sample mean differs from a (Normal) population mean by one or more probable errors (in either direction) 50 percent of the time. (It differs from the population mean by one or more standard errors in either direction about 32 percent of the time.)

20. The probable errors are, respectively, .12 and .30.

pothesis found to be in error, is the Newtonian "half-deflection," that $\lambda = .87''$. The hypothesis H that "passes" is

H: the half-deflection is in error, $\lambda > .87''$.

Consider passing H with the Sobral result of $1.98'' \pm 0.18''$. We ask: What is the probability of "such a passing result"—that is, one as far or farther from $.87''$ than the observed result—given that λ is the Newtonian (half-deflection) $.87''$? The answer is that this probability is practically 0. (The result is more than 6 standard deviations in excess of $.87$.) So π, in construction rule R-π, is nearly 1. The Príncipe result, being around 1.6 standard deviations in excess of $.87''$, is only a reasonably severe passing result. That is, with reasonably high probability, around $.95$, a result more in accordance with $\lambda = .87''$ would be expected, if λ were equal to the Newtonian value ($.87''$). ($\pi = .95$.)[21]

The probabilities, it must always be remembered, are not assigned to the hypotheses about λ. Universes are not as plenty as blackberries—to recall Peirce (from chapter 3). There is one universe, this one blackberry, within which a hypothesized value for λ either does or does not hold true. We know, however, that there are a variety of sources of error that produce differences between actual and estimated deflections. Making use of this knowledge of error we can argue as follows: were the experimental differences from the half-deflection due to the variety of known sources of error and not to a genuine discrepancy from $.87''$, they would practically never, or extremely rarely, be expected to occur in a series of (hypothetical) eclipse experiments at the two sites. This is our standard canonical argument for inferring that a discrepancy from a parameter value is real.

If one were filling out the hierarchy of models, one would explore how at stage i a single question is split off from the primary one. The possible hypotheses at stage i are values for λ. These are the possible answers to this one subquestion. One would describe the link between an observed mean deflection L (itself a model of the data) and hypotheses about λ within the experimental model. The severity criterion warrants accepting the use-constructed hypothesis

$H(L)$: λ exceeds $L - 2$ standard errors,

where the standard error is the estimated standard deviation of L. To see why $H(L)$ is warranted, notice that "$H(L)$ is false" asserts that λ

---

21. I do not mean that this is the only work in weighing these two inferences. Detailed checks to affirm the assumptions of the experimental data models are also needed and would have to be incorporated in a full-blown discussion of the experimental inquiry.

does *not* exceed $L - 2$ standard errors. To calculate severity one calculates the probability of *not* observing a deflection as large as L, given that λ is any of the values included under "$H(L)$ is false." The value of this probability is high (at least .97).[22]

A Result in "All Too Good Agreement" with Newton

There was, however, a third result also obtained from the Sobral expedition. In contrast with the other two this third result pointed not to Einstein's prediction, but, as Eddington ([1920] 1987) declares, "with all too good agreement to the 'half-deflection,' that is to say, the Newtonian value" (p. 117). It also differed from the other two in being discounted due to systematic errors! The instrument used, an astrographic telescope, was of the same type as that used in the counted Príncipe result. Nevertheless, upon examining these Sobral astrographic plates the researchers constructed a hypothesis *not* among the three set down in advance. Because this new hypothesis incorporates the alleged exception into Einstein's hypothesis (1), we may denote it by 1*:

> 1*: The results of these (Sobral astrographic) plates are due to systematic distortion by the sun and not to the deflection of light.

Popper held up this test as a model of severity, unlike the tests of psychological theories of the day, because the Einstein prediction dared to stick its neck out: a deflection far from the predicted value and near .87", Eddington (1918) declared, "would overthrow Einstein's theory" (p. 36). So what is to be made of this discounting of one set of results from Sobral?

Certainly this violates UN. It exemplifies the second entry in our list of ways that such a violation can occur: *exception barring*, or what Worrall calls *exception incorporation*. Here, when confronted with an apparent piece of counterevidence, one constructs a new hypothesis to account for the exception while still saving the threatened hypothesis—in this case, Einstein's. Moreover, while the Einstein hypothesis can accommodate the Sobral astrographics with the help of 1*, Newton's hypothesis accommodates them without any such contrivance. According to the UN requirement, it seems, the result used to construct 1* would count more for Newton than Einstein—contrary to the actual appraisal.

Now, the proponent of UN may deny that this really counts as exception incorporation (because there is a violation of "initial condi-

---

22. Note that although this includes infinitely many alternative values of λ, the high severity requirement is met for each. This instantiates my point in section 6.4.

tions"), but what cannot be denied is that constructing 1* violates UN. Still, it might be objected: the UN requirement never intended to condemn this kind of violation of UN. Here the data are being used to arrive at (and affirm) some low-level auxiliary hypothesis, one which, in this case, indicates that the data may be discounted in appraising the primary hypothesis. Are we to understand the UN theorist as allowing use-constructions in the case of low-level auxiliary hypotheses? Surely not. Otherwise the kind of UN violation (exception incorporation) that started all the fuss in the first place would pass muster. All of this underscores the main thesis of this chapter: the UN requirement fails to discriminate between problematic and unproblematic use-constructions (or double-countings).

Let us return to Eddington and the mirror hypothesis (1*). Consider the actual notes penned by Sobral researchers as reported in Dyson et al. 1920:

> May 30, 3 a.m., four of the astrographic plates were developed. . . . It was found that there had been a serious change of focus, so that, while the stars were shown, the definition was spoilt. *This change of focus can only be attributed to the unequal expansion of the mirror through the sun's heat.* . . . It seems doubtful whether much can be got from these plates. (P. 309; emphasis added)

This is not to say that it was an obvious explanation that could be seen to be warranted right off. It called for a fair amount of (initially unplanned) data analysis, and gave rise to some debate—all of which of course depended on using the suspect data themselves. However, the dispute surrounding this inference was soon settled, and because of that it sufficed for most official reports to announce that the astrographics, despite appearing to support the Newtonian value, were discounted due to distortions of the mirror. Such a report, not surprisingly, raises the antennae of philosophers looking into this historical episode.

John Earman and Clark Glymour (1980) point a finger at Eddington precisely because he "claimed the superiority of the qualitatively inferior Príncipe data, and suppressed reference to the negative Sobral results" (p. 84)—the Sobral astrographics. According to Earman and Glymour, "Dyson and Eddington, who presented the results to the scientific world, threw out a good part of the data and ignored the discrepancies" (p. 85). They question his suppression of these results because, in their view, "these sets of measurements seem of about equal weight, and it is hard to see decisive grounds for dismissing one set but not the other" (p. 75).

There were, however, good grounds for dismissing the Sobral

astrographics—one could not obtain a usable standard error of the estimate. Moreover, as the journals of the period make plain, the numerous staunch Newtonian defenders would hardly have overlooked the discounting of an apparently pro-Newtonian result if they could have mustered any grounds for deeming it biased. And the reason they could not fault Eddington's "exception incorporation" in hypothesizing 1* is that it involved well-understood principles for using this type of data to test, and in this case reject, a key assumption of the experiment. Results were deemed usable for estimating the deflection λ, we said, only if the statistical method (least squares) was applicable; that is, when there was sufficiently precise knowledge of the change of focus (scale effect) between the eclipse and night plates (within .03 mm)— precisely what was *absent* from the suspect Sobral plates.[23]

The discussion of this mirror distortion hypothesis brings out an interesting feature of the analysis. A. C. D. Crommelin (*Nature,* 13 November 1919) remarked that "there is reason to suspect systematic error, owing to the very different character of the star images on the eclipse and check plates" (p. 281). This is contrasted with the Sobral results taken with the 4-inch lens, whose use was allowed:

> The small coelostat used with the 4-inch lens did not suffer from deformation, the images of stars during totality being of the same character as those on the check-plates; this increased the weight of the determination with that instrument. (Crommelin 1919, 392)

Even small systematic errors of focus are of crucial importance because the resulting scale effect (from this alteration of focus) quickly becomes as large as the Einsteinian-predicted deflection effect. After an analysis of the distortion it was evident that the Sobral astrographic results pointed to only one hypothesis: 1*, systematic error.[24]

Over and over again, the discussions reveal that what was of central importance in declaring results usable was the similarity of the pattern of error in the eclipse and check plates. This similarity, or the lack of it, was discerned statistically; and this discernment is a formal

23. To see how important even small systematic errors of focus are, one need only look at how the resulting scale effect (from this alteration of focus) quickly becomes as large as the Einsteinian-predicted deflection effect of interest. The effect of 1.75″ refers to the deflection of the light of a star just at the limb of the sun; but the researchers only observed stars whose distance from the sun is at least 2 times the solar radius. Here the predicted deflection is about 1″ of arc or .015 millimeters on the photographic plate. See von Klüber 1960, 50.

24. Eddington was a specialist in techniques of data analysis, and his notes offer, in effect, rules for legitimately using eclipse evidence to "use-construct" hypotheses about λ (that is, instances of rule R-π).

analog of the informal detection of error patterns noted in chapter 1.[25] The mirror problem of 1919 became what I have in mind by a canonical model of error, and it was used in subsequent eclipse experiments.

Whereas this first stage was relatively uncontroversial, the second stage was anything but.

Stage ii: Can Other Hypotheses Be Constructed to Explain the Observed Deflection?

While even staunch defenders of Newton felt compelled to accept that the eclipse evidence passed the hypothesis that the deflection effect $\lambda = 1.75''$, they did not blithely accept that Einstein's law of gravitation had thereby also passed a good test. As the scientist H. F. Newall put it, "I feel that the Einstein effect holds the day, but I do not yet feel that I can give up my freedom of mind in favour of another interpretation of the effects obtained" (Newall 1919, 395–96). Such skeptical challenges revolved around stage ii, determining the *cause* of the observed eclipse deflection. At issue was the possibility of a mistake about a causal factor. The question, in particular, was whether the test adequately discriminated between the effect due to the sun's gravitational field and others that might explain the eclipse effect. A "yes" answer boiled down to accepting the following hypothesis:

(ii)(1): The observed deflection is due to gravitational effects (as given in Einstein's law), *not* to some other factor N.

The many Newtonian defenders adduced any number of factors to explain the eclipse effect so as to save Newton's law of gravity: Ross's lens effect, Newall's corona effect, Anderson's shadow effect, Lodge's ether effect, and several others. Their plausibility was not denied on the grounds that they were deliberately constructed to account for the evidence (while saving Newton)—as the UN requirement would suggest. On the contrary, as Harold Jeffreys wrote,

25. Dyson, Eddington, and Davidson (1920) say this about the discounted Sobral results:

The images were diffused and apparently out of focus. . . . Worse still, this change was temporary, for without any change in the adjustments, the instrument had returned to focus when the comparison plates were taken in July. (P. 309)

Interestingly, Crommelin explained that if we assume that the bad focus left the scale unaltered, then the value of the shift from these results is 1.54", thereby no longer pointing to the Newtonian value.

before the numerical agreements found are accepted as confirmations
of the theory, it is necessary to consider whether there are any other
causes that could produce effects of the same character and greater in
magnitude than the admissible error. (Jeffreys 1919b, 138)

Were *any* other cause to exist that was capable of producing (a considerable fraction of) the deflection effect, Jeffreys stressed, that alone
would be enough to invalidate the Einstein hypothesis (which asserts
that *all* of the 1.74″ are due to gravity).

Everything New Is Made Old Again

The historian of science Stephen Brush (1989) found, in apparent
violation of the rule to prefer (temporal) novel predictions, that the
ability to explain the known fact about Mercury's orbit provided
stronger support for Einstein's theory of gravitation than did the theory's ability to predict the new fact (in the 1920s) about the deflection
of light. Getting Mercury's orbit correct counted more in favor of Einstein's theory than light bending did, not despite the fact that the former was known and the latter new, but because of that very fact. Severity considerations explain why. The known fact about Mercury—
being an anomaly for Newton—was sufficiently important to have led
many to propose and test Newtonian explanations. These proposed
hypotheses, however, failed to pass reliable tests. In contrast, when
light bending first became known to exist "one might expect that
another equally or more satisfactory explanation would be found"
(Brush 1989, 1126). It is as if before this novel effect could count as
an impressive success for Einstein's theory, scientists had to render
it old and unsatisfactorily explained by alternative accounts (much
like Mercury). I think Brush is right on the money in declaring
that

> the eclipse results . . . provoked other scientists to try to give plausible
> alternative explanations. But *light bending could not become reliable evi-
> dence for Einstein's theory until those alternatives failed, and then its weight
> was independent of the history of its discovery.* (Brush 1989, 1127; empha-
> sis added)

Let us now consider how the new light-bending effect was made appropriately old.

Using the Eclipse Results at Stage ii

The challenges at stage ii to the pro-Einstein interpretation of the
observed deflection were conjectures that the effect was due to some

factor other than the Einstein one (gravity in the sun's field). They were hypotheses of the form

(ii)(2): The observed deflection is due to factor N, other than gravitational effects of the sun,

where N is a factor that at the same time saved the Newtonian law from refutation. Each such hypothesis was criticized in a two-pronged attack: the effect of the conjectured N-factor is too small to account for the eclipse effect; and were it large enough to account for the eclipse effect, it would have other false or contradictory implications.

Stage ii exemplifies several UN violations: the road to hypothesis construction was constrained to account for evidence e, and e also counted in support of that hypothesis. Once the deflection effect was affirmed at stage i it *had* to be a constraint on hypothesizing its cause at stage ii; at the same time, the eclipse results had to be used a second time in appraising these hypotheses. (A similar eclipse would not occur for many years.) Typically they were used to fix a parameter, the extent to which a hypothesized factor N could have been responsible for the observed deflection effect. When explicitly used to save the Newtonian law, they also violated UN by exception incorporation. Note also that the alternative hypotheses were at the same level as the primary hypothesis here.

The arguments and counterarguments (scattered through the relevant journals from 1919 to around 1921) on both sides involved violating UN. What made the debate possible, and finally resolvable, was that all who entered the debate were held to shared standards for reliable experimental arguments. They were held to shared criteria for acceptable and unacceptable use-constructions. It was acceptable to use any evidence to construct and test a hypothesis H (about the deflection effect) so long as it could be shown that the argument procedure was reliable or severe—that it would very rarely yield so favorable a result erroneously. Examples abound in the literature. They supply a useful sampling of canonical arguments for ruling out hypothesized causes of an effect of this sort. I will briefly cite a few.

The shadow effect. Alexander Anderson (1919, 1920) argued that the light deflection could be the result of the cooling effect of the moon's shadow. Eddington responded that were the deflection due to this shadow effect there would have had to be a much larger drop in temperature than was actually observed. (It might have been responsible for the high value of the deflection found at Sobral.) Anderson did not give up, but attempted other hypotheses about how the moon's shadow could adjust conditions just enough to explain the effect and save the Newtonian law. These attempts were found wanting, but only after being seriously considered by several scientists (e.g., by Arthur Schuster [1920]). The problem, in each case, was not that Anderson repeatedly use-constructed his hypotheses, but that in so doing he was forced into classically unreliable arguments. The problem, well put by Donald Moyer (1979), was this:

> The available adjustments are adjustments of parameters of trustworthy laws and these adjustments are tightly constrained by the connections among these laws of phenomena. Temperatures, or air currents, or density gradients cannot be adjusted in one law without also adjusting all the other laws where these terms occur as well and this must not introduce consequences not observed. (P. 84)

A test procedure that relies on inconsistent parameter adjustments to get a hypothesis to pass would frequently pass hypotheses erroneously. The test is highly unreliable.

Newall's corona lens. Another N-factor seriously entertained was put forward by H. F. Newall (1919, 1920), that of the intervention of a corona lens. Again, there was a two-fold response, here by the scientist F. A. Lindemann and others. The refraction required to cause the eclipse result, Lindemann (1919) argued, would require an amount of matter many orders of magnitude higher than is consistent with the corona's brightness, and were there enough matter to have caused it, comets passing through the region should have burned up.

Ether modifications. Sir Oliver Lodge (e.g., Lodge 1919) promised that if the Einstein effect was obtained he would save Newton by modifying conditions of the ether with special mechanical and electrical properties; after the results were in, he did just that. (Lodge, a proponent of spiritualism, held that the ether effected contact with departed souls, in particular his son, Raymond.) Strictly speaking, since these hypotheses were constructed by Lodge before the results, it seems that the case satisfies temporal novelty, and so use-novelty. This hardly made Lodge's arguments more impressive. The problem was not *when* Lodge formulated his hypotheses, but that his procedure for passing them required inconsistent parameter adjustments. Consistent adjustments showed that each hypothesized factor N could not have caused the observed deflection. As Lindemann (1919) put it:

> Sir Oliver Lodge has suggested that the deflection of light might be explained by assuming a change in the effective dielectric constant near a gravitating body. This way of looking at it had occurred to me. . . . It sounds quite promising at first since it explains . . . the shift of the perihelion of Mercury as well as the . . . shift of the spectral

lines, if this exists. *The difficulty is that one has in each case to adopt a different constant in the law,* giving the dielectric constant as a function of the gravitational field, *unless some other effect intervenes.* (P. 114; emphasis added)

The kinds of tactics Lodge employed lead many to insist on the UN requirement. Far from striving to steer clear of classic unreliable use-construction procedures, he employed (whether deliberately or not) precisely the kind of rigging that would allow hypotheses to pass, whether or not they were true.

Not that one can see immediately which use-constructions are kosher and which are not—even the constructors themselves cannot do this. This is because one cannot see immediately which ones have arguably passed severe tests. By indiscriminately prohibiting all tests that violate UN, the UN requirement cannot provide an epistemological ground for the reasoning in this dispute, nor for the way it was settled.

What finally settled the matter (around 1921) was not the prediction of novel evidence, but the extent to which known evidence warranted only a construction of the Einstein gravitational hypothesis. This was argued by Harold Jeffreys (1919a, 1919b) (despite his having initially assigned an extremely low Bayesian prior probability to Einstein's law). Jeffreys—one of the last holdouts—explains:

> It so happens that the three known facts, the truth of Kepler's third law, the motion of the perihelion of Mercury, and the displacement of star images, give different equations for the constants, and *the only solution that satisfies those three conditions happens to be Einstein's theory.* . . . It must be accepted as the only theory that will satisfactorily coordinate these facts. (Jeffreys 1919a, 116; emphasis added)

What he is saying is that in order to use the known results (the eclipse effect together with Kepler's law and the Mercury perihelion) to construct a hypothesis, and do so reliably, one is led to Einstein's law of gravity! After reviewing the tests and all the rival explanations, Dyson and Crommelin concluded in the February 1921 issue of *Nature*,[26] which was entirely devoted to the eclipse tests: "Hence we seem to be driven by exhaustion to the Einstein law as the only satisfactory explanation" (p. 788).

What about other alternative hypotheses that may be dreamt up that will not disagree with *H* on any experimental results (either of a given test or of any conceivable test)? What about, say, possible alternative conceptions of space and time that would agree experimentally

26. Dyson and Crommelin, *Nature* 106 (1920–21): 781–820.

with Einstein's law? We already took up this issue in discussing under-determination (chapter 6). It is readily admitted that the 1919 eclipse tests were not severe tests of these alternative conceptions. The eclipse tests were not even considered tests of Einstein's full theory. As Eddington remarked:

> When a result that has been forecasted is obtained, we naturally ask what part of the theory exactly does it confirm. In this case it is Einstein's *law* of gravitation. (Eddington 1919, 398).[27]

It is important to stress, however, that the existence (or logical possibility) of alternative hypotheses that are not themselves tested by a given experiment leaves unaltered the assessment of hypotheses that *are* severely tested. The severity calculation is unchanged. On the informal side this means that we can learn things one piece at a time and do not have to test everything at once. That is why Jeffreys (and others) could laud the eclipse results as finally putting the Einstein law on firm experimental footing, apart from any metaphysical concepts (e.g., about space and time). (See, for example, Jeffreys 1919b, 146.) However Einstein's full theory is modified, the knowledge gained in the severely tested experimental law remains:

> In this form the [Einstein] law appears to be firmly based on experiment, and the revision or even the complete abandonment of the general ideas of Einstein's theory would scarcely affect it. (Eddington [1920] 1987, 126)[28]

Summary and Next Step

Our critique of use-novelty showed it to be neither necessary nor sufficient for severity. This finding discredits the rule of novelty (RN) when viewed as a policy always to be followed to satisfy severity. It also teaches us how UN may be violated yet avoid a possible threat to severity, namely, with a reliable rule for use-constructing hypotheses such as rule R-π. Proponents of UN err by taking a handful of cases in which UN is violated and where the test lacks severity, and then generalizing to eschew all violations of UN.

There are, however, contexts of inquiry where the methodological rules that have been developed to ensure reliability are invalidated

27. One reason for this is that the redshift prediction had thus far not passed a severe test.

28. It is interesting to consider in this connection the recent progress in partitioning theories of gravity and determining which theories are consistent with given experimental results, as reported in Earman 1992. See section 6.3.

when UN is violated. Their reliability guarantees break down when use-constructing is allowed. These contexts comprise an important subset of standard Neyman-Pearson tests. In other contexts, however, Neyman-Pearson methods seem happy to violate rules against use-constructing. For a long time this has caused a good deal of confusion. Let us see if we cannot dispel it once and for all.

CHAPTER NINE

Hunting and Snooping: Understanding the Neyman-Pearson Predesignationist Stance

If in sampling any class, say the M's, we first decide what the character P is for which we propose to sample that class, and also how many instances we propose to draw, our inference is really made before these latter are drawn, that the proportion of P's in the whole class is probably about the same as among the instances that are to be drawn. . . . But suppose we were to draw our inferences without the predesignation of the character P; then we might in every case find some recondite character in which those instances would all agree. That, by the exercise of sufficient ingenuity, we should be sure to be able to do this, even if not a single other object of the class M possessed that character, is a matter of demonstration. For in geometry a curve may be drawn through any given series of points.

—C. S. Peirce, *Collected Papers*, vol. 2, par. 737

To base the choice of the test of a statistical hypothesis upon an inspection of the observations is a dangerous practice; a study of the configuration of a sample is almost certain to reveal some feature, or features, which are exceptional if the [chance] hypothesis is true.

—E. S. Pearson, *The Selected Papers of E. S. Pearson*, p. 127

9.1 Introduction and Overview

The debate about the novelty requirement in the arena of philosophy of science parallels an ongoing methodological debate in actual scientific practice, and the preceding results have direct ramifications for that dispute. That this dispute is alive and well is brought home by the program put forward in Glymour, Scheines, Spirtes, and Kelly 1987 and its subsequent extensions. The dispute concerns a principle often adhered to in statistical testing based on Neyman-Pearson (NP) methods. It is commonly supposed that the NP account, from which my error-statistical account derives, prohibits all after-trial constructions of hypotheses. Indeed, it is typically thought to mandate an even stricter requirement. It is supposed that a key feature of the NP model of tests is that all aspects of the tests, the hypotheses, the sample size, the significance level, and so on must be laid out in advance of running the experiment. All must be *predesignated*. If the predesignation of tests is always required, then in particular the temporal novelty of hypotheses is required. It is never OK to snoop at the data before formulating a hypothesis—at least not if the same data are to be used in testing that hypothesis. So if NP statistics requires predesignation, then it has all the problems of temporal novel accounts. Known data fail to provide good tests of hypotheses—NP theory has what is called an "old evidence" problem.

But is predesignation part and parcel of the NP methodology? Predesignation is part of what might be called "folk NP statistics." In practice it is often good advice. But violating predesignation does not necessarily conflict with NP principles. In fact, many of its own methods violate this predesignationist stance.

Examples of NP procedures that violate predesignation—by violating use-novelty—are those involved in checking the assumptions of an experimental test. The same data may lead to constructing a hypothesis—say, that the trials are not independent—and at the same time may be used to test that hypothesis. The rationale is analogous to the posttrial scrutiny of the eclipse data discussed in the last chapter. In checking if a particular set of data satisfies assumptions, such a double use of data is likely to offer a better test than looking to the data of some new experiment.

Roger Rosenkrantz puts his finger on this apparent dilemma for NP or "orthodox" statistics:

> It is difficult to live within the confines of a predesignationist methodology. Actual orthodox practice fully bears this out; indeed, standard orthodox texts are all replete with post-designated tests. . . . In analysis of variance, for example, upon rejecting the hypothesis that all means are equal, orthodox texts show you how to go on to test other more particular hypotheses about the means suggested by the data. . . . Those same texts—and their number is legion—also show how to test the underlying assumptions of the usual analysis of variance model . . . again using the same data. Similarly, they show how to test underlying assumptions of randomness, independence and stationarity, where none of these was the predesignated object of the test (the "tested hypothesis"). And yet, astoundingly in the face of all

this, orthodox statisticians are one in their condemnation of "shopping for significance," picking out significant correlations in data *post
hoc*, or "hunting for trends in a table of random digits." . . . It is little
wonder that Orthodox texts tend to be highly ambivalent on the matter of predesignation. (Rosenkrantz 1977, 204–5)

Is the NP statistician being inconsistent in banning postdesignation
in the form of shopping or hunting for significance, while condoning
it in certain other tests? Are NP statisticians justified in insisting on
predesignation with respect to certain kinds of tests and estimation
procedures? And what exactly is the NP statistician condemning in
condemning postdesignation? Answering these questions, which have
been at the center of considerable controversy, is the goal of this
chapter.

In chapter 8 I argued that the real aim of novelty is severity and
that the novelty requirement was justified only to the extent that violating novelty precluded severity. Can an analogous move disentangle
the predesignationist puzzle? The natural suggestion would be to propose that condoning violations of predesignation in one set of cases
while condemning such violations in others may be perfectly justified
if it turns out that severity is only problematic in the latter set of cases.
Will this natural suggestion hold up?

I claim that it will. One must be careful, however, in understanding
what it means for "severity to be problematic." Certainly severity
would be problematic if violating predesignation led to a procedure of
passing hypotheses where the severity was low. For then there would
be a high probability of passing hypotheses erroneously, violating the
NP low error probability requirement. But is predesignation necessary
for severity? We already have our answer to that question as well. If
predesignation is necessary for a good test, so is use-novelty. Hence the
contexts in which one can have a severe test while violating UN are
also contexts that yield a severe test despite violating predesignation.
So having shown that UN is not necessary, we have shown predesignation is not.

However, the real NP argument for predesignation differs from the
argument given by the UN proponents we have considered. Those proponents thought UN necessary because its violation was thought to
lead to zero- or low-severity tests. This is not what is going on in the
contexts where NP statisticians insist upon predesignation—although
it is often thought to be. The real argument is that in certain testing
contexts violating predesignation alters the test procedure in such a
way as to require that it be taken account of in assessing its severity.
What is really being condemned, or so I shall argue, *is treating both
predesignated and postdesignated tests alike.*

My focus in this chapter is on an important class of cases in which violating predesignation is condemned or deemed inadmissible by the error-probability statistician. Disagreement about whether predesignation should be required in this class of cases often masks a disagreement about whether error probability requirements matter altogether. The Bayesian (or other holder of the likelihood principle) says no, while the NP theorist says yes. But a debate also persists among users of the standard significance tests whose position on this matter is much less clear. The individual I have in mind does not deny the importance of error probabilities outright but views the predesignationist stance as unnecessarily hamstringing the researcher in actual practice.

This is how I view the position in the work of Glymour, Scheines, Spirtes, and Kelly (1987). Their project is a rare example of a joint effort by philosophers to articulate methods intended for scientific practice—in particular, for discovering causal hypotheses. But my focus is not on their computer algorithm TETRAD; it is on their discussion of what I take to be the main NP stance against postdesignation. This form of postdesignation falls under the class of cases *hunting for statistically significant correlations*. Glymour, Scheines, Spirtes, and Kelly maintain that despite the general adherence to the inadmissibility of postdesignation procedures, the arguments purporting to show their inadmissibility will not hold up. Admittedly, the real argument against postdesignation (to my knowledge) has never been articulated clearly. With the machinery I have been developing it will be seen that the NP statistician *does* have a legitimate objection to postdesignation, at least with respect to the class of cases of interest.

The general outline of the NP objection emerges naturally from the underlying aim of NP statistics. Despite the different interpretations NP procedures are open to, the following error probability principle stands:

> (EPP): An NP procedure of inference is inadmissible if its error probability characteristics are inconsistently reported or if it prevents the determination of valid error probabilities (even approximately).

In the contexts I will be considering, a test's error probabilities are altered when hypotheses are not predesignated. At the same time there are other tests of hypotheses and models that despite being postdesignated do not violate any NP principles. Several tests are even intended for use in cases that would ordinarily be thought to violate predesignation—Rosenkrantz is right. If my analysis is correct, however, then there is nothing inconsistent in this apparent schizophrenia as regards predesignation.

It is important to get clear on the real NP argument regarding postdesignation procedures, because supposing that NP methods rule them out altogether has opened those methods to unjust criticism. Furthermore, dismissing the NP argument as unsound has only made it easier to ignore the very special constraints that have to be satisfied for validly applying NP tests in such cases.

My strategy in appraising the rule of predesignation follows the general one that I advocate for a philosophy of experiment: the value of a methodological rule is determined by an analysis of how its application allows one to avoid particular types of experimental mistakes. But methodological rules are made to be broken.[1] By understanding the function that a rule of procedure has, we can identify the conditions under which it may be violated—even in testing contexts where it is normally of concern. In the opening epigraph of this chapter, Peirce explains that by prespecifying a test, "our inference is really made before" the data are collected. The "inference" that the NP statistician really wants to make before the trial is about the test's error probabilities. What is really of concern, we shall see, is the validity of the "before-trial" probabilistic guarantees.

The NP argument against violating predesignation says simply that if your worry is ensuring tests with high severity, then you must recognize that the manner of data generation can influence the probability of erroneously passing a hypothesis (the severity of the test). We must look at the entire experimental testing context to correctly assess severity. From this point of view, the NP admonishment against violating predesignation may be regarded as a kind of warning to the error statistician that additional arguments—possibly outside the simple significance test—may be required to rule out the error at hand. The tests themselves cannot be expected to provide the usual guarantees. This, I suggest, should be seen as an invitation for the error statistician to articulate *other* formal and informal considerations to arrive at a reliable experimental argument.

On my treatment of this issue it may be objected that I am assuming that it is unproblematic to determine what a test's error probabilities are, thereby discounting the ambiguities in describing a given test procedure. While I do not wish to minimize the problem of how to describe a given experiment—some might identify it with the *reference class* problem—I regard that as a problem of choosing an appropriate

---

1. I am not just playing on a cliché. While obviously methodological rules are made to be followed, an equally important service they perform is to call attention to an assumption or goal that should be met, if not by following the rule, then (at least approximately) by an alternative route.

test procedure for a given inquiry, a problem distinct from the current issue. The basis for choosing a test will depend upon experimental background knowledge acquired from comparable cases, knowledge that would comprise a kind of repertoire of canonical inquiries into errors. Once one has specified the kind of argument from error one wishes to sustain, this background knowledge directs one to appropriate choices of test statistics and corresponding error characteristics. (I return to this issue in chapter 11.)

9.2 HUNTING FOR STATISTICALLY SIGNIFICANT DIFFERENCES

As is so often the case with contemporary statistical disputes, a history of this debate already exists from early applications of significance tests. A good source for the debates that took place nearly forty years ago is Morrison and Henkel's classic volume *The Significance Test Controversy.* Much of what I would wish to say emerged in these early discussions. I will focus on two contributors to the volume, both social scientists.

Leslie Kish's Example

Leslie Kish (1970, 138) refers to a study regarding infant training. The study, done in the late 1940s, sought to investigate a variety of infant training experiences regarding nursing, weaning, and toilet training that according to Freudian psychological theories of the day were thought advantageous for personality adjustment. In records on children spanning several years, a number of high (statistically significant) correlations were found between children exposed to a certain kind of infant training T and various personality traits. Among those subjected to training T "gradual weaning," for example, there was a significantly higher proportion of children with "high social standards" than among those abruptly weaned. Among the other high correlations found were between "no punishment for toilet accidents" and "good school relations," and between "late bladder training" and "little nail biting." The question is whether the evidence on correlations provides good evidence for the existence of a real effect in the population of children or whether the observed correlations are spurious or "due to chance."

The Key Question. The fact that the researchers searched among many factors for large correlations adds a new twist to this question. The question, which I pose to the reader, is this: Is it relevant to the assessment of the evidential import of the observed correlations that they are the outcome of a procedure of searching for effects large enough to be statistically significant? The Neyman-Pearson error statistician says yes. The question is why?

Let us use Kish's example to describe a procedure sometimes called "hunting with a shotgun." Here the sample size is fixed ahead of time, and even the cutoff for rejecting a test hypothesis may be preset, say at the .05 level. What varies is the hypothesis chosen for testing and reporting. (The hypotheses themselves may or may not have been thought up in advance.) Each such hypothesis asserts that some property or "treatment" T is genuinely correlated with some factor F in some population. Each factor is dichotomous, each subject has it or not. Before getting to the "hunting" aspect, let us recall how a test of such a hypothesis might go in nonhunting cases.

The situation shares several features with example 5.1 on birth control pills. The "null" or test hypothesis H is that the observed correlation is merely due to chance—that in fact the incidence of the effect (cancer, high feeling of belonging) is no different among those treated (with the pill, with gradual weaning in infancy) and those not so treated.[2] (In Kish's example the difference is sought in either a positive or a negative direction.) The null hypothesis is rejected when the observed correlation is sufficiently statistically significant (e.g., at the .05 level). The hypothesis that passes—let us abbreviate it as H^*—asserts that there is a genuine correlation in the population between the factors of interest (in either or in one direction).

As in example 5.1, a simple difference measure can be used for a statistic measuring observed correlations, and to avoid complexities I will stick to that case in this chapter. We observe the proportion of Ts that are Fs and the proportion of not-Ts that are Fs and record the difference D. We can compute the statistical significance of D by considering the number of standard deviations by which it differs from 0 (0 being the expected difference, given the truth of the null hypothesis). The 2-standard-deviation cutoff corresponds to a .05 level of statistical significance (see note 2).

The same mathematical procedure for calculating statistical significance is available even if the particular factor F is specified after hunting for large correlations. At issue is whether it is misleading or

2. In example 5.1, only the existence of a positive difference was being tested. It was a *one-sided* test. Kish's example is a *two-sided* test, because it looks for differences in either direction. A 2-standard-deviation difference in one direction corresponds to a significance level of approximately .025 (I elsewhere round it to .03), so in two directions it corresponds to a significance level of about .05. That is why the two-sided significance tests discussed here have .05 significance levels.

fallacious to report the statistical significance of a difference *in the same way* as in the case where the hypothesis to be tested is prespecified. The "hunter," as Kish calls him, thinks not, and Kish alludes to the infant training example, as discussed by William Sewell (1952), to illustrate how the hunter would get into trouble.

The researchers in the infant training study conducted 460 statistical significance tests! Out of these they found that 18 were statistically significant at the .05 level (or beyond), 11 of these were in the direction expected by the popular psychological account. Sewell (1952) denies that we should be just as impressed with the 11 statistically significant results as we would be if they were the only 11 hypotheses to be tested. Kish agrees and explains why.

Note that the hunting procedure is an example of what I called a "use-constructing" test procedure in chapter 8. Introducing some abbreviations for this simple example will make it easy to characterize the more general argument later. For each factor F_j, calculate the difference statistic D_j, between the proportions with F_j among those given infant training T and those not given T. Finding a factor, F_j, on which the experimental subjects show a statistically significant difference would lead to testing the postdesignated null hypothesis:

Null hypothesis H_j: In the population of children, treatment T is not correlated with factor F_j.

Let us focus on just one type of infant training T—gradual weaning. Suppose, for example, that a statistically significant difference is observed between gradual weaning and factor F_6, say, "a strong feeling of belonging." In that case, the procedure directs one to test the corresponding null hypothesis, H_6:

H_6: Gradual weaning in babyhood is not correlated with a strong feeling of belonging in older children.

Next, the null hypothesis H_j (in this case, H_6) is then rejected if the observed difference is statistically significant at the .05 level. The hypothesis that passes when H_j is rejected is the non-null hypothesis H_j^*:

Non-null hypothesis H_j^*: There is a genuine correlation between gradual weaning and factor (or personality trait) F_j.

But the only time H_j is tested according to this procedure is when the observed difference is statistically significant! So on this procedure, whenever H_j is tested, it is rejected.

What is wrong with this? In the actual study, out of 460 attempts to hunt for statistically significant correlations, 18 were found significant at the .05 level or beyond (11 in the expected direction "on the basis of psychoanalytic writings"). Kish (1970) remarks:

> Note that by chance alone one would expect 23 "significant" differences at the 5 percent level. A "hunter" would report either the 11 or the 18 and not the hundreds of "misses." . . . After finding a result improbable under the null hypothesis the researcher must not accept blindly the hypothesis of "significance" due to a presumed cause. Among the several alternative hypotheses is that of having discovered an improbable random event through sheer diligence. (P. 138)

Keep in mind Kish's statement about what the "hunter" would report. It is not just that the hunter postspecifies (and tests) hypotheses to fit samples, but that the hunter, or one who endorses being a hunter, is saying that *doing so calls for no difference in interpretation*. And because of that, the hunter reports the statistically significant cases *just as if* the successful cases had been predesignated.

However, if one's answer to the "key question" I posed at the start of this section is yes, then one does *not* think that the postdesignated cases should be reported just as if they had been predesignated. One thinks they should be distinguished. The NP error statistician distinguishes between the two on the very grounds Kish cites. But why, one might ask, should the import of the evidence depend upon whether the hypothesis is set out in advance? If the hypothesis, say H_6, had been set out in advance, and a 2-standard-deviation difference observed, we would have computed the statistical significance level in the usual way (.05). Why this change because it was one of the successfully hunted ones?

For the NP or error statistician, the altered interpretation is called for because the test procedure in the postdesignated case is very different from the case in which H_6 is preset as the hypothesis to test. What is allowed to vary—and hence the set of possible outcomes—is very different. With H_6 predesignated, the possible results are the possible differences in F_6 rates between the two differently trained groups— that is, the different values of statistic D_6. In the postdesignated case the possible results are the possible statistically significant F factors that one might hunt down. This difference is reflected in the difference in error probabilities. Let us turn to a second contributor to Morrison and Henkel (1970), Hanan Selvin.

Hanan Selvin's Example

Selvin, in an article first published in 1958, gives a very useful capsule statement of the problem:

When the hypotheses are tested on the same data that suggested them and when tests of significance are based on such data, then a spurious impression of validity may result. The computed level of significance may have almost no relation to the true level. . . . Suppose that twenty sets of differences have been examined, that one difference seems large enough to test and that this difference turns out to be "significant at the 5 percent level." Does this mean that differences as large as the one tested would occur by chance only 5 percent of the time when the true difference is zero? The answer is *no*, because the difference tested has been *selected* from the twenty differences that were examined. The actual level of significance is not 5 percent, but 64 percent! (Selvin 1970, 104)[3]

So more than half the time one will be designating an observed difference (or correlation) unlikely to have been the result of mere chance error when in fact it is a result that easily (commonly) results from chance.

Selvin's distinction between "the computed" and "the true or actual" significance levels is a useful way of making out the NP argument, and it merits some additional clarification.

Computed versus Actual or True Significance Levels. The *computed level* of significance of a difference is the usual one: the improbability of observing such a large difference in the proportion with trait F_j given that in fact there is no real correlation, that is, the null hypothesis H_j is true. The computed level would be .05 if the observed difference were at least 2 standard deviations (the chance or null hypothesis being that the difference is 0). In the case of a prespecified test of null hypothesis H_j, the computed level equals the actual error probability of the procedure—the actual significance level. But the actual significance level differs if H_j arose from a procedure of searching through 20 factors on which the groups might be correlated. In this case, the *actual significance level* would be the probability of observing at least one such 2-standard-

3. As Selvin notes, this can be calculated approximately by considering the probability of finding at least one statistically significant difference at the .05 level when 20 independent samples are drawn from populations having true differences of zero, $1 - P$ (no such difference). This is $1 - (.95)^{20} = 1 - .36$. In general, the probability of obtaining at least one statistically significant outcome (in either direction) with N independent tests and a 2α (computed) significance level is $1 - (1 - 2\alpha)^N$. This would give the *actual significance level*, that is, the actual probability of erroneously affirming a genuine correlation. The assumption of independent samples is made here for simplicity. With real data on a single population, Selvin remarks, this independence assumption does not hold "and the computation of the true level of significance would be extremely difficult" (ibid.).

deviation difference, given that there is no genuine correlation (in the population) on any of the 20 factors. Using various assumptions, Selvin calculates this probability to be .64.[4]

More generally, the probability of error in the postdesignated case is the probability of finding some such α-significant correlation *or other*, given that no real correlation exists. In both cases, one minus the probability of error is the severity of the test (for passing the hypothesized correlation). Thus, while $1 - \alpha$ is the severity (for passing H_j) when H_j is prespecified, severity is no longer $1 - \alpha$ in the postspecified case. In the postspecified case the actual significance level, the actual probability of erroneously finding some such α-significant correlation, is not generally equal to α. (Recall the altered severity criterion for hypotheses constructed from the data in chapter 6, section 6.6.)

It may be objected that in calling this the "actual" significance level I am taking sides in favor of one description of the "actual" test procedure—one that takes into account the fact that searching has occurred. I am, but maintain that this aspect of the procedure cannot be ignored given the aim of the statistical significance test chosen. Remember, I am distinguishing the appropriateness of the test chosen (for a given inquiry) from the error probabilities, *given* that that test is chosen. One chooses a type of test corresponding to the type of argument from error that one wishes to sustain. In the present case, the interest is in arguing from error to infer a genuine correlation. "Hunting" raises a problem because it may invalidate the desired argument.

Ronald Giere's Example

Ronald Giere (1969) generalizes this kind of argument against hunting for a corresponding procedure for estimating a population proportion. Here it is imagined that we hunt through random samples for a property shared by all of the n members in the sample. Finding such a property, we construct a confidence interval estimate with some high confidence level $1 - \alpha$, say, .95 (Giere uses q). As we know, a standard .95 confidence interval estimation procedure includes the true population proportion 95 percent of the time—whatever its value might be. It is possible, as Giere shows, to describe a series of applications of the estimation procedure such that the probability (or the ex-

4. Selvin's calculation, discussed in note 3, is just an application of the Binomial model that we have already considered. As always, each outcome is either a "success" or not. Here, however, a successful outcome is a test result statistically significant at level .05. The probability of getting no successes in 20 independent trials is the probability of not getting a significant difference in one trial, namely, .95 raised to the twentieth power, giving .36. One minus this probability is .64.

pected relative frequency) of successful estimates is not 95 percent but 0! As Giere remarks,

> This will be sufficient to prove [the inadmissibility of this method] because Neyman's theory asserts that the average ratio of success is independent of the constitutions of the population examined. (Giere 1969, 375)

Giere shows how to construct populations that, in effect, illustrate Peirce's point at the outset. Take a population of *A*s and to each set of *n* members from this population assign some shared property. The full population has *U* members where $U > 2n$ members. Then arbitrarily assign this same property to exactly $U/2 - n$ additional members.

> Given a sufficient store of logically independent properties, this can be done for all possible combinations of *n A*'s. The result is a population so constructed that while every possible *n*-membered sample contains at least one apparent regularity, every independent property has an actual ratio of exactly one-half in the total population. (Ibid., 376)

More generally, for any postdesignated selection of the property to be estimated whose population frequency is statistically dependent on its frequency in the observed sample used to arrive at the estimate, "one can always construct a possible series of populations leading to an expected ratio of successful estimates differing from [the predesignated confidence level $1 - \alpha$]" (p. 376). (In Mayo 1980, I give an analogous argument for tests.) This difference corresponds, in the case of postdesignated tests, to the difference between the actual significance level and the computed (or predesignated) level.

Summary

To sum up this section, the dependency of the correlation to be tested, or the proportion estimated, on the correlation or proportion that is observed results in changing the experiment—at least, so far as the NP error statistician is concerned. In the standard statistical significance test (on difference in proportions) where the factors whose correlation will be tested are predesignated, the possible outcomes are the possible different degrees of significance that might be observed with respect to that single predesignated correlation of interest. In the case of hunting for a statistically significant difference, in contrast, what is fixed is the particular level of statistical significance for which one is going to hunt. What varies now are the possible factors or possible correlations that might turn out to be statistically significant at that level.

This formal difference in error probabilities corresponds to an informal difference in the way the procedure can err. The hunting procedure has more and different ways of erring than a procedure of testing a predesignated hypothesis. The ability of a test to protect against the errors of one kind of procedure may have no relation to its ability to protect against errors in some other procedure that has many more ways of going wrong. This is what the NP stance against violating predesignation amounts to. Why, then, would anyone who answered yes to the question at the start of section 9.2 object to the NP stance?

9.3 "NO PEEKING!": GLYMOUR, SCHEINES, SPIRTES, AND KELLY

Glymour, Scheines, Spirtes, and Kelly, henceforth abbreviated as GSSK (1987), develop a computerized procedure for using observed correlations in data to construct a (linear) model that fits the observed correlation or difference. I will use their term "model" interchangeably with my "hypothesis." Although their approach is largely intended as a (computerized) method for *finding* models, presumably to be tested on other evidence, the authors claim they "also believe that there is often nothing wrong with using one and the same body of data to discover a theory and to confirm it or test it" (p. 46). The discovery procedure they provide is essentially a computerized program for carrying out the postdesignated searching procedure described above. (The correlations looked for will typically involve many more variables, but this will not alter the points that they or I wish to make.) The bulk of social scientists and statisticians, steeped as they are in NP methodology, object to such a hunting procedure on the grounds that it entails using the same data both to find as well as to test statistically significant correlations. Because they comprise a key group for whom their method of causal modeling is intended, GSSK are led to examine the basis of such objections.

Under the apt subtitle "No Peeking," GSSK (1987) consider what they take to be the best arguments for prohibiting this kind of double use of data, and find them wanting (p. 45). My focus is on the argument that they regard as most promising. They dub it the "worst case argument" and model it on Giere's argument above. We can use it to elicit what I view to be "the real NP stance on predesignation." Toward this end it is sufficient to keep to the type of correlation hypothesis described above, leaving to one side the additional difficulties of warranting causal hypotheses (which we might imagine to be higher up in the hierarchy of models).

The Worst Case Argument

GSSK (1987) spell out what they view as the NP argument against hunting for statistically significant correlations in terms of an argument against their computerized search procedure. The NP argument would begin with a fact that they recognize, namely, that the procedure of hunting for statistically significant correlations "will produce *some* model, even for data that are in fact randomly generated from independent variables" (p. 55). That is, the hunting procedure would find some statistically significant correlation even if there were none. Moreover, they recognize, it would do so with high probability. Nevertheless, GSSK want to reject the NP objection to such a procedure by denying the soundness of what they regard as the NP argument. Let us quote directly from their gloss on the NP argument against computerized hunting procedures:

> 1. Computer-aided heuristic searches for statistical models must examine the data for statistical dependencies among the variables, search for the model or models that best explain [fit] those dependencies, subject the models thus obtained to statistical tests based on the data, and output those models that survive the tests.
> 2. No procedure for searching for hypotheses is acceptable if there are circumstances in which it is very probable that that procedure will yield a false conclusion.
> 3. For any procedure as in 1, a number r of independent random variables and a sample size n can be found such that it is very probable that a sample of size n will show k statistically significant correlations (or other statistic) among h of the r variables, for some number h and for some number k. . . .
> 4. In the circumstance described in 3, it is very probable that a procedure, such as described in 1, will output false hypotheses.
> 5. Therefore, by 4 and 2, a computer-aided heuristic search procedure is unacceptable. (GSSK 1987, 55–56)

Does this argument capture the NP objection? On a first reading it appears to distort the NP argument, but this is mainly because it is not in the language of NP criteria for judging *inference* procedures. The conclusion, for example, might make it sound as if the NP statistician disallows searching for hypotheses not specified beforehand. In actuality, the only prohibition relates to a procedure of first using data to arrive at a correlation that is statistically significant at some level α, and then using that same data to test the corresponding null hypothesis H, that there is no real correlation. However, that is precisely what GSSK mean by a computer-aided heuristic search procedure. We can avoid misunderstandings by spelling out, in the conclusion of the argument (clause 5), the procedure that is being declared NP inadmissible. That change calls for corresponding adjustments to the premises as well. Once these alterations are made, this argument turns out to be a plausible rendering of the NP objection; but contrary to what GSSK suppose, the resulting argument is sound.

Let us turn to the adjustments. To begin with premise 1, we need to be clear about their phrase that the searches "output those models that survive the tests." First, the output of any NP procedure is never just a hypothesis or model or estimate by itself, it is always accompanied by a statement of the error characteristics of the test or estimation procedure. Here the error characteristic is the significance level of the test, α, such as .05. Second, it is rather curious to say that the outputs here are hypotheses that *survive* statistical tests. The outputs are those hypotheses of form H_j^*: treatment T is correlated with F_j. And this output occurs when the corresponding *null hypothesis* H_j does not survive, that is, when H_j is rejected by the statistical significance test. Accordingly we may rewrite premise 1 as follows:

> 1. Computer aided heuristic searches . . . search for statistical dependencies among the variables, reject those null hypotheses H_j if the data show a difference d_j that is statistically significant at some small level α (e.g., .05) and output the non-null hypothesis H_j^* (at level α).

(In more abbreviated form, the procedure is to search for variables for which observed differences d_j satisfy $P(D_j \geq d_j|\ H_j) \leq \alpha$, declare d_j statistically significant at level α, reject H_j and pass H_j^*.)

In premise 2, once again, "procedure for searching" has to be filled out as the procedure just described. Premise 2 also talks about yielding "a false conclusion," which could be ambiguous were it not for premises 3 and 4. Those premises make it clear that yielding a false conclusion means outputting a statistically significant correlation (at some small level α) even though the variables in question are *not* correlated (in the population) but are independent. (The false conclusion here is a type I error: rejecting the null hypothesis even though it is true.) Premise 2 becomes

> 2. No procedure for significance testing is acceptable if there are circumstances in which it is very probable that that procedure will reject the null hypothesis H_j at a low significance level α (and pass the non-null hypothesis H_j^*) even though the null hypothesis is actually true.

Premise 3, while stated a bit confusingly, simply generalizes the situation we discussed in section 9.2. Since the population imagined is one where all the variables are independent, to assert a statistically significant correlation, that is, to reject the chance hypothesis, is to commit an error. Premise 3 describes a circumstance such that the searching procedure has a high probability of committing such a type I error. This is affirmed in premise 4. It then follows, in 5, that the significance test consisting of the procedure of searching described in premise 1 is unacceptable or inadmissible, on NP grounds.

So, fleshing out the argument given by GSSK turns out to give a reasonable rendition of how an NP argument might go. Will it hold up? GSSK think not.

Their Response and an NP Rejoinder

The main objection of GSSK to the argument concerns premise 2. Premise 2, they object, assumes that a procedure ought to be judged by the worst imaginable case, namely, the circumstances described in premise 3.[5] Why consider the worst possible case, they ask, where all of the variables searched are actually independent?

> In the majority of cases researchers are pretty confident that the statistical dependencies they find are due to some causal structure or other. . . . If the investigator were not strongly inclined to think that there is some explanation other than chance (or bad measurement design) for the patterns found in the data, a causal model would not be sought in the first place. Unless the researcher thinks there is a large probability that the dependencies in the data are spurious, there is no sufficient reason not to use the data to search for the best explanation of it. (GSSK 1987, 56–57)

They continue that,

> of course, some of the correlations found may be due to chance, and that is the more likely the smaller the sample size in proportion to the number of variables considered. The investigator should certainly

---

5. GSSK (1987) also object that premise 2 "puts all of the weight in judging a procedure on the desirability of avoiding *false* theories" (p. 56). But NP criteria also consider the probability that a true hypothesis is accepted—by seeking a small type II error probability. It is just that the situation to which this NP argument is referring is one where it is given as part of the procedure that some hypothesis (or estimate) is going to be accepted. Thus the error of concern is the probability of erroneous acceptance. Of course nothing in the NP argument says there is anything wrong with searching for statistically significant results for the sake of getting some hypotheses to consider, and then testing them on other data.

take account of that fact and, where appropriate, test a model on new
samples. (P. 57)

I will return later to this continuation, which suggests a different
tack than the earlier parts of the passage. What about the first part of
their response? For starters, merely being ignorant of spuriousness is
not sufficient for most researchers to be "pretty confident that the statistical dependencies they find are due to some causal structure." If,
alternatively, this confidence is well founded from background knowledge, then why are the researchers running statistical significance
tests? Perhaps what GSSK mean is that, most of the time, the researchers have a strong belief that any dependencies found to be "statistically
significant" are real: they assume there's a causal story to be had; the
function of the test is to tell them which factors are really connected.

For the test to tell them which are really connected, however, it
has to be able to calculate the actual statistical significance—the actual
error probability of the test procedure. That is, the researchers' reliance
on an analysis of statistical significance to arrive at a reliable argument
from error depends critically on that analysis being done correctly. Significance tests are useful only to the extent that they can be relied on
to alert us to the lack of statistical significance when there is none.
Tests *should* declare the statistical significance level high and *not* low,
when the correlation observed is of the sort that would, very probably,
disappear in subsequent trials. This ability to make reference to what
would probably occur in additional trials is altogether central to NP
principles.

The error statistician might put the rejoinder this way. This type of
statistical significance test is designed for a case in which one is concerned with ruling out the error that an observed correlation is merely
due to chance. It is designed for a case in which one wants to know
what it would be like if this were a population in which it *would* be
an error to declare correlations genuine. As we have seen in previous
chapters, one can then use the sample to see whether this error can be
ruled out.[6] If your case is not one in which there is a need of such
information, then you are not in need of this particular NP test. But
premise 2 does not rest on any claims about the *relevance* of significance
tests for a given research situation. Claims about researchers' confi-

6. What is wanted in such a situation is a standard signal or warning that an
observed correlation is of the sort that can often occur by chance: that it is the sort
of result frequently generated even in working with a population in which the
factors are independent. The actual significance level is a measure of this frequency,
so its being high indicates that such a correlation is highly probable by chance
alone.

dence, warranted or not, are quite beside the point of the NP argument, which is really just an argument, a demonstrable one, about the *properties of these testing tools*. The argument shows that if you change the test procedure the error probabilities change, and if you report significance levels in the usual way—if you are a "hunter" in Kish's sense—then you are going to get your error probabilities wrong.

The upshot of premise 2, we can imagine the NP test saying, is this: "I cannot do my job if you are a hunter. My logic breaks down. Don't blame me if you declare correlations real far more often than the computed level arrived at in searching for significance." Or as Selvin (1970) put it, if you apply statistical significance tests to a hunted hypothesis, "a spurious impression of validity may result. The computed level of significance may have almost no relation to the true level" (p. 104).

What GSSK call the "worst case" is precisely the case that the NP statistical test *must* consider here, because that is just what the type I error would be. It is often said that no one really thinks (point) null hypotheses are exactly true. But it is a mistake to regard this as a criticism of their use in tests. We use them in getting the probability of a type I error, or the significance level, because we seek an objective way of learning how far from true they are. Hunting, on the other hand, allows correlations to be described as improbably far from what would be expected due to chance, when in fact they are quite typical of what would be expected even if chance alone were responsible.

Using Computed Levels of Significance as Fit Measures. Granted, one may only be interested in giving a kind of summary measure of the observed correlation, and the computed significance level could be used for this. (The smaller the computed level, the larger the deviation from the corresponding null hypothesis H_j. This corresponds to a greater "fit" with the alternative hypothesis H_j^*.) However, the process of giving a summary measure of fit is no longer an NP test process. An NP test process always asks about the error probability of the observed correlation or fit measure—it would ask about the actual significance level. Without such an error probability, *there is at most a data summary and not a statistical inference* from the observed correlation to the population. To such a data summary the whole question of NP inadmissibility in premise 2 would not apply. Nevertheless, the soundness of this premise, which after all pertains to NP *tests*, still stands.

The Honest Hunter: Defeating the Worst Case

We have justified the argument of the NP inadmissibility of postdesignated or searching procedures of significance testing. The objection of GSSK exposed no unsoundness in the NP argument—once that argument is properly understood. However, the later part of GSSK's statement quoted earlier, as well as their subsequent research efforts, leads me to suspect that GSSK agree. For there they allude to the advisability of (a) ensuring a large enough sample size relative to the number of variables to be hunted or (b) testing on new samples. Both (a) and (b) presuppose taking seriously the threat posed by hunting for statistical significance. Avenue b is tantamount to setting out a new test and not violating predesignation altogether. Avenue a, however, may be seen as a way of defeating the worst case. I want to pursue this avenue a bit.

Let us use the designation "hunters" to refer to those who engage in "hunting with a shotgun" and allow postdesignated hypotheses about correlations to be reported *just as if* they were predesignated. One might agree that the NP argument above is a sound argument against being a hunter, but deny that this bars all postdesignated tests that double count data. It does not bar the practices of what we might call "honest hunters." Honest hunters report, as far as possible, the true or actual significance level, taking into account the way this is altered by the fact of hunting. In suggesting avenue a, however briefly, GSSK seem to be taking the line of the honest hunter.

NP statisticians can have no *in principle* objection to hunting and reporting the actual significance level[7]—although they may have a practical one, which I will return to in a moment. Quite the contrary, that is what the NP statistician would recommend in cases that violate predesignation. To state it more generally, avenue a advises that by appropriately specifying the test (number of variables searched, sample size, the significance level required), the actual significance level, even in postdesignated tests, may be sufficiently small.

One way that can happen is if many of the hypotheses tried turn out to be statistically significant. Let us return to Selvin. He explains how "curiously enough" the same argument against improperly using significance tests in postdesignation cases

> can be extended to show how the tests might legitimately be used on such hypotheses. Consider once more the twenty differences drawn from populations where the true differences are zero. We have seen that the probability of *at least one* difference "significant" at the 5 percent level is 0.64. By similar calculation it can be shown that the prob-

7. That is so whether or not it is high or low. It cannot bar an entire procedure because it is possible for it to be applied in a case with large error probabilities, for that is true for all NP procedures. It may say it is not a particularly good test, but not that it is inadmissible.

ability of at least *two* "significant" differences is 0.26, that the probability of at least *three* is 0.07, and that the probability of at least *four* is
0.01. In other words, if one examines twenty differences and finds
four or more "significant" at the 5 percent level, then the *set* of differences is significant at the 1 percent level, since this combined result
would have happened only one time in a hundred if the true differences were zero. (Selvin 1970, 104–5)

Several caveats remain. First, while one can reject the worst case
here, one still cannot say any particular hypothesized correlation has
passed a severe test. Second, as Selvin notes, sustaining this argument
would require carefully considering correlated biases and the lack of
independence of the 20 factors (p. 105). For ease of computation, Selvin does not take these into account. Third, here the number of differences looked at was fixed at 20. If the number of variables sought could
be open-ended—in a diligent hunting expedition—then it would be
far more difficult to get a low error probability. It may not even be clear
how to determine what the error probability is in such an open-ended
case. Nevertheless, the honest hunter could argue that particular contexts impose a limit on the possible variables and corresponding
hypotheses.[8]

So the task for honest hunters might be put as that of finding ways
of showing that the overall error probability is fairly low. I take this to
be the idea pursued in subsequent work by GSSK. By means of Monte
Carlo simulations, and considerations of background constraints
(based on knowledge of which variables are or are not related), they
have made progress in investigating the reliability of their search procedures. The impetus of these investigations, in my view, is to show
that their use-construction rules are of the sort that I called R-π rules,
with fairly high severity π in chapter 8, section 8.5. The NP philosophy

8. One should not overlook a related and very serious problem that can arise.
Here we are imagining that the analyzer knows all of the tests attempted. The situation is entirely different if one must resort to rounding up positive and negative
results from the literature. The problem, often called the "file drawer problem,"
is that nonstatistically significant results may remain in file drawers, never to be
published. Because these negative results would not be counted, a much higher
proportion of statistically significant results would be found than actually exist. This
is a good example of a canonical mistake.

Robert Rosenthal, a leader in the relatively recent area of "meta-analysis," discusses how one might "estimate the degree of damage to any research conclusion
that could be done by the file drawer problem" (1987, 223). This attempt to estimate and subtract out the effect of studies remaining in file drawers is, in its intent,
very much in the spirit of the error statistical program.

has no valid argument against postdesignation in such cases. (But see note 9.)

C. S. Peirce, that arch-predesignationist, actually anticipated the gist of the responses of the honest hunter. Perhaps this is not really surprising because Peirce in many ways seemed to have anticipated the appropriate construal of NP methods—a thesis to be pursued in chapter 12. Peirce discusses an important modification of the rule of predesignation, namely, when it is not necessary.

> Without any voluntary predesignation, the limitation of our imagina-
> tion and experience amounts to a predesignation far within those
> limits; . . . thus . . . if the number of instances be very great . . . the
> failure to predesignate is not an important fault. . . . So that if a large
> number of samples of a class are found to have some very striking
> character in common, or if a large number of characters of one object
> are found to be possessed by a very familiar object, we need not hesi-
> tate to infer, in the first case, that the same characters belong to the
> whole class, or, in the second case, that the two objects are practically
> identical. (Peirce 2.740)

A Tactical NP Objection

One can articulate another kind of NP objection to hunting for correlations. Although it too can be seen to have its basis in what I called the error-probability principle (EPP), it is not an argument about the inadmissibility in principle of violating predesignation. The objection now is more practical: it is so difficult to figure out what the actual error probabilities are in postdesignated cases, this objection goes, that they are not recommended by this school of inference. Even adjusting the test specifications to get reasonable significance levels rests on very slippery assumptions or requires too-large sample sizes to be practic-able. Nevertheless, new uses of computer-driven Monte Carlo simula-tions might get around these tactical criticisms. To their credit, GSSK, in their most recent work, appear to be heading in that direction.

9.4 THE CREATIVE ERROR-THEORIST

Some practitioners may feel dissatisfied at what I have provided thus far. They may grant that the NP theorist is correct to charge that the error probability guarantees of NP tests break down in hunting proce-dures. They may likewise grant the validity of calling for elaborate ad-justments to significance levels, sample sizes, and so on to ameliorate the problem of high error probabilities in hunting procedures. But in point of fact they are still confronted with the realities of their inquiries, and being an honest hunter by the straight and narrow path still prevents them from doing the kinds of things that it seems they ought to be able to do with the evidence they have. If one keeps in mind that the ultimate goal is a severe argument from error, then it is possible to go further, while remaining within the NP school. The previous chapters give us a head start on this problem.

This takes us into the realm of informal error probability arguments, and hence from formal NP statistics into the broader realm of the error statistician. Consider a case in which a violation of predesignation results in the actual significance level being high (and thus, the severity being low). The honest hunter must report the actual significance level. So the statistical significance test analysis does not provide a severe test of the reality of some correlation. But that is not the only kind of analysis possible, even without getting more data. (Of course, if it were feasible to get more data, this might be desirable.) Even if none of the quantitative NP tools has anything more to offer (at least not at present), the error-statistician's tool kit certainly might. What is needed is an argument from error, an argument to rule out error. If it is remembered that that is the underlying rationale for NP inferences in the first place, then arriving at or approximating some such argument must be countenanced on error-severity grounds.

Attention to the inadmissibility argument spelled out above alerts the researcher that the significance test does not license a certain inference. It says nothing about other arguments that might be put forth. The error statistician requires only that it be able to be shown that the argument used is reliable in our sense. (The researcher cannot just say that he or she feels very strongly about the conclusion.) Nothing in what we have said precludes assessing the reliability of some other method that a researcher might custom-design. It might well be shown that the custom-designed method constitutes a reliable method for inferring a type of correlation. It might be shown to be a high severity use-construction rule (R-π), as defined in chapter 8.

Recall our discussion of the informal arguments from coincidence in chapter 3—how they justified inferring that an effect is real, that it will not go away. Hacking, for example, presented such an argument for dense bodies. The same was seen in Perrin's argument for coincidence for the causes of Brownian motion. At this stage of their arguments there was no need to formally define a test statistic with quantitative error probabilities. They could arrive at arguments that fairly well rule out the error of mistaking an artifact for a real effect. The informal calculation of severity mimics the formal one.

Of course, whether one is relying on a formal or an informal argument from error, one must be careful not to infer more than the argument warrants. Having knowledge of a real effect is not the same as having knowledge of an important effect; much less is it the same as knowing about its specific cause. The need for a piecemeal breakdown into arguments from error, and the need to limit the inference to what is strictly warranted by each argument, must be respected (chapters 5 and 6).[9]

The quantitative NP test procedures serve as canonical models of error—so do NP inadmissibility arguments. Canonical models of error, recall, are exemplars, of both admirably high and infamously low reliability. The "worst case" scenarios, as in the examples of Kish, Selvin, and Giere, are examples of the latter. They demonstrate how hunting expeditions can lead you terribly astray. The lesson is that the onus is on you to show how you are getting around that possibility. The situation is similar to what happens in retrospective analyses of correlations. Knowledge of how that can lead one astray has given rise to a host of procedures for reliably inferring correlations retrospectively. In a much less systematic way, individual researchers, especially in medicine, develop informal procedures for learning from correlations that are only noticed from "peeking" at the results of medical trials. Some oncologists hold that the major advances in cancer chemotherapy were made based on retrospective studies of small groups of patients (Greenspan 1982, 8). However difficult it may be to arrive at these approximate or qualitative error probabilities, if they are arrived at and are valid from a frequentist point of view, there are no error statistical grounds for condemning them.

9.5 CONCLUSION

The real NP argument against postdesignated significance tests is that if the same data are used both to construct as well as to test statistically

9. Even if one manages to obtain a low actual level of significance and thereby pass the nonchance hypotheses severely, this severity does not carry over into particular interpretations of the correlation found. These distinct inferences (we may locate them in a model above the experimental hypothesis of a real correlation) introduce a distinct set of possible errors.

For example, in the social sciences, as Paul Meehl has steadily warned, genuine correlations can be found among nearly all variables "since in social science everything correlates with everything to some extent" (1990, 207). This fact—Meehl calls it the "crud factor"—introduces an important canonical error: when the crud factor is high, a test that takes evidence of a real correlation as evidence of a particular causal hypothesis would often pass causal hypotheses erroneously—the test would have poor severity. Meehl makes the intriguing proposal of estimat-

hypothesized correlations, the actual probability of erroneously declaring a correlation to be genuine—that is, the actual significance level—differs from, and may be much greater than, the computed significance level. Using the computed significance level in postdesignation cases forces one to adopt a different interpretation of a significance level, one that conflicts with the intended interpretation and use of significance levels (as error probabilities).

The upshot of the real NP argument is a warning: since violating predesignation may alter the actual significance level (by altering the test procedure), it is invalid to report the results *in the same way* as if hypotheses were predesignated. Now the high actual significance level corresponds to low severity, for it means that there is a high probability of affirming the existence of a real correlation erroneously. Hence the justification for the NP warning is that if one fails to heed it, tests will be construed erroneously as having high severity.

One could suggest, as I believe GSSK do, that the computed significance level be calculated simply as a kind of measure of fit and then give some other report of the overall error probability. In getting these other reports, they also allude to various background considerations or "constraints" regarding which variables are or are not connected. If these other reports of reliability are valid from a frequentist point of view, then they may certainly be sanctioned and even welcomed by the error probability statistician. It is important to see, however, that this is not a defense of hunting in standard significance tests, quite the opposite. It is to grant that the NP inadmissibility argument necessitates some wholly other kind of test or analysis. I take it that this is what GSSK are really hunting for.

The examples we have considered of hunting for statistical significance should be viewed as canonical models of error—as classic ways of being led into clearly unreliable tests. Rather than viewing them as part of an utter prohibition of violating predesignation, they should be viewed as invitations to articulate creative arguments to substantiate reliability by other means. Such developments are quite in keeping with the statistical philosophy of E. S. Pearson:

> There is perhaps in current literature a tendency to speak of the Neyman-Pearson contributions as some static system, rather than as part of the historical process of development of thought on statistical theory which is and will always go on. (Pearson 1966d, 276)

---

ing the crud factor for given domains—something I hope will be pursued. Meehl's work provides an excellent source for building a tool kit of errors for social science research.

Considerations of the creative postdesignationist provide the groundwork for justifying a break with overly narrow construals of NP methodology. This break is not new but reflects sound uses of those procedures in much of scientific practice. What is still needed is a clear articulation of the associated error-statistical arguments. This is part of the larger task of setting out an adequate methodology of experiment, a task that requires domain-specific considerations and is beyond the scope of this book. Despite wide latitude for such a program, the one thing retained is the constraint—formal or informal—of error statistics or severity. This stands in marked contrast to the alternative program represented by the Bayesian Way. Deliberate disregard for this constraint, as will be seen in the next chapter, "frees" the Bayesian to view hunting and data snooping as irrelevant to the import of the evidence in hand.

CHAPTER TEN

Why You Cannot Be Just a
Little Bit Bayesian

To understand how radical the likelihood principle must appear to many objectivists, note first that in accepting this principle one renounces all desire to make his estimates unbiased. An even more radical consequence of the likelihood principle is the thesis of the innocuousness of [rules to stop the experiment].

—Bruno de Finetti, *Probability, Induction and Statistics: The Art of Guessing*, p. 170

The likelihood principle emphasized in Bayesian statistics implies, among other things, that the rules governing when data collection stops are irrelevant to data interpretation. It is entirely appropriate to collect data until a point has been proved or disproven.

—Ward Edwards, Harold Lindman, and Leonard Savage, "Bayesian Statistical Inference for Psychological Research," p. 193

It would indeed be strange if the information to be extracted from a body of data concerning the relative merits of two hypotheses should depend not only on the data and the hypotheses, but also on the purely external question of the generation of the hypotheses.

—A. W. F. Edwards, *Likelihood: An Account of the Statistical Concept of* Likelihood *and Its Application to Scientific Inference*, p. 30

The likelihood principle implies . . . the irrelevance of predesignation, of whether an hypothesis was thought of beforehand or was introduced to explain known effects.

—Roger Rosenkrantz, *Inference, Method and Decision: Towards a Bayesian Philosophy of Science*, p. 122

IN APPRAISING METHODOLOGICAL RULES for scientific inference the normative epistemologist needs to assess how well the rules promote a given experimental aim. It is entirely reasonable to expect that philosophical accounts of hypothesis testing or confirmation should have something to say in such a metamethodological assessment. I listed this task in chapter 3 as the third way in which accounts of hypothesis testing may be applied in philosophy of science. A danger persists, however, that the view of testing appealed to in such a metamethodological appraisal already embodies principles at cross-purposes with the aim underlying the account to be appraised. This situation exists, I have already alleged, in Bayesian appraisals of the use-novelty (UN) requirement. In this chapter I shall give a full-blown justification for my allegation—but that will be only my first stopping point on the way to a further destination. Catching the Bayesians in this misdemeanor uncovers a pervasive illicitness in the Bayesian Way of performing a methodological critique. The problem, in a nutshell, is this: the underlying rationale of a number of methodological rules is the aim of reliability or severity in the sense I have been advocating, yet that aim runs counter to the aim reflected in Bayesian principles. In section 10.3 I will explicitly take up an even more far reaching outgrowth of this recognition, which explains the title of this chapter.

The intent of the title is not to suggest that all Bayesians are radical subjectivists or strict Bayesians—but somewhat the opposite. What I am arguing is that insofar as one accepts inference according to Bayes's theorem, one is also buying into distinctive principles of relevant evidence, hence criteria for inferences, hence grounds for judging methodological rules. The key issue is the question of the relevance of error probabilities. Accepting minimal Bayesian principles compels renouncing standard error probability principles and their informal counterparts (e.g., severity and reliability in our sense).

The conflict between these two sets of principles is familiar to philosophers of statistics:

> It seems that the divisions in statistics result almost completely from differences in attitude to the question of whether operating characteristics of data analysis procedures are important or not. (Kempthorne 1972, 190)

Error probabilities are examples of operating characteristics of procedures, and they are the linchpin of error statistics. The difference in attitude reflects different principles for interpreting data. Given an observed outcome x the error statistician finds it relevant—indeed essential—to consider the other outcomes that could have resulted from the procedure that issued in data *x*. Those considerations are needed to calculate error probabilities. Bayesian inference—although it comes in many different forms—must hold to the likelihood principle, and this leads to the irrelevance of such calculations. James Berger and Robert Wolpert, in their monograph *The Likelihood Principle* (which they abbreviate as LP), assert that

> the philosophical incompatibility of the LP and the frequentist viewpoint is clear, since the LP deals only with the observed x, while frequentist analyses involve averages over possible observations. . . . Enough direct conflicts have been . . . seen to justify viewing the LP as revolutionary from a frequentist perspective. (Berger and Wolpert 1988, 65–66)

I will be making use of their work in section 10.3.

Despite these direct conflicts, particular error statistical procedures often correspond to procedures Bayesians would countenance, albeit with differences in interpretation and in justification. This apparent overlapping of procedures is regarded by some as belittling the significance of the "philosophical" differences between Bayesians and error statisticians on matters of interpretation and justification. Even in the apparent eclecticism of statistical practice, however, the issues of interpretation and justification do not go away; and when it comes to utilizing statistical ideas in philosophy of science, these issues are paramount—although they have generally been overlooked. Since philosophy of statistics, in its formal guise, tends to occupy a rather separate niche in philosophy of science, it is not surprising to find that most philosophers of science are unaware that there are two major conflicting principles of confirmation, support, or testing. Nor is it obvious that this conflict should be of any particular concern to philosophers.

Is it possible that a conflict that, strictly speaking, emanates from two opposed formal statistical schools could shed any light upon the problems still facing philosophers of science? Is it possible that even as the logical empiricist ways are being replaced with "postpositivist" ones that a fundamental principle of evidence and evidential appraisal is still, unknowingly, retained? Is it possible that a good deal of the debate about methodological principles is rooted in the opposition between two principles, made explicit in theories of statistics? The answer to all these questions, I believe, is yes.

First I will illustrate how the novelty debate is skewed when seen through Bayesian glasses. Then we will arrive, finally, at the heart of the conflict between error probability principles and the likelihood principle.

10.1 NOVELTY AND SEVERITY THROUGH BAYESIAN GLASSES

Let us begin with the flaw in Bayesian critiques of arguments for the UN requirement. While not immediately obvious—at least it does not seem to have been recognized—the flaw is not difficult to spot, having the results from the previous two chapters under our belts.

In chapter 8, recall, the UN requirement was found to reflect the desire to ensure that evidence counts as good grounds for H only to the extent that it may be seen to constitute a good test of H, meaning that the evidence stems from a procedure with a low probability of erroneously passing hypothesis H—that is, one with high severity. I then set out to evaluate how well the UN requirement accorded with the aim of severe tests. I showed that while a test that violates the UN requirement is assured of passing the hypothesis it tests, it does not follow that it was assured of doing so *whether or not that hypothesis is false*. In short, I showed the argument for requiring UN to be unsound by showing that violating UN need not lead to violating severity, despite the fact that the *intended* aim of use-novelty is severity.

The Bayesian appraisal of accounts of novelty takes a very different tack. To the Bayesian, it has been said, all things are Bayesian, and the Bayesian appraisal of the UN requirement is a perfect illustration of this. The Bayesian appraises the UN requirement according to whether it has a rationale from *its own* vantage point of what counts as good support for a hypothesis. Running the UN requirement through the Bayesian machinery means asking whether satisfying UN is necessary for Bayesian support.[1] Howson and Urbach (1989) make this Bayesian strategy very clear. They note that although "the Bayesian theory of support is certainly inconsistent with" the UN requirement,

> there are arguments for the view, and these both sound convincing and also number among their subscribers many if not most contemporary philosophers of science. We shall examine these arguments now and show that their plausibility vanishes on closer inspection. (Howson and Urbach 1989, 276)

They get the plausibility to vanish only by changing the argument—at least as it is offered by the non-Bayesians they consider (e.g., Giere, Glymour, Worrall). They change the argument by making the "closer inspection" consist of an examination through a Bayesian magnifying glass—through the Bayesian rule for support. I am not criticizing their

1. That the Bayesian asks about support rather than about good and bad tests does not impede this analysis.

Bayesian scrutiny of the arguments offered by other Bayesians (e.g., Redhead). I concur that attempts at Bayesian justifications of UN will not wash. But aside from these few Bayesian exceptions, the "many if not most contemporary philosophers of science" to whom Howson and Urbach refer are not giving Bayesian arguments for requiring UN. Here is where the inappropriateness comes in.

Whose Rule of Support?

According to Howson and Urbach,

> attempts to show that data which hypotheses have been deliberately designed to entail, as opposed to independently predicting, do not support those hypotheses fail. On the contrary, *the condition for support,* that $\dfrac{P(e \mid \text{not-}H)}{P(e \mid H)}$ be small, may be perfectly well satisfied in many such cases. (Howson and Urbach, 1989, 279; emphasis added; I replace their *h* with *H*)

The condition for support? So confidently do Bayesians speak of "the condition for support" that the UN proponent may forget to ask whether this was the intended condition when thinking that UN is required for a good test. If it is not, then the Bayesian criticism fails to make a dent in the argument for UN. In fact, it is not.

First, let us be clear about the origin of this condition for support. It comes from the Bayesian condition that for evidence *e* to provide support for hypothesis *H* the posterior probability of *H* given *e* must be higher than the probability of *H* prior to *e*. That is, the posterior probability of *H* must exceed the prior probability of *H*:

> *Bayesian rule for support (first form): e* supports *H* if $P(H \mid e)$ is greater than $P(H)$.

Although it is not immediately obvious, this rule is equivalent to the requirement that *e* be more probable under *H* than under not-*H*. That is,

> *Bayesian rule for support (second form): e* supports hypothesis *H* if *e* is more probable under *H* than under not-*H*.

Equivalently,[2]

2. To see how the second form of the Bayesian rule of support falls out from its first form, consider Bayes's theorem and then calculate the ratio of $P(H \mid e)$ and $P(H)$:

$$P(H \mid e) = \frac{P(e \mid H)\, P(H)}{P(e \mid H)\, P(H) + P(e \mid \text{not-}H)\, P(\text{not-}H)}.$$

e supports *H* if $P(e \mid H)$ is greater than $P(e \mid \text{not-}H)$.

We have arrived at the rule for support to which Howson and Urbach refer in the above passage. Let us get a little fancier. Let us abbreviate the *Bayesian ratio of support*[3] for *H* as *BR:*

$$BR: \frac{P(e \mid \text{not-}H)}{P(e \mid H)}.$$

We can then write what Howson and Urbach refer to as "the condition for support" as the condition: *e* supports *H* if the Bayesian ratio BR is less than 1. The smaller the BR is, the greater the support for *H*. This condition does not require the posterior probability to be high, just that it be higher than the prior.

For the case where *H* is constructed to fit *e*, Howson and Urbach suppose that *H* entails *e* (hence $P(e \mid H) = 1$), so in considering their discussion I will too. In that case, the Bayesian rule of support becomes extremely simple:

Bayesian rule of support where $P(e \mid H) = 1$: *e* supports hypothesis *H* if $P(e \mid \text{not-}H)$ is less than 1.

It is now easy to see, when the argument for the UN requirement is scrutinized through Bayesian glasses, why the proponent of the UN requirement *appears* to be claiming that violating UN leads to $P(e \mid \text{not-}H)$ being 1. Why? Because that is what it would mean for *a Bayesian* to assert that no support accrues (when *H* entails *e*).

Recall that $P(e \mid \text{not-}H)$ is our friend the Bayesian catchall factor (section 4.3). Calculating the Bayesian catchall factor (except where not-*H* is a point hypothesis) requires prior probability assignments to

---

Then the ratio,

$$\frac{P(H \mid e)}{P(H)} = \frac{P(e \mid H)}{P(e \mid H)P(H) + P(e \mid \text{not-}H)\,P(\text{not-}H)}$$

$$= \frac{1}{P(H) + \dfrac{P(e \mid \text{not-}H)}{P(e \mid H)}\,P(\text{not-}H)}.$$

This exceeds 1 just so long as the denominator is less than 1. And remembering that $P(H) = 1 - P(\text{not-}H)$, it is seen that this occurs whenever $P(e \mid \text{not-}H) < P(e \mid H)$.

3. The BR is often called the likelihood ratio but this is misleading since the hypotheses involved can be disjunctions requiring averaging over prior probabilities. What I am calling the Bayesian ratio of support is also called the Bayes's factor against *H*, but I did not want to confuse the BR with what I call the Bayesian factor on the catchall.

all alternatives to *H*. But the non-Bayesian refuses to employ prior probabilities. Nevertheless, this does not stop Howson and Urbach from using this critique against non-Bayesian arguments for use-novelty. From their Bayesian perspective, all that is needed to vitiate arguments for requiring UN is that UN is not required for Bayesian support. For this it suffices to show that even when *H* is use-constructed, an agent can assign the Bayesian catchall factor a value less than 1. And that is what they do.

Howson and Urbach most specifically consider Giere's position, which we are already familiar with, that evidence used to construct a hypothesis has "no chance of refuting it." In this connection, they consider Giere's discussion of Gregor Mendel. A constraint on Mendel's model, says Giere, was to fit the evidence of the two-to-one ratio of tall to dwarf plants. Thus, following the same pattern of argument articulated in chapter 8, he considers that such a fit was assured even if Mendel's hypothesis had been false. Through Bayesian glasses, Giere looks to be claiming that the Bayesian catchall factor is assured to be 1, and this Howson and Urbach deny:

> It is far from self-evident that Mendel's data would not be improbable were his own explanation of them to be false; indeed, as that was the *only* explanation which seemed plausible to Mendel, its falsity would presumably render those data, were they assumed to be still conjectural, relatively improbable as far as he was concerned. And this, as we have seen, is sufficient for a Bayesian to be able to explain the undoubted fact that Mendel himself took his data to be strongly confirmatory of his model. *Giere has not justified his thesis—nor indeed could he—that* $P(e \mid not\text{-}H) = 1$ *when* H *has been designed to explain* e. (Howson and Urbach 1989, 277; emphasis added. To be consistent with my notation, I capitalize their lowercase *h* and use "not-*H*" rather than ~*h*.)

So it suffices to vitiate Giere's argument, according to Howson and Urbach, that *as far as a given agent is concerned* there is no other plausible explanation of the evidence. In fact, Bayesian support will accrue (however small) so long as the subjective Bayesian catchall factor is less than 1.[4] Thus it suffices for support that the agent believes there to be at least one alternative to *H* that does not make the evidence certain! Since there is vast latitude to what agents can believe, their dismissal of Giere's argument is simplicity itself.

---

4. There are really two mistaken assumptions in this Bayesian critique of Giere: that the probability of concern is the Bayesian factor on the catchall and that subjective probabilities are relevant.

But it has nothing to do with Giere's argument. Admittedly, Giere is careless in stating what he regards as the intended rationale for UN, but, having acknowledged that he is a non-Bayesian, it is odd that Howson and Urbach suppose that a subjective Bayesian analysis has relevance. They disregard Giere's warning that

> it is crucial to remember that the probabilities involved are physical probabilities inherent in the actual scientific process itself. If one slips into thinking in terms of probability relations among hypotheses, or between evidence and hypotheses, one will necessarily misunderstand this account of the nature of empirical testing. In particular, one must not imagine that to estimate the probability of [a failing result] one must be able to calculate the probability of this result as a weighted average of its probabilities relative to all possible alternative theories. No such probabilities are involved. (Giere 1983, 282–83)

And yet that is precisely how Howson and Urbach construe the probabilistic claim in Giere's argument.

Now it is true that Giere's argument for the necessity of UN is unsound (as I argued at length in chapter 8), but not for the reason the Bayesian alleges. The difference is altogether crucial. The proponent of the UN requirement is not at all claiming that violating UN precludes Bayesian support (whether subjectively or objectively interpreted)! The concern, rather, is with violating the severity requirement. And Bayesian support is easy to obtain even where severity is violated.

Low or Minimally Severe Tests Can Satisfy the Bayesian Requirement for Support

One way of making this point clearer is to consider a restricted version of the UN requirement, which *does* hold:

> *Restricted UN requirement:* Data used to arrive at and test a use-constructed hypothesis cannot count as a good test of that hypothesis if there is a high probability for passing some such hypothesis, even if it is false.

This is, of course, just an instance of the severity requirement: any test that lacks severity is a poor one, and in some cases violating UN makes it easier to carry out strategies that hinder severity. Equivalently, the restricted UN requirement says that a use-constructed hypothesis is poorly supported or poorly tested by evidence if the use-construction rule is one of the unreliable ones. In the extreme case of a gellerized rule, there is no test, and so no genuine support at all.

The existence of highly unreliable use-construction procedures is the reason that many are led to uphold the UN requirement and eschew "double counting" of data. Admittedly, as we saw in chapter 8, it is really only the restricted UN requirement that is warranted. Nevertheless, Howson and Urbach's Bayesian argument against the general UN requirement goes through just as well for this restricted UN requirement. Another way to put this is that even when severity is low or zero, the condition for Bayesian support can be satisfied. Hence finding Bayesian support still available simply cuts no ice with an error-severity person. Satisfying Bayesian support is not sufficient for severity.

This is obvious where calculating the Bayesian ratio (BR) is subjective, that is, where the reason the Bayesian support ratio is small is that the agent simply believes evidence e to be incredible under alternatives to H. However, minimally severe tests can muster Bayesian support even if the BR is determined by objective likelihoods from a probability model.

Maximally Likely Alternatives Again

We have already seen several examples that would show this. Recall our discussion of maximally likely alternatives and the problem of underdetermination in section 6.5. Let H_0 be the null or test hypothesis, and H an alternative hypothesis. The Bayesian condition of support for H is satisfied so long as H makes e more probable than does H_0—so long as the Bayesian ratio BR is less than 1. Here the BR equals

$$\frac{P(e \mid H_0)}{P(e \mid H)}.$$

But one can always find such an H (so long as H_0 does not give e probability 1). One simply uses evidence e to construct or select an H that perfectly fits the evidence e. Support would thereby accrue to H, even when the restricted UN requirement would be violated. The extreme example of gellerization could illustrate, but so could less artificial examples such as those from "hunting for statistical significance," discussed in chapter 9 (especially in section 9.2).[5]

---

5. To see how the gellerized process in example 6.1 would do to make this point, note that the severe testing theorist would describe the process this way: Observe the outcome e, find a hypothesis $G(e)$ that makes e maximally likely, and then deem $G(e)$ supported by e. (Whether one goes on to measure the support is irrelevant.) Although the particular hypothesis erected to perfectly fit the data will vary in different trials, for every data set some such alternative may be found. Therefore, supporting the maximally likely hypothesis constructed is assured, even if that hypothesis is false.

Teddy Seidenfeld clued me in to a nifty real-life example from sports that offers a different kind of illustration. A person who wants to show he has a system for

Hunting Again

To see how the Bayesian rule accords support to a hypothesis that
an error theorist would consider poorly tested, we can recall the strategy of hunting used in the study of infant training discussed by Kish
(1970) and detailed in section 9.2. For simplicity, consider that a single
infant training experience is of interest, early weaning. Suppose that
the procedure is to search among 100 factors for 1 that is highly correlated with having been subjected to early weaning. Say that such a
correlation is found between early weaning and a tendency toward
shyness in older children. Outcome e is the difference between the proportions of the early weaners and the late weaners who are or claim
to be shy. Hypothesis H_0 asserts that the observed correlation between
early weaning and shyness is spurious, or due to chance. The procedure will test hypothesis H_0 only upon finding an e that is very improbable given that H_0 is true—so the numerator of the Bayesian ratio is
small. For example, the difference sought may be required to be at
least 2 standard deviations (corresponding to the .05 "computed" level
of significance). The alternative H may assert that the correlation is real
and in the direction observed:

H: Early weaning is correlated with shyness in young children.

Hypothesis H is deliberately chosen or constructed so that the denominator of the BR, $P(e \mid H)$, is high (perhaps maximal). The evidence is far
more probable given hypothesis H than given the null hypothesis H_0.

Notice the similarity between the improbability of the particular e
given the chance hypothesis H_0 and the small "computed significance
level" in the last chapter. The probability of this particular outcome e—
a high correlation between early weaning and shyness—is very low
given the null hypothesis H_0. The alternative—*by design*—makes this
observed correlation highly probable. So the Bayesian rule of support
is satisfied—indeed, it is well satisfied, since the ratio is not just less
than 1 but very small.

Such an example reveals in no uncertain terms the mistake that
has gone unchecked in Bayesian reconstructions of non-Bayesian arguments. The mistake stems from the fact that satisfying Bayesian sup-

---

predicting the winning slate in a football series offers this "test": If the list of winners he sends you at the start of the season turns out to be correct, then it is regarded as good support that his system really works. However, each possible per-
mutation of winners is sent to a different sports fan, so his system is assured of
acquiring high support (by someone) even if his system has no predictive ability
beyond mere guessing.

port is not sufficient for satisfying severity. That a methodological rule is not required for Bayesian support does not license inferring that the rule is not required for non-Bayesian measures of support—such as those based on error probabilities.

Satisfying Severity versus Satisfying Bayesian Support

It will be useful to recapitulate the distinguishing feature of a severity calculation by way of the Bayesian rule of support (that e supports H if the BR is small). In appraising such a rule, the error statistician is concerned with its behavior under repetitions. With respect to the case being discussed, the severity criterion requires asking how often the rule would award support to hypotheses about the effects of infant training on personality, even if they are false and there is no real difference in personality traits among those subjected to the different infant training.[6] We can answer this question by viewing the Bayesian ratio BR as a statistic, a function of the data. For each trial of the experiment the BR takes some value. So long as the BR is less than 1, support accrues to the use-constructed alternative H, but we are imagining that the rule is even more demanding: the BR must be very small. Severe testers want to know how often such strong support would accrue for some nonchance hypothesis (between infant training and personality) or other, even if H_0 is true. And they want to know this even after the result is in and a particular value for the BR has been calculated. With 100 different tests, it is highly probable that at least one 2-standard-deviation difference would be found, even if all the null hypotheses were true.[7] The severity of the test that H passes is 1 minus this, and thus is very small, practically 0. Although the probability of *any particular* statistically significant result is low, the probability of some high correlation *or other* is high.

To answer the severity question, the error statistician needs to consider something that from the Bayesian standpoint is irrelevant—the behavior of the statistic (in this case the BR) in a series of (real or hypothetical) repetitions. This is the experimental distribution of statistic BR. (Precisely why it is irrelevant for the Bayesian will be taken up in section 10.3.) Such considerations, in the view of the error statistician, are necessary to scrutinize the Bayesian procedure of assessing support by calculating the BR.

---

6. Since this is an example of the type where the hypothesis selected for testing can vary, the severity criterion becomes SC with hypothesis construction (defined in section 6.6).

7. Applying the calculation discussed in chapter 9, the probability is about .99.

Sum-Up

We have shown in this section that the Bayesian rule of support is unreliable in the sense that it allows support to accrue to hypotheses, with high probability, even if the hypotheses are false. True, merely satisfying the rule of support does not say that the posterior probability of *H* is high, nor that the increment in the posterior is large. It says merely that *some* support accrues to H; it may, depending on the prior probabilities, be tiny. *But that is the Bayesian condition for support that Howson and Urbach use to denounce arguments for the UN requirement.* Hence it is appropriate for us to consider it in questioning their denunciation. Moreover, it makes sense for them to consider the minimal requirement for support because that is what they regard as being challenged by the claim that UN is necessary. What I have shown is that their criticism is unsound because satisfying Bayesian support does not entail satisfying severity. And all I claim to be doing just now is discounting this Bayesian criticism.

It may be objected that I am evaluating the Bayesian criterion of support from an error probability (e.g., severity) stance. That is exactly right. It is entirely appropriate to do this in answering Bayesian critiques of the novelty requirement, because the aim of novelty is severity.

Novelty through Bayesian Glasses   .

Since the UN requirement reflects a concern about error-severity, it is easy to see why the Bayesian concludes that the novelty requirement will not hold up. The point is made informally by remembering what was said about Bayesian philosophers being descendants of holders of "logical theories" of confirmation. Like the logical theorists before them, those who assess the import of evidence by way of the Bayesian ratio regard as irrelevant how the hypothesis was generated. (There are also prior probabilities, of course, but these are separate from the import of the evidence for support.) The Bayesian support ratio BR, which is how the import of the evidence comes through for a Bayesian, is unaffected by the manner of hypothesis generation. Hence Howson and Urbach's (1989) assertion that "the Bayesian theory of support is certainly inconsistent with" the novelty requirement (p. 276).

Not all Bayesians deny the UN requirement, but both supporters and detractors share the assumption that what has to be shown or denied is its bearing on the Bayesian measure of support (the Bayesian support ratio). To be fair, the reason for this is not just the Bayesian tendency to appraise all principles of evidence according to the Bayesian formula. In this case, non-Bayesian UN proponents have unconsciously opened up their argument to Bayesian scrutiny by making ambiguous statements about the grounds for requiring UN. (We saw this in discussing Worrall and Giere in section 8.3.) What UN proponents have failed to see or failed to state unequivocally is that the raison d'être for the use-novelty principle has to do with severity—in our non-Bayesian sense. To think that severity matters is to think that test results cannot be appraised without considering the error properties of the entire procedure from which the results arose—the very properties that Bayesians are happy to declare irrelevant.

Bayesian Ways to Make Novelty Matter

If violating novelty is not relevant to Bayesians, then they are faced with the problem of accounting for scientific cases where it does seem to have mattered. It is open to the Bayesian to propose the classic Bayesian move. Any scientific appraisal that you think turned on the mode of hypothesis construction (and a concern with the corresponding lack of severity or reliability), the Bayesian may allege, can be reconstructed as having turned on some difference in prior probabilities. Indeed, any time there is a difference in the appraisal of two hypotheses that entail (or equally fit) evidence, the Bayesian *must* locate the source of the difference in the priors. There is no place else to locate the difference in the Bayesian algorithm. Granted, with given assumptions about one's prior probabilities in hypotheses, some, though not all, violations of severity can be made to correspond to tests that are poor or comparatively poor on Bayesian grounds. Even so, the Bayesian reconstruction incorrectly locates the actual rationale for disparaging these tests.

Various attempts in which, through just the right assumptions and prior probabilities, use-novel hypotheses receive higher Bayesian support than use-constructed ones include Campbell and Vinci 1983; Howson and Urbach 1989; Maher 1988, 1993c; Redhead 1986; and Rosenkrantz 1977. Regrettably, my remarks on them must be brief. These attempts, if they are not guilty of mistakes about probabilities, err in one or both of two ways: either (1) they violate the likelihood principle[8] or (2) they fail to capture the actual epistemological rationale for why certain violations of use-novelty are taken as problematic and why they should be.

---

8. The arguments between Maher and his critics Howson and Franklin (1991) really turn on this issue, although the debate has not been framed in these terms. Maher needs to appeal to the idea of the reliability of the method by which hypotheses are generated. Such an idea finds its home in error-statistical not Bayesian accounts.

My basis for (2) is that the problem caused by unreliable use-construction procedures is not a problem about prior degrees of belief in hypotheses. To underscore this point, consider a single hypothesis so that the problem cannot be traced to prior probability assignments. A hypothesis that might do is one we saw in the example with which I began our discussion of novelty—the bombing of the World Trade Center. Let hypothesis H assert that group X drives into the garage at the given time and explodes the bomb. When advanced before the bombing, H passes a (relatively) severe test. The probability that the before-the-fact description of the bombing would have fit the actual facts so well if H were false is low. In contrast, hypothesis H does not pass a severe test when advanced after the details of the bombing have been reported. Being able to come up with a hypothesis that fits the reported occurrence is precisely the sort of move that is open to anyone who wants to assign credit for the bombing, even though the alleged group had nothing to do with it. At the stage that we disparage the after-the-fact calls claiming responsibility for the bombing, nothing has been done to rule out this error.

More generally, the difference in the evidential import of two pieces of evidence, both of which fit hypotheses equally well, is located in a difference in the reliability of test processes. This is where the error statistician locates it. The Bayesian desiring to make out a difference locates its source elsewhere (depending on which attempt one considers). For this reason, error statistical criteria match our intuitions about differential support better than Bayesian ones.

10.2 THE OLD EVIDENCE PROBLEM

The position that how or when a hypothesis is generated is irrelevant to a Bayesian may seem puzzling in the light of what has been said about Bayesians having an "old evidence" problem. The puzzle results because, whereas for most Bayesians novelty *never* matters to support, for a few others it *always* does! These other Bayesians are said to have an old evidence problem.

An account is said to have an old evidence problem if it has the consequence that old or known data fail to count as evidence in support of a hypothesis (it requires temporal novelty). As critics of temporal novelty show, this conflicts with many cases where known evidence is regarded as providing excellent support for hypotheses.[9]

9. We have shown (in chapters 8 and 9) that contrary to what some have claimed, Neyman-Pearson statistics does not have an old evidence problem.

Why is the subjective Bayesian supposed to have an old evidence problem?

The allegation, brought to the forefront by Glymour (1980), goes like this: if probability is a measure of degree of belief, then if an agent already knows that e has occurred, the agent must assign $P(e)$ the value 1. Hence $P(e \mid H)$ is assigned a value of 1. But this means no Bayesian support accrues from e. For if $P(e) = P(e \mid H) = 1$, then $P(H \mid e) = P(H)$. The Bayesian condition for support is not met.

Another way of phrasing the problem is that if evidence e is known and so assigned a probability of 1 by an agent, then the agent also assigns a probability of 1 to the Bayesian catchall factor, that is, $P(e \mid \text{not-}H) = 1$. So, the BR equals 1, and no Bayesian support accrues to H.

How do subjective Bayesians respond to the charge that they have an old evidence problem? The standard subjective Bayesian response is given by Howson and Urbach (1989) and by Howson (1989).

> It has been argued (e.g., by Glymour . . .) that since e is, by assumption, known at the time h is formulated, its probability must be 1, so that P(e | ~h) = 1 also. . . . But is it? If Glymour is right then P(e | ~h) would be 1 even if h had not been constructed to explain e, if e is a known fact, and so e would not support h in this case either. But Glymour is not right. . . . The Bayesian interprets P(e | ~h) as how likely you think e *would be* were h to be false. . . . On this construction the value of P(e | ~h) is *independent* of whether h was or was not constructed to explain e. (Howson 1989, 386)

But many people—Bayesians included—are not too clear about how this "would be" probability is supposed to work.

Consider the known evidence in the Brownian motion example (chapter 7). Brownian motion was known before formulation of the Einstein-Smoluchowski theory. To assess the support that this phenomenon affords this theory, the subjectivist imagines Perrin or some other agent asking something like this: How probable would I regard the phenomenon of Brownian motion, were the Einstein-Smoluchowski theory false? If the agent thinks that H is the only plausible explanation of Brownian motion (e), he or she may well think the occurrence of the phenomenon very improbable in a world in which H were false. So the agent may well assign a small, and not a maximal, value to $P(e \mid \text{not-}H)$. If we ask how the agent figures out this probability, Howson and Urbach will say (recall chapter 3), *it is not our job*. "We are under no obligation to legislate concerning the methods people adopt for assigning prior probabilities" (Howson and Urbach 1989, 271).

It is pretty clear that an agent would not want to assign a probability of 1 to known evidence because the evidence is *always* known by the time Bayes's theorem is applied. Glymour's "mistake," according to Howson and Urbach, is in supposing that probability assignments are to be relative to the totality of current knowledge. In fact, "they should have been relativised to current knowledge minus *e*" (p. 271):

> Once *e* has become known . . . the probabilities $P(e \mid h)$, $P(h)$, and $P(e)$ . . . are relativised to the counter-factual knowledge state in which you still do not know *e*. (P. 271)

In their view, the support that *e* gives to *H* is to be computed by considering how knowledge of *e* would alter one's degree of belief in *H*, on the supposition that one did not yet know *e*.[10]

Counterfactual Degrees of Belief

The need to consider an agent's counterfactual knowledge state while still keeping assignments coherent is fraught with problems, many of which Glymour discusses (e.g., Glymour 1980, 87–91). Does one try to go back in time, and if so how far back? Were the counterfactual knowledge view taken seriously, says Glymour,

> we should have to condemn a great mass of scientific judgments on the grounds that those making them had not studied the history of science with sufficient closeness to make a judgment as to what their degrees of belief would have been in relevant historical periods. (Glymour 1980, 91)

What if we stay in the present but imagine subtracting out the knowledge of *e*? That seems silly and hardly an easier feat. In this view, for Einstein to assess the evidential bearing of the perihelion of Mercury on the relativistic theory of the gravitational law, he needs to imagine whether there would be an increase in his assignment of probability to the law had he not already known of the perihelion phenomenon. However, as Earman (1992) nicely points out, Einstein developed the theory hoping to account for just this phenomenon; had he not known about it, perhaps he would not have developed the theory. "And if someone else had formulated the theory, Einstein might not have taken it seriously enough to assign it a nonzero prior" (p. 123). In any

---

10. Paul Horwich has a somewhat different tactic for dealing with this problem. He seems to allow that the probability of known evidence *e* equals 1, but maintains that a Bayesian should assess how much *e* would alter our degree of belief assignment to *H* relative to "our epistemic state prior to the discovery" of *e*, when its probability was not yet 1 (Horwich 1982, 53).

event, a scientist does not actually assess the import of the evidence *e* in hand by imagining the probability change that would obtain if *e* were unknown to him.

The problem of assigning counterfactual degrees of belief also makes it difficult to carry out Bayesian reconstructions of scientific episodes, upon which Bayesianism's usefulness to the philosophy of science depends. The Bayesian "solution" to Duhem's problem (recall chapter 4) requires the philosopher to assign probabilities so as to reflect the beliefs actually held by given scientists. But are philosophers in a position to follow Dorling's suggestion that we consider the betting odds a typical scientist would have been willing to place on *e*, assuming that *e* had not yet been discovered? (Dorling 1979, 182). Even if one could arrive at such a counterfactual probability assignment, the deeper question remains: why is it relevant to either making or scrutinizing a scientific inference from evidence *e*?

Take the example of the World Trade Center bombing. In the "imagine the evidence is not known yet" view, to assess the support to give a hypothesis *H* about the group responsible for the bombing—where this hypothesis is called in after the fact—I must consider how the evidence of the explosion would have altered my belief in *H* if I did not yet know of the explosion. I am to be just as impressed when nonnovel evidence fits a hypothesis as I am when novel evidence fits it. The reliability of the use-construction method does not enter. At the end of 1994 I enjoyed fitting the data on profitable stocks over the past twelve months into a highly profitable strategy for buying and selling during that time. Would the Bayesian assess how well these data support my system in the *same way* had I advanced it at the end of 1993? Yet isn't that what pretending you didn't yet know the data would seem to countenance?

The most satisfactory Bayesian way around the old evidence problem, in my view, is for the Bayesian to restrict the probability assignments to specific statistical models of experiments and generic outcomes of those experiments. For example, the probability of heads on a toss of a fair coin is one-half, independent of anyone knowing the outcome. Presumably, this is the "objective" Bayesian solution. This would considerably limit the scope of the Bayesian account in philosophy of science, and the problem of prior probabilities of hypotheses would remain.

Although there seems to be no single agreed-upon solution to the old evidence problem, some such solution is assumed by the generally accepted Bayesian position on novelty described in section 10.1: novelty (temporal and use-novelty)—in and of itself—never matters to support.

Garber's and Jeffrey's Way

The Bayesian rejection of the UN requirement may be put this way: an argument showing that violating UN precludes Bayesian support would also show that known evidence in general precludes Bayesian support (leading to the old evidence problem). There are a few Bayesians, however, who want to accept the UN requirement, yet get around the old evidence problem—even if this requires changing the probability axioms.

This way around the old evidence problem, developed by Daniel Garber, Clark Glymour, Richard Jeffrey, and others, reflects the idea that even old evidence can be made to support H, as long as the fact that H entails e is new. This attempt requires altering the axioms of the probability calculus relative to one's knowledge. For example, instead of stipulating that all tautologies have probability 1, the assignment would be relativized to *knowing* it to be a tautology. (Essentially the same idea was originally developed by I. J. Good, but for a different reason.) The problems that such attempts raise for Bayesians will not be gone into here.[11] They make confirmation even more a matter of subjective, psychological beliefs than traditional subjective Bayesianism.

Note mainly that accepting the Garber-Jeffrey way out of the old evidence problem takes us right back to the problem of use-novelty. With a use-constructed hypothesis H it *is* known that H entails or otherwise fits e, so it is not saved from the chopping block that spares cases in which the entailment is unknown. Those who appeal to such accounts to solve the old evidence problem *start out assuming* the position that violating UN always precludes support. Take Garber:

> Suppose that S constructed h *specifically* to account for e, and knew, from the start, that it would. It should not *add* anything to the credibility of h that it accounts for the evidence that S knew all along it would account for. (Garber 1983, 104)

There is no attempt to justify this on Bayesian principles. It is hard to see how it could be so justified since, as Bayesians are quick to show, the UN requirement violates the likelihood principle. Moreover, as I argued in chapter 8, the position that violating UN precludes support is untenable: it would rule out many cases with excellent and even maximal support (as in estimation techniques). It seems hardly worth changing the probability axioms only to be left upholding this position.

Regardless of how Bayesians settle this family quarrel about how

11. See for example, Miller 1987, 305–8 and Earman 1992.

best to deal with old evidence, the problem only furthers my criticism of Bayesian critiques of methodological principles: methodological principles are often based on non-Bayesian ideas about error probabilities, and these ideas run counter to a fundamental Bayesian principle, the likelihood principle. If we are to use ideas from statistics to obtain a philosophical understanding of reasoning in science—something I heartily endorse—then we need to be very clear on the fundamental differences between Bayesian and error-probability approaches. That is the purpose of the next section.

10.3 THE LIKELIHOOD PRINCIPLE (LP) AND STOPPING RULES

One of the claims [of the Bayesian approach] is that the experiment matters little, what matters is the likelihood function after experimentation. . . . It tends to undo what classical statisticians have been preaching for many years: think about your experiment, design it as best you can to answer specific questions, take all sorts of precautions against selection bias and your subconscious prejudices. (LeCam 1977, 158)

Why does embracing the Bayesian position tend to undo what classical statisticians have been preaching? Because Bayesian and classical statisticians view the task of statistical inference very differently.

In chapter 3 I contrasted these two conceptions of statistical inference by distinguishing evidential-relationship or E-R approaches from testing approaches, and explained why E-R approaches in general, and the Bayesian Way in particular, have appealed more to philosophers than classical testing approaches. The E-R view is modeled on deductive logic, only with probabilities. In the E-R view, the task of a theory of statistics is to say, for given evidence and hypotheses, how well the evidence confirms or supports hypotheses (whether absolutely or comparatively). There is, I suppose, a certain confidence and cleanness to this conception that is absent from the error-statistician's view of things. Error statisticians eschew grand and unified schemes for relating their beliefs, preferring a hodgepodge of methods that are truly ampliative. Error statisticians appeal to statistical tools as protection from the many ways they know they can be misled by data as well as by their own beliefs and desires. The value of statistical tools for them is to develop strategies that capitalize on their knowledge of mistakes: strategies for collecting data, for efficiently checking an assortment of errors, and for communicating results in a form that promotes their extension by others.

Given the difference in aims, it is not surprising that information relevant to the Bayesian task is very different from that relevant to the task of the error statistician. In this section I want to sharpen and make more rigorous what I have already said about this distinction.

The different positions staked out by error statisticians and by those who accept the likelihood principle (e.g., Bayesians), it should by now be clear, are not only of concern to philosophers of statistics. This opposition, I have been urging, while crystallized in formal statistical principles, is implicated, if only informally or implicitly, in a cluster of disputes in philosophy of science. Overlooking this distinction in underlying principles, we saw in section 10.1, has permitted what are essentially question-begging appraisals of methodological rules to go unchallenged. Further, the secret to solving a number of problems about evidence, I hold, lies in utilizing—formally or informally—the error probabilities of the procedures generating the evidence. It was the appeal to severity (an error probability), for example, that allowed distinguishing among the well-testedness of hypotheses that fit the data equally well (the alternative hypothesis objection, chapter 6).

Having been reminded of the philosophy of science ramifications of the dispute between opposing statistical philosophies, I want to revisit that dispute, but this time I want to go deeper into its core. Whereas in thinking of the key difference between Bayesians and error statisticians one most often thinks of the former's willingness to assign prior probabilities to hypotheses and the latter's insistence upon methods that do not require such assignments, the difference I now want to concentrate on is more fundamental. In this place, more than any other, one can see the chasm that divides the Bayesian from the error-statistician.

Stopping Rules

Let me begin with a question, as I did with hunting for statistical significance in chapter 9. The situation now resembles that case, but instead of hunting for a statistically significant property, we will imagine that the researchers have an effect they would like to demonstrate, and that they plan to keep experimenting until the data differ statistically significantly, say at the .05 level, from the null hypothesis of "no effect." (In other words, the researchers keep going until they get a 2-standard-deviation difference. .05 is the *computed* or "nominal" level of significance.) We can call this a "try and try again" procedure. The effect may be anything one would like to consider—that a subject can do better than chance at guessing an ESP card, that one treatment does better than some other (with regard to some symptom), that the discrepancy from some parameter value is real—or any of the other kinds of examples we have considered. You are presented, say, with the statistically significant data ultimately arrived at. The question is whether it is relevant to your appraisal of the effect that the data resulted from the try and try again procedure.

For a simple example, imagine a subject of an ESP experiment, Zoltan. During each trial of the experiment Zoltan must predict ahead of time the next ESP card in a deck of cards. Suppose that after a long series of trials Zoltan scores a relative frequency of successful predictions that exceeds the relative frequency expected by chance alone by an amount sufficient to attain a .05 significance level. Would it be relevant to your evaluation of the evidence if you learned that Zoltan had planned all along to keep running trials for as long as it took to reach the (computed) significance level of .05? Would you find it relevant to learn that after, say, 10 trials, having failed to rack up enough successes to reach the .05 level of statistical significance, Zoltan went on to 20 trials, and failing yet again he went to 30 trials, and then 10 more, and on and on until, say on trial 1,007, he finally attained a statistically significant result?

A plan for when to stop an experiment is called a *stopping rule*. So my question is whether you would find knowledge of the stopping rule relevant in assessing the evidence from a statistical test. If you would you are in good company, for that is how standard error statistics answers the question. From the Bayesian point of view, however, you are incoherent!

The Likelihood Principle

Having alluded more than once to the likelihood principle (LP), I will now say more specifically what it asserts.[12] The LP is regarded as having been articulated by non-Bayesian statisticians, principally George Barnard (1947) and R. A. Fisher (1956). But, as it is *their* principle now, I will let the Bayesians do the talking.[13]

In their classic piece, Edwards, Lindman, and Savage (1963) spell out the LP as follows. They consider two experiments involving the same set of hypotheses H_1 up to H_n. Let D be an outcome from the first

12. There is also something called the "weak likelihood principle," but since that is not in dispute between Bayesians and error statisticians I will not discuss it. Richard Miller (1987) uses the term to mean something different. What he has in mind is a principle sometimes called the law of likelihood (e.g., by Hacking; noted in section 6.6). The formal likelihood principle should not be confused with these other notions.

13. I do not think there are more than a handful of non-Bayesian ("likelihoodists") who still accept the LP.

experiment and D' from the second. They ask, "Just when are D and D' thus evidentially equivalent, or of the same import?" (p. 237). Their answer is when, for some positive constant k,

$$P(D' \mid H_i) = kP(D \mid H_i)$$

for each i. That is, D and D' are evidentially equivalent when the likelihood of H_i given D is a multiple of the likelihood of H_i given D'. That is because the posterior probabilities of the hypotheses come out the same, as the interested reader can check.

A reminder: $P(D \mid H)$ is called the likelihood of H, but for a non-Bayesian this can be calculated only where H is a simple statistical hypothesis—not a disjunction. That is why, for example, where we considered $P(D \mid \text{not-}H)$, we gave it a different name (the Bayesian catchall factor). In discussing the LP, however, the Bayesian often wishes to identify a conflict between Bayesian and non-Bayesian treatments of evidence. To demonstrate this conflict the Bayesian has to consider only examples in which $P(D \mid H_i)$ is calculable for a non-Bayesian, that is, where these likelihoods are calculated the same way for Bayesians and non-Bayesians. I will also maintain this restriction.

To this end, the LP is often stated with reference to hypotheses about a particular parameter μ, such as the probability of success (on a Binomial trial) or the mean value of some characteristic. L. J. Savage (1962) states it this way, where x and y (rather than D' and D) now refer to the two results:

> According to Bayes's theorem, $P(x \mid \mu)$ . . . constitutes the entire evidence of the experiment, that is, it tells all that the experiment has to tell. More fully and more precisely, if y is the datum of some other experiment, and if it happens that $P(x \mid \mu)$ and $P(y \mid \mu)$ are proportional functions of μ (that is, constant multiples of each other), then each of the two data x and y have exactly the same thing to say about the values of μ. (P. 17; I substitute P for his Pr and μ for his λ)

It is a short step from this reasoning to see the conflict with classical or "orthodox" theory. As Lindley (1976) puts it,

> we see that in calculating [the posterior], our inference about μ, the only contribution of the data is through the likelihood function. . . . In particular, if we have two pieces of data x_1 and x_2 with the same likelihood function . . . the inferences about μ from the two data sets should be the same. *This is not usually true in the orthodox theory, and its falsity in that theory is an example of its incoherence.* (P. 361; emphasis added. I replace his q with μ)[14]

14. Where the likelihoods are proportional for the hypotheses under consideration they are sometimes said to be the same likelihood function. That is how

Savage's Message at the 1959 Forum

It was this conflict that was uppermost in Savage's mind when in 1959 he led a forum attended by several leaders in statistics. He declared:

> In view of the likelihood principle, all of these classical statistical ideas come under new scrutiny, and must, I believe, be abandoned or seriously modified. (Savage 1962, 18)

Attendees at this forum included P. Armitage, I. J. Good, G. Barnard, M. S. Bartlett, E. S. Pearson, D. Lindley, D. R. Cox, and others, representing a mixture of statistical schools. Savage announced to this distinguished group that all the classical statistical notions—all the notions under "error statistics"—significance levels and tests, confidence levels and interval estimates, criteria based on error probabilities—all are suspect. They are suspect because they come into conflict with the LP.

The conflict is most pronounced, Savage explains, on the relevance of stopping rules. While it is widely held that the import of the evidence depends on the stopping rule in examples like the one above, in fact, Savage warns, this violates the LP. The LP tells you that it can make no difference to the import of evidence whether the experimenter had planned to "try and try again" until a (computed) .05 significant result is achieved, or whether the experimenter had planned to run just one experiment, with some fixed sample size, and let the chips fall where they may. Let us refer to the former, try and try again plan, as the *optional stopping plan,* and the latter, prespecified plan as the *fixed sample size plan.* Why, according to the LP, does a result have exactly the same thing to say about μ when generated through optional stopping as when generated through a fixed sample size plan? Because the probabilities of the results from the two experiments (given μ)—i.e., the likelihoods—are proportional to each other.

Zoltan's 1,007 trials—whether by optional stopping or fixed sample size—consist of a string of k successes and $1,007 - k$ failures. It can be pictured as a string of 1,007 ss and fs, such as

$$s,s,f,s,f,f,f,s,f,f,s,f,s,\ldots\ldots\ldots\ldots$$

This string is the outcome x. The hypothesis of interest is a hypothesized value for μ—the probability of success on each trial. The posterior probability accorded to μ with *either* experimental plan is a function of the prior probability and the likelihood, $P(x \mid \mu)$. And in both cases,

---

Lindley is using "same likelihood function" here. It will be less confusing to just say that their likelihoods are proportional.

$P(x \mid \mu) = \mu^k(1 - \mu)^{1007-k}$. That is, the data x enter into the Bayesian computation the same way whether they arose from the optional stopping plan or the fixed sample size plan.

> In general, suppose that you collect data of any kind whatsoever—
> not necessarily Bernoullian, nor identically distributed, nor indepen-
> dent of each other . . . —stopping only when the data thus far col-
> lected satisfy some criterion of a sort that is sure to be satisfied sooner
> or later, then the import of the sequence of n data actually observed
> will be exactly the same as it would be had you planned to take ex-
> actly n observations in the first place. (Edwards, Lindman, and Savage
> 1963, 238–39)

This is called the *irrelevance of the stopping rule.* Those who accept the LP hold to the irrelevance of the stopping rule.[15]

How then, in contrast, do error statisticians render stopping rules relevant? By operating with a different notion of relevant evidence. In their view, it *is* relevant to what the data are saying about the population parameter μ to learn that the result in front of them—x—came from the try and try again (optional stopping) method. Mathemati-cally, this corresponds to the fact that x does not enter the error statistician's computations by itself but always by considering error properties of the experimental procedure from which x arose. Information about stopping rules does not show up in likelihoods, but it sure shows up in a procedure's error probabilities.

Edwards, Lindman, and Savage, quite rightly, regard this differ-ence in attitude on the relevance of stopping rules as a central point of incompatibility between the two approaches. That is why it is so im-portant for us. To the holder of the LP, the irrelevance of the stopping rule is a point in its favor, but to the error statistician the situation is exactly the reverse. P. Armitage (1962), the most forthright error statistician at the 1959 Savage forum, puts it plainly:

> I think it is quite clear that likelihood ratios, and therefore posterior
> probabilities, do not depend on a stopping rule. Professor Savage, Dr
> Cox and Mr Lindley take this necessarily as a point in favour of the
> use of Bayesian methods. My own feeling goes the other way. I feel
> that if a man deliberately stopped an investigation when he had de-
> parted sufficiently far from his particular hypothesis, then "Thou shalt
> be misled if thou dost not know that." If so, prior probability methods

15. There are certain exceptions where the stopping rule may be "informa-tive," but I keep to examples that Bayesians do not regard as falling under this qual-ification.

seem to appear in a less attractive light than frequency methods, where one can take into account the method of sampling. (P. 72)

The error statistician wants to take the method of sampling into account because, as was known in 1959, the try and try again method allows experimenters to attain as small a level of significance as they choose (and thereby reject the null hypothesis at that level), even though the null hypothesis is true.[16] If allowed to go on long enough, the probability of such an erroneous rejection is one! So the *actual* or *overall* significance level is not .05 but 1!

Optional Stopping Leads to High or Maximal Overall Significance Levels

Just as, in chapter 9, we calculated the actual significance level to be the probability of hunting down *some statistically significant factor or other* given that none are really correlated, here we calculate the actual or overall significance level as the probability of finding a statistically significant difference from a fixed null hypothesis *at some stopping point or other* up to the point at which one is actually found. The overall significance level accumulates.

We need to be extra careful with the term *statistically significant difference* in the optional stopping case. Here, one keeps taking more and more samples until the observed difference is *computed* to be statistically significant, until it is, say, 2 standard deviations away from the null hypothesis. The computed significance level with an optional stopping plan refers to the significance level that would be calculated under a fixed sample size plan—.05. Say it took k tries to achieve a difference computed to be .05 statistically significant. The *actual* or overall significance level is the probability that out of k tries at least one would be computed to be .05 statistically significant, even if the null hypothesis is true.

Unlike the case of hunting, there is a substantial literature on how to run and calculate overall significance levels—error probabilities— for tests with different stopping rules. These kinds of tests are called *sequential*. Sequential tests have long been part of the error-statistician's tool box. One reason they are so useful is that often it is estimated that a smaller number of samples is required with a sequential than with a fixed sample size test. Medical trials, especially, are often deliberately designed as sequential. Armitage, as it happens, is a leader in the devel-

16. Feller (1940) is the first to show this explicitly. Other early discussions of this result include Anscombe 1954 and Robbins 1952. The result is also implicit in Good 1956 and Lindley 1957.

opment of sequential trials, having devoted whole books to their use and interpretation within the error statistical framework.

Example 10.1: Armitage: The Effect of Repeated Significance Tests

In his *Sequential Medical Trials,* Armitage (1975) discusses the effect of repeated significance tests. He illustrates with a common example from medicine. Suppose that each patient scores in some numerical fashion the effectiveness of two different treatments, say two types of painkillers A and B. Drug A is administered one week, drug B on a different week. The recorded observation on each patient is the difference between the two scores. Imagine that a new significance test is performed after each patient's scores are obtained, with a view toward finding a difference (in either direction) computed to be statistically significant at the .05 level. The null hypothesis assumes that drugs A and B are equally effective. By the time 30 patients are sampled, the probability of computing a statistically significant difference even though the null hypothesis is true is around .3—not .05. So with 30 patients, the actual probability of rejecting the null hypothesis erroneously is not .05 but around .3. We would say that the *calculated* (or fixed sample size) significance level is .05, but that the actual or *overall* level is .3. Armitage gives the overall significance level for each number of patients (based on the standard test of the difference between means called the *t*-test), as shown in table 10.1.

Before we rule out the null (or "mere chance") hypothesis and argue that the result is indicative of a genuine difference, we want to be able to sustain a reliable argument from error. We want to be able to say that our procedure would probably have ruled in favor of the "mere chance" explanation, were that the case. But the procedure of trying and trying again cannot be said to have a good chance of ruling in favor of the null hypothesis—even if the null is true. With enough significance tests, the try and try again procedure will almost never pass the null hypothesis even if it is true. In their useful booklet on statistics for doctors, Bjorn Andersen and Per Holm (1984) provide a humorous analogy for this unfairness toward the null hypothesis:

> The procedure might be compared with new rules for determination of the world championship in heavy-weight boxing: Only the reigning champion is allowed to strike. The fight is over, whenever the contender is out for the count of 10. The contender (like H_0) has little chance of winning, no matter how "good" he is. (P. 57)

A Funny Thing Happened at the 1959 Savage Forum

Now Savage knows all about the effect of optional stopping—he knows all about how the try and try again method ensures reaching

TABLE 10.1 The Effect of Repeated Significance Tests (the "Try and Try Again" Method) (Armitage 1975, p. 29)

Number of patients (differences in scores) n	Probability of a 0.05 "significant" result at or before this stage, given the null hypothesis is true
2	0.05
10	0.09
20	0.26
30	0.29
40	0.31
50	0.33
100	0.39
infinity	1.00

statistical significance. In his opening remarks at the 1959 forum, Savage rehearses how "the persistent experimenter can arrive at data that nominally reject any null hypothesis at any significance level, when the null hypothesis is in fact true" (Savage 1962, 18). Because the persistent experimenter is thereby assured of rejecting a perfectly true null hypothesis, the standard error statistician denies that such a rejection provides genuine evidence against the null. But Savage audaciously declares that the lesson to draw from the optional stopping effect is just *the reverse* of the one the error statistician draws. The problem is not with the data arrived at by a procedure of trying and trying again, the problem is with significance levels!

> These truths [about the optional stopping effect] are usually misinterpreted to suggest that the data of such a persistent experimenter are worthless or at least need special interpretation. . . . The likelihood principle, however, affirms that the experimenter's intention to persist does not change the import of his experience. (Savage 1962, 18)

I shall come to the business of the relevance of "intentions" in a moment. Savage's argument is this: if calculating the significance level is altered (by the stopping rule), then there must be something wrong with significance levels because likelihoods are unaffected. According to the LP, says Savage, "optional stopping is no sin," so the problem must lie with the use of significance levels (1964, 185). But why should we accept the likelihood principle?

Yes, the LP follows from Bayes's theorem, but significance tests are non-Bayesian techniques. Apparently, the LP is regarded by some as so intrinsically plausible that it seems any sensible account of inference should obey it. Bayesians do not seem to think any argument is necessary for this principle, and rest content with echoing Savage's declaration in 1959: "I can scarcely believe that some people resist an idea so patently right" (1962, 76). However much Savage deserves reverence, that is still no argument. Ironically, what prompted Savage's famous declaration as to the patent rightness of the LP was a heretical confession by George Barnard. Barnard—the statistician whose arguments Savage claims (p. 76) convinced *him* (in 1952) of the *irrelevance* of optional stopping—had just announced to the forum that he had changed his mind!

Explaining why he now thinks that stopping rules do matter, Barnard describes an example quite like the one with which I began this section:[17]

> Suppose somebody sets out to demonstrate the existence of extrasensory perception and says "I am going to go on until I get a one in ten thousand significance level." Knowing that this is what he is setting out to do would lead you to adopt a different test criterion. What you would look at would not be the ratio of successes obtained, but how long it took him to obtain it. And you would have a very simple test of significance which said if it took you so long to achieve this increase in the score above the chance fraction, this is not at all strong evidence for E.S.P., it is very weak evidence. (Barnard 1962, 75)

By altering the test criteria accordingly, Barnard continues, one would avoid misinterpreting the evidence.[18] That is just what error statisticians recommend—thereby making them incoherent from the Bayesian standpoint.

The Argument from Intentions

Startled by this turnabout, Savage reminds Barnard of the persuasive argument he himself urged (in 1952) against the relevance of stopping rules.

> The argument then was this: The design of a sequential experiment is, in the last analysis, what the experimenter actually intended to do. His intention is locked up inside his head. (Savage 1962, 76)

The experimenter's intentions about when to stop sampling are locked up in his head, and it seems absurd for intentions to influence what

17. The suggestion Barnard made to the forum was that stopping rules matter when you do not have explicit alternatives. He himself was a likelihoodist and not a Bayesian, although he came to give that up as well. (See, for example, Barnard 1972.)

18. In practice, the alteration is generally to lower the computed or nominal significance level sufficiently so that the overall significance level is still .05. Armitage and others have done extensive work on this for a variety of types of sequential trials.

the data have to say. Since significance levels take stopping rules into account, significance levels let experimenter's intentions count. In their joint paper, Edwards, Lindman, and Savage remark:

> The irrelevance of stopping rules is one respect in which Bayesian procedures are more objective than classical ones. Classical procedures . . . insist that the intentions of the experimenter are crucial to the interpretation of data. (Edwards, Lindman, and Savage 1963, 239)

Although Savage (1962, 76) declared himself uncomfortable with the argument from intentions, it is repeated again and again by followers of Savage. Howson and Urbach think it substantiates some rather dire conclusions about significance tests:

> A significance test inference, therefore, depends not only on the outcome that a trial produced, but also on the outcomes that it could have produced but did not. And the latter are determined by certain private intentions of the experimenter, embodying his stopping rule. It seems to us that this fact precludes a significance test delivering any kind of judgment about empirical support. . . . For scientists would not normally regard such personal intentions as proper influences on the support which data give to a hypothesis. (Howson and Urbach 1989, 171)

In their view, apparently, to take account of the experimenter's sampling plan is to take personal intentions into account and is unscientific, while the properly scientific way of assessing evidential support is to ask for the agent's personal degrees of belief in hypotheses.

In fact, the whole insinuation that to regard optional stopping as relevant is to make private intentions relevant is fallacious. Any and all aspects of what goes into specifying an experiment could be said to reflect intentions—sample size, space of hypotheses, prediction to test, and so on—but it does not mean that paying attention to those specifications is tantamount to paying attention to the experimenter's intentions. Yet Howson and Urbach are pretty plainly arguing that since a significance test's error probabilities are determined by the experimenter's personal intentions and since intentions should not matter to support, a test's error probabilities (e.g., significance levels) do not or should not be relevant to support. Are Bayesians just committing a gross fallacy here?

They are, but they cannot see it. They have got Bayesian glasses on, and they will not take them off. Through Bayesian glasses, there is no place in the inference scheme to record the effect of the sampling plan—at least not once the data are in hand. So, they view it as locked inside someone's head.

Recall hunting for a statistically significant difference in the infant training example again (chapter 9). Suppose the hunter reports the single factor found to be statistically significant out of 20 that are checked (e.g., late weaning and left-handedness). We have this one statistically significant result before us, but where, one might ask, is the fact that it was the single factor found significant in a hunting expedition through 20 factors? Is it locked up in the experimenter's head? Not if he or she is an honest hunter, nor if the one scrutinizing the result is an error statistician. But that means having an eye for error probabilities—being able to see, in particular, that the actual probability of erroneously declaring statistical significance in this case is not .05 but over .6. The Bayesian glasses have a substantial blind spot here.

Likewise with optional stopping. If one is wearing Bayesian glasses, that is, if one adheres to the LP, then two experiments that give the same (i.e., proportional) likelihoods to a hypothesis have the same evidential bearing on the hypothesis. If one is wearing Bayesian glasses, then, once the data are available, one cannot make out any difference between that data having arisen from a try and try again method or from a (nonsequential) experiment where the subject declares ahead of time, "If I have not shown statistical significance in exactly n trials then conclude I have not shown the effect." One cannot see the difference because the likelihoods are unchanged. One may well know there is a difference in sampling plans, but that just means one knows they had different intentions, and that cannot possibly make a difference to the meaning of evidence. That, at any rate, is the way things look through Bayesian glasses. That is the way things look to anyone peering at evidence through the LP. Ian Hacking, in his Likelihood testing period, also gives the argument from intentions:[19]

> Can testing depend on hidden intentions? Surely not; hence optional stopping should not matter after all. (Hacking 1965, 109)

Examples of philosophers espousing the argument from intentions could easily be multiplied.

Notice a certain similarity with justifying why novelty should matter. If a violation of novelty is nothing more than that the experimenter intended to find a way to account for the data, then it looks as if propo-

19. That account, developed in Hacking 1965, was based on Hacking's likelihood rule of support noted in section 6.6.

nents of novelty appeal to the psychological intentions of the investigator. Once the aim of novelty is recognized to be severity, violating novelty shows up as a problem (when it *is* one) with a test's severity; and the effect on severity, whether formally or informally calculated, shows up in a procedure's error probabilities. In precisely the same way, the error statistician has a perfectly nonpsychologistic way of taking account of the impact of stopping rules, as well as other aspects of experimental plans. The impact is on the error probabilities (operating characteristics) of a procedure.[20]

In the optional stopping plan, the difference in the test procedure clearly shows up in the difference in the set of possible experimental outcomes. Certain outcomes possible in the fixed sample size (nonsequential) version of the test are no longer possible.[21] If the stopping rule is open-ended, then the possible outcomes do not contain any that fail to reject the null hypothesis!

It might be asked: But does the difference in error probabilities corresponding to a difference in sampling plans correspond to any real difference in the experiments? Absolutely. The researchers really did something different in the try and try again scheme and, quoting Armitage, "thou shalt be misled" if you do not know this. It is not just that incorrectly reporting a test's error probabilities incorrectly reports what happened in obtaining a result, it also incorrectly reports *what should be expected to happen* (with various probabilities) in subsequent experiments on the phenomenon of interest. It must be remembered that every error statistical inference includes a statement about future experiments, whether or not they will be carried out. With an incorrect report of a test's error probabilities, an experimenter seeking to check or repeat the previous results would be misled. The reported error probabilities would not be close to those that would actually be found in such repetitions.

I think enough has been said to banish the common allegation that letting stopping plans matter is tantamount to letting intentions matter. As such, we can reject the argument that the LP must be embraced on pain of unjustly letting intentions enter into the appraisal of evidence.

20. The only time it seems unwarranted to draw a distinction is if the experimenter stops after the first test because a statistically significant result is achieved on the first try. But in that case the difference between the computed and the overall significance level is extremely small, and should make no difference in a "nonautomatic" use of tests. I discuss what is wrong with automatic or recipelike uses of tests in chapter 11.

21. The sample space differs but because the likelihoods are proportional, the difference cancels out for a holder of the LP.

But, if threats will not win us over, the Bayesian tempts with the goodies that await those who accept the LP.

Bayesian Freedom, Bayesian Magic

A big selling point for adopting the LP, and with it the irrelevance of stopping rules, is that it frees us to do things that are sinful and forbidden to an error statistician.

> This irrelevance of stopping rules to statistical inference restores a simplicity and freedom to experimental design that had been lost by classical emphasis on significance levels (in the sense of Neyman and Pearson). . . . Many experimenters would like to feel free to collect data until they have either conclusively proved their point, conclusively disproved it, or run out of time, money or patience. . . . Classical statisticians . . . have frowned on [this]. (Edwards, Lindman, and Savage 1963, 239)

Breaking loose from the grip imposed by error probabilistic requirements returns to us an appealing freedom.

LeCam, a leading error statistician (cited at the start of section 10.3) hits the nail on the head:

> It is characteristic of [Bayesian approaches] . . . that they . . . tend to treat experiments and fortuitous observations alike. In fact, the main reason for their periodic return to fashion seems to be that they claim to hold the magic which permits [us] to draw conclusions from whatever data and whatever features one happens to notice. (LeCam 1977, 145)

In contrast, the error probability assurances go out the window if you are allowed to change the experiment as you go along. Repeated tests of significance (or sequential trials) are permitted, are even desirable for the error statistician; but a penalty must be paid for perseverance—for optional stopping. Before-trial planning stipulates how to select a small enough significance level to be on the lookout for at each trial[22] so that the overall significance level is still low. That is what Armitage's work on sequential clinical trials is all about.

But the Bayesian pays no penalty, or so it seems. I. J. Good, a veteran Bayesian, often puts it this way:

> Given the likelihood, the inferences that can be drawn from the observations would, for example, be unaffected if the statistician arbi-

---

22. This is the level that would be required to be reached on any given significance test so as to stop the trials. Setting it small enough ensures that the probability of an erroneous rejection of the null is still small in sequential trials.

trarily and falsely claimed that he had a train to catch, although he really had decided to stop sampling because his favorite hypothesis was ahead of the game. . . . On the other hand, the "Fisherian" tail-area method for significance testing violates the likelihood principle because the statistician who is prepared to pretend he has a train to catch (optional stopping of sampling) can reach arbitrarily high significance levels, given enough time, even when the null hypothesis is true. (Good 1983, 36)

"Arbitrarily high significance levels" means significance levels as *small* as one wants. Elsewhere in Good's *Good Thinking:*

The way I usually express this "paradox" is that a Fisherian [but not a Bayesian] can cheat by pretending he has a train to catch like a gambler who leaves the table when he is ahead. (Good 1983, 135)

As often as my distinguished colleague presents this point, I remain baffled as to its lesson about who is allowed to cheat. The significance tester—as Good well knows—does not allow reaching arbitrarily high (meaning small) significance levels through optional stopping. The significance tester is not allowed to change the sample size at will, stopping just because he is ahead. When error statisticians perform sequential tests, the *overall* (and not the computed) significance level must be reported. To the error statistician, what would be cheating would be to report the significance level you persevered to attain, say .05, *just as if* the test were the ordinary nonsequential sort.

Good's point seems to be this: Error statisticians are forced to fret about a consideration the Bayesian is free to ignore. Wearing our error probability glasses—glasses that compel us to see how certain procedures alter error probability characteristics of tests—we are forced to say, with Armitage, that "Thou shalt be misled if thou dost not know that" the data resulted from the try and try again stopping rule. To avoid having a high probability of following false leads, the error statistician must scrupulously follow a specified experimental plan. But that is because we hold that error probabilities of the procedure alter what the data are saying—whereas Bayesians do not. The Bayesian is permitted the luxury of optional stopping and has nothing to worry about. The Bayesians hold the magic.

Or is it voodoo statistics?

Armitage's Example

To some, the magic is accomplished by smoke and mirrors and wearing Bayesian glasses. At the 1959 forum, Armitage, building on his earlier remarks, went on to say that

[Savage] remarked that, using conventional significance tests, if you go on long enough you can be sure of achieving any level of significance; does not the same sort of result happen with Bayesian methods? The departure of the mean by two standard errors corresponds to the ordinary five per cent level. It also corresponds to the null hypothesis being at the five per cent point of the posterior distribution. Does it not follow that by going on sufficiently long one can be sure of getting the null value arbitrarily far into the tail of the posterior distribution? (Armitage 1962, 72)

Armitage's point can be simply put as follows. In many cases, rejecting a null hypothesis H_0, say at level of significance .05, corresponds to a result that would lead a Bayesian to assign a low (e.g., .05) posterior probability to H_0. This occurs with so-called uniform or uninformative priors. (That this is so is often touted by Bayesians as a point in their favor: Whereas the most an error statistician can say is that this procedure has a low [.05] probability of erroneously rejecting the null, the Bayesian, thanks to his prior probability assignment, can assign the low probability to the specific hypothesis H_0.)[23] Hence, Armitage reasons, if error statisticians—if they go on sampling long enough—are assured of reaching a .05 significant result, even though H_0 is true, then Bayesians—if they go on sampling long enough—would be assured of reaching a low (.05) posterior probability in H_0, even though H_0 is true. (The assurance here is with high probability or, in the limit, with probability one.) That is:

1. In certain cases, rejecting a null hypothesis H_0, say at level of significance .05, corresponds to a result that would lead a Bayesian to assign a low (e.g., .05) posterior probability to H_0.

2. If one is allowed to go on sampling long enough (i.e., the try and try again procedure), then, even if H_0 is true, one is assured of achieving a .05 statistically significant difference from the null hypothesis H_0.

3. Therefore, if one is allowed to go on sampling long enough, then, in the cases described in (1), one is assured of reaching a low posterior probability in H_0, even though H_0 is true.

Now the error statistician is not allowed to go on trying and trying, at least not without paying a penalty. The penalty, we said, is that the overall significance level—in the extreme case 1—must be reported. The stopping rule matters. But Bayesians are free! They are allowed to go on sampling and the stopping rule does not alter the likelihoods,

23. See, for example, DeGroot 1973.

hence the posterior is just the same as if the case were nonsequential. It follows that, in going on long enough, a Bayesian is assured of assigning a low posterior probability to H_0 even though H_0 is true.

So, who is allowed to mislead?

Although Savage wants to deny Armitage's implication, he appears to grant it, though fuzzily, and skips to a different sort of example. While I think there are problems with this different example as well (see Rosenkrantz 1977, 199), I want to keep to Armitage's kind of example.[24]

Armitage's example goes like this. The null hypothesis H_0 is an assertion about a population parameter μ. As in example 10.1, μ might measure the mean difference in the effectiveness of two drug treatments. H_0 asserts that the treatments are equally effective, that is, that μ equals 0.

H_0: The treatments are equally effective: μ equals 0.

The experiment records $\overline{X}$ which, in this case, is the mean difference in scores accorded to the two drug treatments in a sample of n patients. The sample size n, however, is not fixed but is determined by a stopping rule. The stopping rule—an example of a try and try again plan—is to keep taking more samples until H_0 is rejected at the .05 level, *by the usual prespecified significance test:*

(1) *Stopping rule:* Keep sampling until H_0 can be rejected at the .05 level.

That is, the stopping rule is to keep sampling until $\overline{X}$ is 2 standard deviations away from 0 (the hypothesized value of μ in H_0) in either direction. The standard deviation here is the standard deviation of the statistic $\overline{X}$, but for simplicity, let us just abbreviate it as s.d.[25] So we have (letting $|\overline{X}|$ be the absolute value of $\overline{X}$)

(1) *Stopping rule:* Keep sampling until $|\overline{X}| \geq 2$ s.d.

Following this stopping rule, one is assured of achieving a .05 significant difference even if H_0 is true. But with a so-called uninformative

---

24. The example Savage skips to involves comparing two simple hypotheses. Rather than lending plausibility to Savage's cause, Rosenkrantz (while himself a Bayesian) thinks it shows that Savage goes too far in ignoring the stopping rule. Rosenkrantz's analysis has been questioned by Seidenfeld (1979b, n. 4).

25. This would more properly be written as s.d.$(\overline{x})$. When the standard deviation is estimated, as is most often the case, it is called the standard error, but it seems simpler to stick with a single term. Armitage takes the random variable X to be Normally distributed with mean μ and standard deviation 1. In that case, s.d.$(\overline{x})$ is $n^{-1/2}$.

or diffuse prior probability assignment to μ, such an occurrence would correspond to assigning a low posterior probability to H_0. Hence, following this stopping rule, the Bayesian would be assured of assigning a low probability to H_0 even though H_0 is true. This is Armitage's argument. No satisfactory answer has been forthcoming, nor is there one. Armitage is right.

Berger and Wolpert

To my knowledge, there are only a handful of Bayesians (or other holders of the LP) who have specifically addressed Armitage's example: most just accept Savage's dismissal of it.[26] Berger and Wolpert (1988), in their interesting monograph, show themselves to be as ardent a pair of proponents of the LP as it is likely to have. Still, even they concede Armitage's point. But, as they want to retain the LP, some defensive moves are called for.[27]

Since Berger and Wolpert treat Armitage's example in terms of confidence intervals, I will too. Recall example 8.2. Given some result, one forms an interval within which the parameter of interest, μ, is hypothesized to lie. As is standard, we can use a lowercase $\bar{x}$ to represent the observed value of the random variable $\bar{X}$. (This is easier to read than $\bar{X}_{obs}$ here.) The standard 95 percent confidence interval takes this form

(2) Estimate that μ is within 2 standard deviations of the observed mean $\bar{x}$.

(I use 2 rather than the exact value of 1.96.) That is, the 95 percent confidence interval is

(2) Estimate that μ equals $\bar{x} \pm 2$ s.d.

Berger and Wolpert agree that a Bayesian would use this interval in the usual fixed sample size case, adding:

> Of course, he would not interpret confidence in the frequency sense, but instead would (probably) use a posterior Bayesian viewpoint with the noninformative prior density. (Berger and Wolpert 1988, 80)

26. No one, to my knowledge, has identified the flaw in Savage's use of "the simple general formula" on page 73 of Savage 1962.
27. In a forthcoming paper, "Reasoning to a Foregone Conclusion," Kadane, Schervish, and Seidenfeld set out mathematical conditions under which Bayesians are and are not allowed to reason "to a foregone conclusion" erroneously.

Whereas the frequentist says only that this particular estimate was generated by a procedure with a 95 percent probability of correctly including the value of μ, the Bayesian can assign the particular estimate a posterior probability.[28] In particular, with the so-called noninformative prior, they would assign it a .95 posterior probability. We can now see how Armitage's argument goes through.

Berger and Wolpert continue: "Suppose now that the experimenter has an interest in seeing that $\mu = 0$ is not in the confidence interval. He could then use the stopping rule" (1) above (ibid.; I replace θ with μ). Let us rewrite the stopping rule to relate directly to confidence intervals. The null hypothesis H_0 asserts that $\mu = 0$. Rejecting H_0—finding $\bar{x}$ statistically significant from 0—is equivalent to 0 *not* being included in the interval estimate formed with $\bar{x}$. Hence assuring that $\mu = 0$ is *not* in the 95 percent confidence interval is equivalent to assuring that the null hypothesis H_0 is rejected at the .05 level. So the stopping rule in (1)—keep sampling until H_0 is rejected—stated in terms of confidence intervals is

(1) Keep sampling until the 95 percent confidence interval formed excludes 0.

So the Bayesian experimenter interested in keeping 0 out of the interval is free to use stopping rule (1). At the same time, Berger and Wolpert concede, "the [Bayesian] conditionalist, being bound to ignore the stopping rule, will still use (2) as his confidence interval, but this can *never* contain zero" (ibid., 81).

(The term "conditionalist" comes from the fact that, for a holder of the LP, inference must be conditional on the actual $\bar{x}$ observed.)

Hence Berger and Wolpert allow that "the frequentist probability" that intervals formed by this procedure would include 0, even when 0 is the true value, equals zero! Equivalently, there is zero probability of accepting the hypothesis that $\mu = 0$, even when that hypothesis is true. In short, they find they cannot get around the conclusion that, despite the fact that μ *does* equal 0,

the experimenter has thus succeeded in getting the conditionalist to perceive that $\mu \neq 0$, and has done so honestly. (Pp. 80–81)

Thus, they concede Armitage's point—the very point that Savage had denied or skirted.

Now for the defensive moves. Berger and Wolpert are at pains to

28. It leads to a posterior distribution for μ equal to a normal distribution with mean $\bar{X}$ and a standard deviation equal to $n^{-1/2}$.

uphold the LP. In examples such as Armitage's, Berger and Wolpert maintain, the LP only "seems to allow the experimenter to mislead a [Bayesian] conditionalist. The 'misleading,' however, is solely from a frequentist viewpoint, and will not be of concern to a conditionalist" (ibid., 81). Bayesians remain unconcerned, presumably, because they are not in the business of calculating error frequencies.

Despite their professed lack of concern, Berger and Wolpert, like Savage, are plainly uncomfortable with Armitage's result. They leave off the example suggesting that in appraising the plausibility of the LP we should trust our intuitions in one of the other examples they offer—one where the LP gives the intuitively correct inference—"rather than in extremely complex situations such as [Armitage's example]" (ibid., 83). But the example we are to trust does not involve optional stopping,[29] and the confidence interval example is rather ordinary. Armitage tells us it is a standard situation in clinical trials.[30]

Examples analogous to Armitage's have been produced by others, notably Alan Birnbaum.[31] To the error statistician, such examples are counterexamples to adopting the LP:

> Thus it seems that the likelihood concept cannot be construed so as to allow useful appraisal, and thereby possible control, of probabilities of erroneous interpretations. (Birnbaum 1969, 128)

I shall come back to Birnbaum in chapter 11.

It should be emphasized that this problem exists even for so-called objective Bayesians (those who strive to determine objective prior probabilities). That is the reason I said it was the outgrowth of a differ-

29. Perhaps the open-endedness of the stopping rule makes the case exotic, but less dramatic and still seriously troubling cases are generated with stopping rules only as high as 100, as Armitage shows in his examples with medical trials.

30. One Bayesian ploy would be to insist that learning of the use of such a stopping rule would make the agent change his prior in such a way that the high posterior would be avoided. Not only is the kind of prior that leads to the trouble a commonly acceptable one, but such an admission would also conflict with the Bayesian insistence that once the evidence is at hand the likelihoods tell all. (As always, we are talking about the cases in which stopping rules are uninformative according to the LP.) Ironically, since error probabilities are not supposed to matter for a Bayesian, this ploy really would seem to appeal to the intentions of the investigator. Moreover, this Bayesian ploy depends on the agent reasoning that an experimenter using such a stopping rule probably thinks the null hypothesis is true, and so revising his prior accordingly. But it seems at least as plausible, if not more so, to suppose that an experimenter planning to go on until the null hypothesis is rejected really believes that the effect is real and that the null hypothesis is false.

31. Birnbaum cites a similar result by Neyman from 1938, collected in Neyman 1952.

ence between Bayesians and error statisticians that runs even deeper than the use or nonuse of prior probabilities in hypotheses. It is the difference between the irrelevance and the relevance of error probabilities of procedures. If one is Bayesian enough to adhere to Bayesian coherency, hence to the LP, one is enough of a conditionalist to reject error probabilities and with them the familiar methods of standard error statistics.

That the standard methods conflict with the LP is readily accepted by leading Bayesians. Take Lindley:

> The most obvious violation of the likelihood principle occurs with the idea of a confidence interval, with its concept of repetition of the experiment. (Lindley 1976, 361)

Savage, discussing the

> "nice properties," exemplified by unbiasedness, stringency, minimum mean squared error, symmetry (or invariance), a given significance level, and so on (Savage 1964, 179)

declares that

> practically none of the "nice properties" respect the likelihood principle. (Ibid., 184)

That is why I think Berger and Wolpert's initial response to Armitage is the honest one from the Bayesian viewpoint; namely, that "the 'misleading' . . . is solely from a frequentist viewpoint." After all, it is only through frequentist considerations of error probabilities that Armitage's case is problematic, and those considerations violate the LP to which Bayesians adhere.

There, then, we have it. The reason Bayesians cannot be misled (in the case of optional stopping) is that they reject (as violating the LP) the frequentist viewpoint on which the error calculation depends! Anguish over a procedure's high probability of being wrong (in Armitage's example, as high as probability 1) is an error statistician's affliction. The Bayesian is not so afflicted. If I never check my bank account (and I always believe the correctness of my statement), then, in a sense, the bank can never mislead me.

The Relevance of Outcomes Other Than the One Observed

Let us explore a bit more why error probabilities violate the LP. The reason, in a nutshell, is that error probabilities ask what would happen for data sets other than the one actually observed. What is wrong with us error statisticians, from the Bayesian conditionalist perspective, is that we keep thinking that considerations of outcomes that could have resulted—outcomes other than the one that did result—are relevant for interpreting the evidential import of the data.

> Those who do not accept the likelihood principle believe that the probabilities of sequences that might have occurred, but did not, somehow affect the import of the sequence that did occur. (Edwards, Lindman, and Savage 1963, 238)

The error statistician has only one way of responding to this allegation. "Guilty as charged!" We remain steadfast no matter how leadingly Bayesians ask (echoing a line made famous by Harold Jeffreys), "What has what might have happened, but did not, got to do with inferences from the experiment?" (Lindley 1976, 361), and no matter how intimidating the rhetoric of prominent Bayesians is (e.g., E. T. Jaynes, an objective Bayesian):

> The question of how often a given situation would arise is utterly irrelevant to the question how we should reason when it *does* arise. I don't know how many times this simple fact will have to be pointed out before statisticians of "frequentist" persuasions will take note of it. (Jaynes 1976, 247)

What we error statisticians must rightly wonder is how many times we will have to point out that to us, reasoning from the result that did arise *is* crucially dependent upon how often it would arise. Lacking such information prevents us from ascertaining which inferences can be reliably drawn.

In criticizing the hunter, the error statistician notes, "But had this one not been statistically significant, it is very probable that you would have unearthed some other factor that was—even if none are really correlated." What would have happened is at the heart of the worry in the try and try again (optional-stopping) plan as well. The severity of a test is a measure of the relative frequency with which the test would lead to correctly failing (or not passing) a hypothesis in some sequence of applications. Virtually all the uses of statistical ideas in learning from error throughout this book depend critically on such considerations of "would have beens." What makes standard error statistical tools so useful for scientific inference is that their formal properties, error probabilities, enable learning about what would be expected if various errors exist—the key to experimental arguments from error. Yet these error properties and test criteria based on them are what the Bayesian is only too happy to declare irrelevant. As Lindley (1971) stresses, unbiased estimates . . . sampling distributions, significance levels, power, all depend on something more [than the likelihood function]—something that is irrelevant in Bayesian inference—namely the sample space. (Lindley 1971, 436)

In his admirably fair-minded work comparing different schools of inference, Vic Barnett explains why. In the Bayesian view,

> inferences are conditional on the realized value x; other values which *may* have occurred are regarded as irrelevant. . . . No consideration of the *sampling distribution* of a statistic is entertained; sample space averaging is ruled out. Thus there can be no consideration of the *bias* of an estimation procedure and this concept is totally disregarded. (Barnett 1982, 226)

Bayesian consistency requires rejecting the foundations of the error statistical methods, despite the widespread use of these methods throughout science.

This, then, is the bottom line. Our aims, our notions of relevant evidence, our criteria for judging satisfactory inference (the "nice properties"), our notions of probability (with a few exceptions) are strikingly different from those of the Bayesians. Quite aside from whether one accepts my position on the value of error statistical ideas, what cannot be denied are these differences. The lesson for metamethodology is this: Every critique of methodology from the Bayesian perspective must be seen as contingent upon accepting their aims in favor of error statistical ones. If the methodological rule in question turns out to concern promoting an error statistical aim (e.g., severity), then a Bayesian critique will be misleading if not just question-begging. The error statistician's conception of "being misled" is very different from that of the Bayesian: perhaps it is a gestalt switch that separates them. The philosopher seeking to apply ideas from statistics to the philosophy of science needs to decide whether to sign up for the LP (e.g., Bayesian) paradigm or the error statistical one, or perhaps something altogether different.

10.4 SOME ANTICIPATED OBJECTIONS

Some might object that I am overlooking the ways in which some manage to be Bayesian while at least a little bit of an error statistician at the same time. A main way would be to use Bayesian ways and yet strive to assess the reliability of these methods in a genuine error probability sense. Such error statistical (or "robust") Bayesians, if I understand their position, seem to me to fall onto the error statistical side.[32] Nothing in the error probability approach prevents using Bayesian measures as measures of "fit" whose operating properties can be investigated. Such developments may well be part of "the historical process of development" of error statistical theory to which E. S. Pearson alludes (Pearson 1966d, 276).

Aside from such new and innovative hybrid approaches, am I not overlooking the eclecticism that exists in statistical practice? No. I began this chapter acknowledging that error statistical inferences often correspond to procedures Bayesians would countenance, albeit with differences in interpretation and justification. It is certainly open to error statisticians to apply Bayes's theorem with well-defined statistical hypotheses where standard prior probabilities have been found to work well. Likewise, Bayesians can and do appropriate standard (error-statistical) methods by giving them Bayesian justifications. One might regard I. J. Good's "Bayes-non-Bayes compromise" as a systematic attempt to appropriate error statistical methods in this way.[33] In practice, dabbling in one or the other of these methodologies even without being too clear on the justification is often innocuous. This is not the case when Bayesian principles are applied to philosophy of science.

I earlier outlined three main ways of applying a theory of statistics to philosophy of science: (1) to solve philosophical problems (e.g., Duhem's problem); (2) to model scientific inference; and (3) to carry out a metamethodological critique (e.g., appraise novelty requirements). For each of these applications, the differences of interpretation and justification called for by the Bayesian and error-statistical philosophies are serious and are ignored at our peril.

32. This is not the case for Bayesians who are only willing to employ error-statistical methods if they can be given a subjective Bayesian interpretation, or who employ error probabilities disingenuously (e.g., because the customer wants or expects them).

33. Good's compromise, as I understand it, remains fully Bayesian (see note 32). However, his program has brought forth a number of important relationships between error probability and Bayesian calculations.

CHAPTER ELEVEN

Why Pearson Rejected the Neyman-Pearson (Behavioristic) Philosophy and a Note on Objectivity in Statistics

The two main attitudes held to-day towards the theory of probability both result from an attempt to define the probability number scale so that it may readily be put in gear with common processes of rational thought. For one school, the degree of confidence in a proposition, a quantity varying with the nature and extent of the evidence, provides the basic notion to which the numerical scale should be adjusted. The other school notes how in ordinary life a knowledge of the relative frequency of occurrence of a particular class of events in a series of repetitions has again and again an influence on conduct; it therefore suggests that it is through its link with relative frequency that a numerical probability measure has the most direct meaning for the human mind.

—E. S. Pearson, "On Questions Raised by the Combination of Tests Based on Discontinuous Distributions," p. 228

11.1 Introduction

The two main attitudes Pearson is speaking of correspond to two views of the task of a theory of statistics: the evidential-relation or E-R view and the error probability view. We have traced the key ways in which disputes about methodological rules reflect this underlying distinction in aims. Philosophers of induction, we said, have typically embraced the first of these two views. My primary aim has not been to settle this question of aims, but rather to show how a number of disputes in philosophy of science reflect this difference in aims, and to build an account of experimental learning based on the error statistics approach. I am also concerned with showing that the error approach is at the heart of the widespread applications of statistical ideas in scientific inquiry, and that it offers a fruitful basis for a philosophy of experiment.

Despite the widespread use of error statistical methods, the official school of inference in which they are formally couched—Neyman and Pearson (NP) statistics—has been the subject of enormous controversy and criticism. From the philosophy of statistics debates of the '70s and early '80s, NP theory emerged with several black eyes, spurring on the popular new Bayesian Way. Fetzer (1981); Hacking (1965); Kyburg (1971, 1974); Levi (1980a); Rosenkrantz (1977); Seidenfeld (1979a); and Spielman (1973); as well as several statisticians have raised doubts about the appropriateness of NP theory for statistical inference in science. In a 1977 issue of *Synthese* devoted to the foundations of probability and statistics, Neyman expressed surprise at the ardor with which subjectivists (e.g., de Finetti 1972) attacked NP tests and confidence interval estimation methods:

> I feel a degree of amusement when reading an exchange between an authority in "subjectivistic statistics" and a practicing statistician, more or less to this effect:
> *The Authority:* "You must not use confidence intervals; they are discredited!"
> *Practicing Statistician:* "I use confidence intervals because they correspond exactly to certain needs of applied work." (Neyman 1977, 97)

Neyman's remarks hold true today. The subjective Bayesian is still regarded, in many philosophy of science circles, as "the authority" in statistical inference, and yet scientists from increasingly diverse fields still regard NP methods (e.g., confidence intervals) as corresponding exactly to their needs.

Howson and Urbach (1989) have attempted to renew the old efforts to cleanse science of NP methods, declaring "that the support enjoyed by classical methods of estimation among statisticians is unwarranted" (p. 198). These, along with the other NP methods, they apparently feel, should be taken to the dump heap and replaced with their brand of subjective Bayesianism. Given the new emphasis philosophers of science have placed on taking cues from actual scientific practice, this disregard if not outright condemnation of procedures that are widely and successfully used across a vast spectrum of science is curious and out of place. I think it is time to remedy the situation. Philosophers of statistics can no longer operate on the image of the philosopher issuing pronouncements on the appropriateness of the scientist's tools—not if they want to contribute to an experimental methodology that will be of relevance to science.

Much of the reason philosophers have rejected NP methods may be traced to the difference in aims just mentioned: these philosophers seek an E-R view and NP does not give them one. To a large extent, such criticisms stem from holding to a certain philosophical image of the "logic" of statistical inference—that it should mirror deductive logic only with degrees—and not at all from finding these methods unproductive in scientific applications. In this view, a theory of statistical inference must provide a quantitative measure of evidential relationship—an E-R measure (whether a measure of support, confirmation, probability, or something else). From this perspective, NP methods will be judged inadequate for statistical inference unless NP error probabilities can be interpreted as E-R measures. Unsurprisingly, as critics show, if error probabilities (e.g., significance levels) are interpreted as E-R measures, misleading and contradictory conclusions are easy to generate. Such criticisms are not really criticisms but flagrant misinterpretations of the quantities in error statistical methods—misinterpretations repeatedly warned against in good textbooks on statistics. I have discussed criticisms based on E-R misinterpretations of error probabilities at length elsewhere (e.g., Mayo 1980, 1981, 1982, 1983, and 1985a), and I will not give them much additional consideration.

A second set of criticisms that can also be seen to follow from the E-R image of statistics is that based on assuming the likelihood principle. Since this assumption, we saw, is tantamount to assuming the irrelevance of outcomes other than the one observed, and therefore to rejecting error probabilities, these criticisms beg the question against error statistical methods. To remind us, recall the criticism of error statistical methods based on the "argument from intentions" discussed in section 10.3. If one adheres to the likelihood principle (as Bayesians do), then it does not matter whether data arose from a try and try again method or from a nonsequential experiment—the stopping rule is irrelevant. To deem stopping rules relevant—as statistical significance tests do—is, from the Bayesian point of view, tantamount to making the experimenter's intentions relevant. All the other error statistical properties are similarly found to be "incoherent" on the likelihood principle. The tables are turned completely, we saw, for an error statistician. Given an observed outcome x, the error statistician finds it essential to consider the other outcomes that could have resulted from the procedure that issued x. Ignoring aspects of the experiment that alter error probabilities (e.g., the stopping rule) violates error statistical reasoning and permits systematically misleading results.

However, we can separate out from the critical literature several legitimate questions of the epistemological basis of the NP methods: How should test results be interpreted in scientific contexts? What is so good about tests that are good or "best" on error-probability criteria? How can any of the seemingly arbitrary choices of tests and error probabilities be justified? I grant that without adequate answers to these questions, the NP prescriptions can appear to license counterintuitive and unsatisfactory results.

The problem stems from the decision-theoretic framework in which NP methods are standardly couched. Although this framework has its uses, it does not adequately reflect most of the reasons that scientists find these methods correspond precisely to their needs. We need a framework that captures the nature and rationale of NP methods in scientific practice.

Happily, we already have it. In the error statistical account, formal statistical methods relate to experimental hypotheses, hypotheses framed in the experimental model of a given inquiry. Relating inferences about experimental hypotheses to severe tests of primary scientific claims is, except in special cases, a distinct step. Standard statistical ideas and tools enter into this picture of experimental inference in a number of ways, all of which are organized around the three chief models of inquiry. Their role is to (i) provide techniques of data generation and modeling along with tests for checking whether the assumptions of data models are met; (ii) provide tests and estimation methods that allow control of error probabilities; and (iii) provide canonical models of local experimental questions with associated tests and data modeling techniques.

Knowing what we want from our statistical theory, and having the elements of our framework at our disposal, it will be easy to cut through the seemingly complex arguments from philosophy of statistics. Getting NP tests to do what we want them to do, however, requires diverging from some of the key tenets that are presumed to be integral to the NP theory. The focus in this chapter is tests. The key tenets of NP testing from which we may be required to diverge are at the same time at the heart of many of the criticisms of NP theory. Accordingly, my reformulation of NP statistics will simultaneously respond to two challenges: how to answer the main criticisms of that approach, and how error statistical methods provide the needed tools for learning from error.

While it seems correct to call my approach a reinterpretation of NP statistics, I want to argue that the appropriate use of NP methods is already to be found—albeit only by hints and examples—in one of the two founders of NP statistics: Egon Pearson (as well as in most of actual practice). Egon S. Pearson (not to be confused with his father, Karl[1]), although one of the two founders of NP methods, rejected the statisti-

---

1. Karl Pearson's subjectivist philosophy contrasts with that of his son Egon.

cal philosophy that ultimately became associated with NP statistics—or so I shall argue. Many contemporary criticisms of NP methods mirror Pearson's own reasons for this rejection. Extricating the view E. S. Pearson *did* hold gives a much deeper and more accurate understanding of NP principles than that which comes out in either statistics textbooks or in the presentations of critics of the NP approach. It is against these caricatures of NP methods that the criticisms of NP are largely directed. Understanding Pearsonian statistics shows how and why actual uses of NP methods generally circumvent the pitfalls without forfeiting what is central to error statistical methods: the fundamental importance of error probabilities.

11.2 NEYMAN-PEARSON THEORY OF STATISTICAL TESTS (NP TESTS)

I want to begin by putting aside for a moment the concepts of our new framework and broaching NP tests in their more formal rubric. I want to get us to consider the tests in their naked mathematical form, the better to see the latitude for their use and interpretation. The highlights of chapter 5—the examples of NP tests, the discussion of probabilistic models, and the hierarchy of models in experimental inquiry—prepare us for each of the ideas we now need. As we proceed, the connection with severity and arguments from error will emerge.

To really get down to the bare bones, the NP testing theory can be seen to define mathematical functions on random variables. The variables may take on different values corresponding to different outcomes of an experiment. Tests are functions that map possible values of these variables (i.e., possible experimental outcomes) to various hypotheses about the population from which outcomes may have originated. Commonly, the hypotheses are assertions about some property of this population, a *parameter*, which governs the statistical distribution of the experimental variable X. As before, I confine myself to cases with only a single unknown parameter, say μ. A test is like a postal system wherein different values of X (different addresses) get sent to different values of μ (different destinations).

An example already considered several times is the Binomial experiment, the common exemplar being coin-tossing. Here the statistical variable might be the proportion of heads in n tosses, written as $\overline{X}$, and the hypotheses, assertions about the (Binomial) parameter p, the probability of heads on each toss. The test is a rule that "sends" the different observed proportions of heads to various values of the parameter p.

The standard NP test splits the possible parameter values into

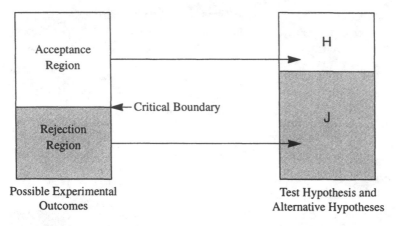

FIGURE 11.1.   NP tests as mapping rules.

two—there are, so to speak, two destinations. One represents the *test
hypothesis H*, the other the set of *alternative hypotheses J*. For example, *H*
might assert that *p* = .5, while *J*, that *p* > .5. Hypothesis *H* here is
simple because it consists of just one value of *p*, while *J* is *composite*.
The test maps each of the possible outcomes—the experimental *sample
space*—into either *H* or *J*; those mapped into *H* (i.e., into "accepting"
H) form the *acceptance region*, those mapped into alternative *J* form the
rejection (of H) region. This partition of the sample space is typically performed by specifying a cutoff point or *critical boundary* $\overline{X}^*$. Any outcome
falling outside bound $\overline{X}^*$ falls into the rejection region.

An example would be to reject *H* whenever the observed propor-
tion of heads, $\overline{X}$, is at least .8. The critical boundary $\overline{X}^*$ is .8. There are
two ways to specify the critical boundary. The critical boundary may
be given by specifying a *distance measure D* between $\overline{X}$ and *H*, and indicating "how far" $\overline{X}$ can be from *H* before slipping into the rejection (of
H) region. Equivalently, the cutoff point may be given by specifying
the significance level α, such that once that level is reached, *H* is rejected. (Recall that the larger the difference *D*, the smaller the significance level.) Leaving these acceptances and rejections uninterpreted
for now, the formalism of the NP model simply describes the partitioning that results from the mapping rules as illustrated above (fig.
11.1).

NP tests focus on the probabilistic properties of these mapping
rules, that is, on the probabilities with which the rule leads to one or
another hypothesis, under varying assumptions about the true hypothesis. Two types of errors are considered: first, the test leads to reject

H (accept *J*) even though *H* is true (the type I error); second, the test leads to accept *H* although *H* is false (the type II error). The test is specified so that the probability of a *type I error*, represented by α, is fixed at some small number, such as .05 or .01. In other words, the test is specified to ensure that it is very improbable for an outcome to fall in the "rejection (of *H*) region" when in fact the hypothesis *H* is correct. Having fixed α, called the *size* or *significance level* of the test, NP principles seek out the test that at the same time has a small probability, β, of committing a *type II error*: accepting *H* when *J* is actually the correct hypothesis. 1 − β is the corresponding *power* of the test. That is:

P(test *T* rejects *H* | *H* is true) ≤ α = probability of type I error.

P(test *T* accepts *H* | *J* is true) ≤ β = probability of type II error.

When, as is quite common, alternative *J* contains more than a single value of the parameter, that is, when it is *composite*, the value of β varies according to which alternative in *J* is true. α and β are the test's formal *error probabilities*. To reemphasize, error probabilities are not probabilities of hypotheses, but the probabilities that certain experimental results occur, were one or another hypothesis true about the experimental system. Consider, for example, the probability of a type I error in testing *H* with test *T*. This is the probability of getting an experimental result that test *T* maps to "reject *H*," when in fact *H* is true.

This leads to the cornerstone of NP tests: their ability to ensure that a test's error probabilities will not exceed some suitably small values, fixed ahead of time by the user of the test, regardless of which hypothesis is correct. These key points about the bare bones of NP tests can be summarized as follows:

> An *NP test* (of hypothesis *H* against alternative *J*) is a rule that maps each of the possible values observed into either Reject *H* (Accept *J*) or Accept *H* in such a way that it is possible to guarantee, *before the trial* is made, that (regardless of the true hypothesis) the rule will erroneously reject *H* and erroneously accept *H* no more than α (100 percent) and β (100 percent) of the time, respectively.

The "best" test of a given size α (if it exists) is the one that at the same time minimizes the value of β (equivalently, maximizes the *power*) for all possible alternatives *J*.

Note that the size of a test is the same as the significance level of the cutoff point beyond which *H* is rejected. That is why tests with size α are often described as tests with significance level α. The relationship between severity and size and power will be discussed explicitly in section 11.6.

11.3. THE BEHAVIORAL DECISION PHILOSOPHY: NP TESTS AS ACCEPT-REJECT ROUTINES

The proof by Neyman and Pearson of the existence of "best" tests encouraged the view that tests (particularly "best" tests) provide the scientist with a kind of *automatic rule* for testing hypotheses. Here tests are formulated as mechanical rules or "recipes" for reaching one of two possible decisions: "accept hypothesis *H*" or "reject *H*" (accept alternative *J*). The justification for using such a rule is its guarantee of specifiably low error rates in some long run.

This interpretation of the function and the rationale of tests was well suited to Neyman's statistical philosophy. For Neyman, "The problem of testing a statistical hypothesis occurs when circumstances force us to make a choice between two courses of action: either take step *A* or take step *B*" (Neyman 1950, 258). These are not decisions to accept or believe that what is hypothesized is (or is not) true, Neyman stresses. Rather, "to accept a hypothesis *H* means only to *decide to take action A rather than action B*" (ibid., 259; emphasis added). On Neyman's view, when evidence is inconclusive, all talk of "inferences" and "reaching conclusions" should be abandoned. Instead, Neyman sees the task of a theory of statistics as providing rules to guide our behavior so that we will avoid making erroneous decisions too often in the long run of experience. A clear statement of such a rule is the following:

> Here, for example, would be such a "rule of behaviour": to decide whether a hypothesis, *H*, of a given type be rejected or not, calculate a specified character, *x*, of the observed facts; if $x > x_0$ reject *H;* if $x \leq x_0$ accept *H*. Such a rule tells us nothing as to whether in a particular case *H* is true when $x \leq x_0$ or false when $x > x_0$. But it may often be proved that if we behave according to such a rule . . . we shall reject *H* when it is true not more, say, than once in a hundred times, and in addition we may have evidence that we shall reject *H* sufficiently often when it is false. (Neyman and Pearson 1967b, 142)

Tests when interpreted as rules of *inductive behavior* make up a key portion of the *behavioristic (or behavioral) model* of tests. Because this model is typically associated with Neyman and Pearson theory, defects of that model are taken as defects of the theory. My position is that there are other, more satisfactory models to direct the use and interpretation of the NP methods, and that they are provided by the present approach to experimental learning. But before getting to that it is important to do battle with a certain, not uncommon, misunderstanding.

What the Behavioral Model Does Not Say

The misunderstanding concerns the construal of "accept" and "reject" on the behavioristic model. Actually, Neyman is quite clear on what he intends. Accept *H*, Neyman says, means *to take action A* rather than *B*. Accept *H* does not mean believe *H* is true. Accept *H* does not mean act as if you knew *H* was true, in the sense of behaving in any and all of the ways you would if you knew that *H* was true. Supposing that the NP model intends this last interpretation of Accept *H*, Howson and Urbach dismiss NP theory as inappropriate for science as well as for practical action. If a scientist were to interpret accepting a statistical hypothesis in the way Howson and Urbach think NP theory intends,

> he would never bother to repeat the experiment. Moreover, he would be happy to stake his entire stock of worldly goods . . . on a wager offered at odds of, say, 10 to 1 against that hypothesis being true. Or suppose a food additive conjectured to be toxic were subjected to a trial involving 10 persons and the conjecture were rejected, then the manufacturer would be prepared to go directly into large-scale production and distribution. This interpretation of acceptance and rejection has merely to be stated to reveal its absurdity. . . . Nevertheless, despite its immense implausibility, this seems to be the way statisticians standardly interpret the notions. (Howson and Urbach, 1989, 162–63)

Their hilarious portrayal of the way they suppose "statisticians standardly interpret" the acceptance of a statistical hypothesis has no relation to any real statistician. This is not just because in reality statisticians do not strictly follow Neyman's behavioristic model, but because no such interpretation is licensed by that model. Howson and Urbach confidently assert, but on what basis I cannot imagine, that

> it is evident that the behaviour Neyman and Pearson had in mind was the acceptance and rejection of hypotheses as being true or false, that is, the adoption of the same attitude towards them as one would take if one had an unqualified belief in their truth or falsehood. (Ibid., 163)

But this is not at all evident, and Neyman and Pearson could not have been clearer in their rejection of anything like the construal that Howson and Urbach pin on the NP approach. (Nor do any of Howson and Urbach's citations offer any evidence otherwise.)

Neyman's behavioristic model literally identifies the acceptance of

H with the adoption of a decision to take some specific action *A* (rather than *B*) where *A* is set out at the start. One cannot choose or even articulate a test of any hypothesis until one identifies the acceptance of *H* with some one action *A*. Only then is it possible to determine the test's error probabilities, the basis upon which the choice of the test depends. The test's error probabilities may be acceptably low as regards one action while unacceptably high as regards some other. (For instance, they might be acceptably low regarding deciding to do further research on a topic, while unacceptably high regarding taking the act of publishing the results.) In the behavioral model of tests, "accept *H*" gets its interpretation from the specific action (pre)designated by the test in question. So the Howson-Urbach reading conflicts with the idea of fixed, predesignated error probabilities, aside from being a perversion of both Neyman's and Pearson's views.

Admittedly, from the fact that "accept *H*" gets its interpretation from the specific action designated by a given test, it follows that its meaning varies in different tests of *H*. In the behavioral model of tests, its meaning will vary according to the action identified with the result "accept *H*." This is just what Neyman intends.

Isaac Levi (1980a, 1984) offers perhaps the clearest depiction of the behavioristic model of NP among contemporary philosophers of statistics. He suggests "that a good approximation to [Neyman and Pearson's] intent is obtained by construing them as recommending the use of programs for using observation reports as inputs into programs designed to select acts" (Levi 1980a, 406). The idea is to have a rule, laid out ahead of time, for which action to take upon the occurrence of each possible experimental result. Such a "routine" procedure contrasts with what Levi calls a "deliberational" procedure. Where Levi and I may disagree, if we do, is on whether NP theory also admits of a nonbehavioristic (and deliberational) interpretation. Neyman himself is quite clear about his philosophy of inductive behavior, and I want to look a little at what he says.

Neyman and His Inductive Behavior

Neyman's idea of a rule of behavior is innocuous enough. Humans notice çertain fairly stable patterns, Neyman begins—for example, that rain or snow storms follow the appearance of heavy clouds—and form various habits in regard to them—for example, taking cover at the sight of dark clouds. A similar kind of regularity is recognized, Neyman says, in the relative frequency with which a result occurs in repeated trials of some game of chance (real random experiments). Mathematical statistics developed as a way of providing systematic rules for how to act with regard to this latter type of regularity. They are like rules for good habits.

Neyman offers the following, very general definition of a rule of inductive behavior:

> Let $E_1, E_2, \ldots, E_n, \ldots$ be all possible different outcomes of an experiment or of observations relating to some phenomena. Let $a_1, a_2, \ldots, a_m, \ldots$ be all the different actions contemplated in connection with these phenomena.
> *If a rule R unambiguously prescribes the selection of action for each possible outcome E_i, then it is a rule of inductive behavior.* (Neyman 1950, 10)

The statistical test, then, is a special case of a rule of behavior, one where the outcomes occur with some probability, that is, the experimental variable E follows some probability distribution. The acts, on Neyman's model, are condensed into two, a_1 and a_2. The hypotheses are assertions about the probabilities of the possible outcomes E_i—they are statistical hypotheses—and the desirability of performing the two actions depends upon which statistical hypothesis is true (Neyman 1950, 258). Correspondingly, the set of (admissible) hypotheses is split up into two, H and J, where J is regarded as not-H. The idea is that if hypothesis H (or any of the hypotheses making up region H) is true, then action A would be preferable to B, while if any of the hypotheses in J are true, action B would be preferable to A. A rule of inductive behavior determining the choice of A or B according to the experimental outcome E is a test of a statistical hypothesis.

Why does Neyman call them rules of "inductive behavior" as opposed to, say, test rules? He is led to this term because of his scruples about the term "inductive inference." Neyman begins *First Course in Probability and Statistics* as follows:

> Claims are occasionally made that mathematical statistics and the theory of probability form the basis of some mental process described as "inductive reasoning." However, in spite of substantial literature on this subject, the term "inductive reasoning" remains obscure and it is uncertain whether or not the term can be conveniently used to denote any clearly defined concept. On the other hand, as was first remarked in 1937, there seems to be room for the term "inductive behavior." This may be used to denote the adjustment of our behavior to limited amounts of observation. (P. 1)[2]

In addition to wanting to highlight the contrast with "inductive inference," Neyman was doubtless influenced by the common parlance

2. Neyman's reference is to Neyman 1967a, 250–90.

during the time NP tests were being developed. As Alan Birnbaum notes, "the 1920's and 1930's were a period of much critical concern with the meanings and possible meaningless[ness] of terms. . . . These concerns were usually pursued in terms of such doctrines as behaviorism, operationalism, or verificationism" (Birnbaum 1977, 33).

The idea of tests as rules of behavior is not all there is to the behavioristic model of tests. The other key features come in when considering how to select which of the many possible test rules to employ. In selecting such a rule one is led to consider that there are four possible situations that can result. To paraphrase Neyman 1950, p. 261:

  I. Hypothesis H is true and action A is taken.
 II. Hypothesis J is true (H is false) and action A is taken.
III. Hypothesis H is true and action B is taken.
 IV. Hypothesis J is true and action B is taken.

These can be represented using the familiar 2 by 2 square:

	Action A	Action B
H	I	III
J	II	IV

It is assumed that A is preferable to B when H is true and that B is preferable to A if J is. As such, when a test results in situations II and III, the test errs by instructing one to take the less preferred action. The test rule is to be selected in such a way as to control the probabilities of these two types of errors. There are, however, several ways of doing this.

Neyman is led by the consideration that "with rare exceptions, the importance of the two errors is different, and this difference must be taken into consideration when selecting the appropriate test" (Neyman 1950, 261). Typically, he finds, one of the two errors is "more serious," more desirable to avoid. The behavioral model instructs one to let H—the test hypothesis—be the one whose erroneous rejection is considered the more serious. (Situation III is worse than situation II.) This error—the one that comes first in importance—is to be made the type I error of the test. The test is selected to fix the probability of the type I error at some low value and then choose the test that does best (or at least reasonably well) as regards the probability of a type II error.

The paradigm example that seems to fit the behavioristic model is acceptance sampling in industrial quality control. Here a sample from some batch of products is observed in order to decide whether or not to reject the batch as containing too many defectives, say, for shipping.

This is a paradigmatic case in which the importance of errors reflects economic values, and the differential weighing of the errors reflects the losses judged affordable. The values can also be ethical, as in one of Neyman's main illustrative examples.

Testing the Toxicity of Drugs: Neyman

In manufacturing drugs, impurities occasionally enter that are sufficiently toxic that minute quantities that escape ordinary chemical analysis can be dangerous. Prior to putting a newly manufactured lot on the market, it is tested. Small doses are injected into experimental animals and the effect recorded. Let X, the variable recorded, be the number of deaths among the n animals injected with a specified dose of the drug. The experiment is modeled as observing this random variable X. The probability of the different possible number X of deaths depends on how toxic the drug is: all or most animals die if the drug is toxic. The different values of X, Neyman supposes, may lead to one of two possible courses of action: (a_1) put the lot of drug on the market, and (a_2) return the lot to the manufacturer:

> The two kinds of error connected with actions a_1 and a_2 are very different. . . . First consider the case where action a_1 is taken when the appropriate action is a_2. This means that the drug is dangerously toxic but declared harmless through the unavoidable inaccuracies of the experiment. . . . Error of this kind may cause death to the patients treated with the drug. Actual cases of this kind are on record. (Neyman 1950, 263)

Neyman contrasts this with the error of taking action a_2—returning the lot—when in fact a_1 is appropriate. Although the consequences of this error are unpleasant, and may result in financial losses to the manufacturer and an increased price of the drug, "the occasional rejection of a perfectly safe drug is clearly much less undesirable than even an infrequent death of a patient" (ibid.). So this type of error is less important than the first, and would be identified as the type II error.

Neyman's test, then, is a rule of inductive behavior with two hypotheses, two corresponding actions, and two associated errors, one (typically) more important than the other; and the basis for selecting among tests is the goal of controlling the probabilities that they would lead to these errors. In Neyman's view, "in many cases the relative importance of the errors is a subjective matter" and "lies outside of the theory of statistics" (Neyman 1950, 263). Such remarks have led to misunderstandings (see section 11.6). To understand Neyman's attitude, I suggest we think back to his view regarding the use of statistical models generally (discussed in chapter 5). His attitude seems to be this: here is a formal statistical technique that seems to reflect certain features of a standard testing context. It is up to you to assign acts to the two hypotheses, to ascertain which of the two errors is more important to avoid in your testing context (making that one the type I error) and to determine how often such an error seems acceptable (which will direct you to fix the α level of your test). That is "subjective". The NP theory can then use its machinery to find the test that at the same time minimizes the probability of a type II error (β). Once the rule is selected (and assuming the assumptions are approximately met), hypothesis testing is on automatic pilot—on the behavioral model. Applying the test just means following the rule. The experimental outcome is observed, and the test tells you whether to take one of two actions, *A* or *B*, according to whether or not the outcome falls in the rejection region of hypothesis *H*. There is no hemming and hawing, no agonizing over the particular case. Your long-run low error rate needs are guaranteed, and they are guaranteed objectively.

All this is fine and dandy, say critics, if your actual needs correspond to the kind of decision-making context envisioned in the behavioristic model; but scientific inquiry does not seem to be such a context. The issue is not just whether science involves decision making or whether inference can be seen as a kind of decision. Many who are happy to regard all of science as decision making—the typical Bayesian decision maker—reject the NP theory, not because of its development along decision-theoretic lines, but because it does not go far enough in its decision-theoretic leanings. (A full decision theory would involve not only the losses captured in error probabilities but explicit loss functions, prior probabilities, and all the rest of the "full dress" Bayesian approach.) The issue now, raised by both Bayesian and non-Bayesian critics of the behavioristic approach, concerns the appropriateness of the particular kinds of decision strategies depicted in the behavioristic model. Letting a decision to accept or reject a hypothesis turn on whether data reaches a cutoff point just seems too, well, too automatic. Many statisticians allege that no one, not even Neyman, ever tests a scientific claim along the strict behavioristic line.

I agree with them. My position all along is that the NP account admits of a nonbehavioral construal that is more satisfactory and more accurately reflects how NP methods are used in experimental learning. By and large, however, NP tests are still formulated along the lines of the behavioristic model, with the probability of a type I error generally set at the conventional levels of .01 or .05. Why are NP methods so productively used in science despite their "rule of behavior" formulation? How, paraphrasing Neyman, do they manage to correspond precisely to the needs of applied research? There seem to be two main reasons: First, many scientific tasks fit the "assembly line" behavioral-decision model. At many junctures in the links between experimental and data models there is a need for standardized approaches to data analysis that allow us to get going with few assumptions, enable results to be communicated uniformly, and help ensure that we will not too often err in declaring "normal puzzles" solved or not. Second, the behavioral decision approach provides canonical models for nonbehavioral and non-decision-theoretic uses. The behavioral concepts simply serve to characterize the key features of NP tools, and *these features* are what enable them to perform the nonbehavioral tasks to which tests are generally put in science.

Although I take the second reason to be weightier, as well as being the more interesting for our purposes, the first reason should not be disregarded altogether. There are uses for statistics in science for which the behavioristic construal is apt and for which NP as a theory for routine decision making has made for real progress. One type of example is discussed by the statistician Irwin Bross.

Controlling the Noise in Communications Networks

The context Bross (1971) discusses concerns decisions to report a given message, say, that a drug is effective or, more generally, that an effect "is real," not spurious. Bross's particular focus is on analgesics. The act of reporting that a drug is effective is not tantamount to taking any and all acts that would be licensed were it known to be effective, a point we have already made. (The act of reporting is distinct from subsequent possible actions, say, for physicians to use the drug or to buy stock in the drug company.) But a decision to report it as effective may have repercussions for subsequent decisions, and tools for routine error control may be called for.

An NP test may be used as a routine for declaring an analgesic effective in the following manner. It may stipulate: report a drug effective only if an observed difference in effect rates is statistically significant at, say, the 1 percent level. Following Bross, the various sources of error that can creep into scientific reports may be seen as sources of noise in a scientific communications network. Noise from random sources—which is inevitable in experimental research—is often called sampling variation.[3] The adoption of a fixed critical level or size, say 1 percent, is useful in "controlling the noise in communication networks." According to Bross, prior to the advent of controlled clinical

---

3. Noises from nonrandom sources are sometimes called biases or extraneous variables.

trials the noise level in analgesic testing was high enough to impede progress seriously (Bross 1971, 503–4).

To illustrate, Bross describes an uncontrolled drug testing network where researchers report favorably on all new drugs tested, noting that "some years ago this would not have been an entirely unrealistic model for certain networks" (p. 504). If four out of five of these drugs are actually ineffective, being no more effective than a standard agent, 80 percent of the favorable reports would be false. With such a high level of noise, no reliance could be placed on the reports. If, on the other hand, each member of the network reports favorably only on those drugs that pass a test with critical level or size of, say, 5 percent, then the proportion of false positive reports is kept low, at 5 percent.

Thus the use of statistical significance tests as accept-reject routines for a thumbs up or down approach on analgesics helped to control the noise in the scientific communication network. True, it would not do to apply such a behavioristic construal of tests to deciding to accept or reject substantive scientific hypotheses directly. Nevertheless, the hierarchy of models in an experimental inquiry may also be seen as a "communication network," and it is plainly desirable to have tools for controlling errors at numerous points in this network of models. Controlling the errors at various segments of the inquiry is what enables the overall reliability and severity to be achieved. One could well imagine, for example, how Jean Perrin might have used a routine test to give a "yes or no" pronouncement to whether a given grain, after undergoing his technique of fractional centrifuging, was sufficiently uniform to be included in the next stage of the Brownian motion analysis (chapter 7). One act would be including the grain into the analysis, a second would be to subject the grain to further centrifuging. Assurance that he would rarely include insufficiently uniform grains as well as rarely carry out unnecessary centrifuging was precisely what Perrin sought.

These points go toward illustrating what I gave as the first reason that the behavioral model of tests has a serviceable role in research, namely, that there are scientific tasks that fit the behavioristic model. Even these uses, however, depend upon designing, interpreting, and combining several tests in a manner that is decidedly *not* automatic. The second and more important reason that NP tests supply needed tools for research is that their methods provide standard or canonical models for nonbehavioral and non-decision-theoretic uses. Undoubtedly, many of the behavioral concepts with which Neyman chose to characterize the key features of NP tests would not have been chosen by Pearson. But these concepts succeed in characterizing the features of the tests well enough, and these features are what enable tests to perform the nonbehavioral tasks to which they are generally put in science. In these nonbehavioristic contexts, tests license not acts, but arguments or inferences as to what is learned from particular experimental results. The arguments are arguments from error.

This, I propose, is why behavioral models of tests provide serviceable canonical models for nonbehavioral tasks. The tests can and should be seen as tools whose distinctive properties enable them to be used to ask a variety of standard questions about errors—quite generally construed. The result of a single statistical test does not license a substantive scientific inference. Instead, each such test, or set of tests, teaches the answer to a specific question, and error control at local points is the key to arriving at substantive severity arguments.

11.4 PEARSON REJECTS THE NEYMAN-PEARSON (BEHAVIORISTIC) PHILOSOPHY

Alan Birnbaum (1969, 1977) argued that NP admits of two types of interpretations: in one, Neyman's behavioral decision view, we saw that the test result is literally a decision *to act* in a certain way; in the other, which Birnbaum called an "evidential" view, the test result is interpreted as providing strong or weak evidential support for one or another hypothesis.[4] While I do not embrace the particular evidential interpretation Birnbaum favored, I think he was quite right that in situations of scientific research the behavioral interpretation of tests is

---

4. Birnbaum called the concept underlying this evidential interpretation of NP the *confidence concept*, which he formulated (1977, 24) as follows:

> (Conf): A concept of statistical evidence is not plausible unless it finds "strong evidence for J as against H" with small probability (α) when H is true, and with much larger probability ($1 - \beta$) when J is true.

Birnbaum argued that scientific applications of NP tests made intuitive use of something like the confidence concept. Birnbaum's approach, incomplete at the time of his death, sought to make explicit the correspondence between an NP result and a statement about the strength of evidence (e.g., conclusive, very strong, weak, or worthless). For example, he interprets reject H against J with error probabilities α, β equal to .01 and .2, respectively, as very strong statistical evidence for H as against J. A main shortcoming, as I see it, is that it interprets a test output—say, reject H—from two tests with the same α, β as finding equally strong evidence for J. Depending upon the particular outcome and the test's sample size, the two rejections may constitute very unequal tests of J—something I take up in later sections. Birnbaum's rules do not seem to reflect such differences. Further criticism along these lines occurs in Pratt 1977. I discuss more generally attempts at "evidential" interpretations of NP methods in Mayo 1985a.

intended to apply "in a way which is heuristic or hypothetical, serving to explain the inevitably abstract theoretical meanings associated with the error probabilities [and] formal 'decisions' such as 'reject H'" (Birnbaum 1977, 32–33). The behavioristic formulation of tests, Birnbaum proposed, should simply be seen as a way of articulating the new statistical ideas of the NP approach. That the behavioral construal of tests is still with us, I suggest, testifies that they still serve the kind of heuristic function that Birnbaum had in mind.

Birnbaum found clues of these nonbehavioral intuitions in the writings of Pearson. One particularly interesting document that Birnbaum (1977, 33) supplies includes an unpublished remark by Pearson in 1974:

> I think you will pick up here and there in my own papers signs of evidentiality, and you can say now that we or I should have stated clearly the difference between the *behavioral* and *evidential* interpretations. Certainly we have suffered since in the way the people have concentrated (*to an absurd extent often*) on behavioral interpretations. (Emphasis added)

Pearson never articulates just what evidential interpretation he supports, and I do not think that Birnbaum's evidential model, so far as he worked it out (in which NP results are reinterpreted in terms of strong or weak evidence for hypotheses), is indicated in Pearson's "signs of evidentiality." Nevertheless, I endorse Birnbaum's proposal that the behavioral model of NP tests be regarded as a device to communicate what the tests could be used for, while requiring reinterpretation in scientific contexts. This, I believe, was also Pearson's view, and that is why I say Pearson rejects what have come to be identified as the key tenets of the NP behavioral philosophy. What Pearson rejects is the philosophy associated with Neyman's inductive-behavior model.

The Rationale of Tests according to the NP Behavioristic Philosophy

Because NP theory developed mathematically in a decision-theoretic framework (along with the work of Abraham Wald), the statistical philosophy generally associated with these tools is Neyman's behavioral decision one. Often it is referred to as the Neyman-Pearson-Wald (NPW) approach.[5] We can identify two closely connected aspects

5. Even that arch opponent of Neyman, Bruno de Finetti, held that the expression "inductive behavior . . . that was for Neyman simply a slogan underlining and explaining the difference between his own, the Bayesian and the Fisherian formulations" became, with Wald's work, "something much more substantial" (de Finetti 1972, 176). He called this "the involuntarily destructive aspect of Wald's work" (ibid.).

of this decision philosophy: first, the justification of tests in terms of low (long-run) error rates, and second, the function of tests as routine decision rules. While these features, taken strictly, give a caricature of tests, even as Neyman intended them, they are at the heart of the philosophical criticisms of NP to which we need to respond.

Long-run (low error-probability) justification. Since the criteria for goodness of a test are its low error probabilities in the frequentist sense, the justification for using tests is (apparently) solely their ability to guarantee low long-run errors in some sequence of applications. This is not a final measure of the degree of support or probability acquired by hypotheses—it is not an E-R measure. For example, to reject H with a test having a low probability of erroneous rejections does not say that the *specific* rejection has a low probability of being in error, but only that it arises from a testing *procedure* that has a low probability of leading to erroneous rejections. Likewise with confidence levels attached to particular interval estimates. Critics of NP theory deny that low error rates in the long run are relevant to justifying any particular inference.

Tests as decision "routines" with prespecified error properties. This feature is associated with two main criticisms. First, there is the fact that the NP decision model does not give an interpretation customized to the specific result. A test result either is or is not in the prespecified rejection region. But intuitively, if a given test rejects H with an outcome several standard deviations beyond the critical boundary (between rejection and acceptance of H), there is an indication of a greater discrepancy from H than if the same test rejects H with an outcome just at the critical boundary. Both, however, are identically reported as "reject H" (and accept some alternative J), and the probability of a type I error (the test's prespecified *size*) is identical for any such rejection.[6] Second, there is the problem of how to interpret test results. Deciding to accept or reject hypotheses, construed as deciding how to act, does not seem to offer the kind of evidential appraisal needed for scientific inference.

A Dialogue between Pearson and Fisher

These features are not only the source of contemporary criticisms of NP theory. They lie at the heart of R. A. Fisher's original attack on

6. The point here is that to do no more than report the error probabilities, while condoned by the strict NP decision model, is not sufficient to discriminate between these two results—one of the sources of the criticisms of NP tests. *Other* uses of error probabilities, however, can make this discrimination along the lines I discuss in sections 11.6 and 11.7.

Neyman and Pearson's reworking of (what Fisher regarded as "his") significance tests. In his forceful style, Fisher declared that followers of the behavioristic approach are like

> Russians (who) are made familiar with the ideal that research in pure science can and should be geared to technological performance, in the comprehensive organized effort of a five-year plan for the nation. (Fisher 1955, 70)

Fisher makes a similar comparison with the United States:

> In the U.S. also the great importance of organized technology has I think made it *easy to confuse the process appropriate for drawing correct conclusions, with those aimed rather at, let us say, speeding production, or saving money.* (Ibid.)

The allegation is essentially the one cited earlier: NP methods seem suitable for industrial acceptance sampling, but not for drawing inferences in science.

Pearson, however, responds to Fisher's attacks—something critics seem to have overlooked. Perhaps this is because it occurs in an obscure, very short (but fascinating) paper, "Statistical Concepts in Their Relation to Reality" (Pearson 1955), that is not included in *The Selected Papers of E. S. Pearson.*

Pearson Responds to Fisher

What one discovers in Pearson's (1955) response to Fisher (and elsewhere in his work) is that for scientific contexts Pearson rejects both the low long-run error probability rationale and the nondeliberational, routine use of tests. These two features are regarded as so integral to the NP model that I think it is fair to say that Pearson rejected the NP philosophy (but not NP methods).[7] Pearson did not publish much on his own statistical philosophy per se, but evidence scattered throughout his statistical papers offers a fairly clear picture of the rationale underlying his rejection of these decision features of NP tests. These are the "signs of evidentiality" to which Pearson alluded.

Pearson's Original Heresy

Let us begin with Pearson's (1955) response to Fisher's main criticism—that the NP model turns tests into a pragmatic, five-year-plan type of a process. Pearson insists that

---

7. Perhaps it is clearest to say that what Pearson rejected was the Neyman-Pearson-Wald (NPW) model of NP methods. See also Note 5.

there was no sudden descent upon British soil of Russian ideas regarding the function of science in relation to technology and to five year plans. It was really much simpler—or worse. *The original heresy, as we shall see, was a Pearson one!* (Pearson 1955, 204; emphasis added)

Interestingly, Fisher directs his attacks at *Neyman's* behavioral approach, leaving Pearson out of it.[8] Nevertheless, Pearson protests here that the "original heresy" was really his!

Pearson does *not* mean it was he who endorsed the behavioral-decision model that Fisher attacks. The "original heresy" refers to the break Pearson made (from Fisher) in insisting that tests explicitly take into account alternative hypotheses, in contrast with Fisherian significance tests, which did not. With just the single hypothesis (the null hypothesis) of Fisherian tests, the result is either reject or fail to reject according to the significance level of the result. However, just the one hypothesis and its attended significance level left too much latitude in specifying the test, rendering the result too arbitrary. With the inclusion of a set of admissible alternatives to H, it was possible to consider type II as well as type I errors, and thereby to constrain the appropriate tests.

In responding to Fisher, Pearson is not merely arguing that NP methods *can* be interpreted in a manner other than a pragmatic behavioral-decision one, he is claiming that their original formulation (admittedly "heretical" in the above sense) was not even intended to capture decision-theoretic aims. Those aims came later, and were not his:

Indeed, to dispel the picture of the Russian technological bogey, I might recall how certain early ideas came into my head as I sat on a gate overlooking an experimental blackcurrant plot. (Ibid., 204)

Having sketched for Fisher this marvelous image of his sitting on a gate (my own sketch being the frontispiece), Pearson goes on to explain that his thoughts had not at all to do with speeding up production or saving money. Rather, Pearson continues,

To the best of my ability I was searching for a way of expressing in mathematical terms what appeared to me to be the requirements of the scientist in applying statistical tests to his data.

8. George Barnard, in a private communication, revealed the part he played in Fisher's reception of NP theory. It was Barnard who alerted Fisher to the consequences of proceeding within the behavioristic model of tests favored by Neyman. At the same time, Barnard told Fisher that Neyman's model was to be distinguished from Pearson's philosophy. Barnard 1985 provides an excellent discussion of historical developments in statistics, as well as comments from a number of statisticians.

After contact was made with Neyman in 1926, the development of a
joint mathematical theory proceeded much more surely; *it was not
till after the main lines of this theory had taken shape* with its necessary
formalization in terms of critical regions, the class of admissible
hypotheses, the two sources of error, the power function, etc., *that the
fact that there was a remarkable parallelism of ideas in the field of acceptance
sampling became apparent. Abraham Wald's contributions to decision theory
of ten to fifteen years later were perhaps strongly influenced by acceptance sampling problems, but that is another story.* (Pearson 1955, 204–5; emphasis added)

So it was only after the main NP theory had taken shape that a "remarkable parallelism" with acceptance sampling problems was discovered. And while the NP methods clearly benefited from the mathematical rigor of the newly developed work in decision theory, the original
application, as Pearson saw it, was to learning from data in science.

Pearson proceeds to "Fisher's next objection": to the terms "acceptance" and "rejection" of hypotheses, and to the type I and type II
errors. His admission is again revealing of his philosophy:

It may be readily agreed that in the first Neyman and Pearson paper
of 1928, more space might have been given to discussing how the
scientific worker's attitude of mind could be related to the formal
structure of the mathematical probability theory. . . . *Nevertheless it
should be clear from the first paragraph of this paper that we were not speaking of the final acceptance or rejection of a scientific hypothesis on the basis of
statistical analysis.* . . . Indeed, from the start we shared Professor Fisher's view that in scientific enquiry, *a statistical test is "a means of learning."* (Ibid., 206; emphasis added)

Thus, says Pearson, the NP framework, with its consideration of
alternative hypotheses, grew out of an attempt to provide tests then in
use with an epistemological rationale—one based on their function as
learning tools. In this role, the test's output was not supposed to be
identified with the final acceptance or rejection of a scientific hypothesis. Instead, the test teaches about a specific aspect of the process that
produced the data. A suitable reformulation of NP tests, I believe,
grows directly out of the distinct roles that statistical tests play in filling
out and linking models in an experimental inquiry. Although Pearson
did not himself propose such a reformulation, Pearson clearly distances
the original learning function of NP methods from the later behavioral-
decision construal to which Fisher is objecting. He declares in the last
line of this paper that

Professor Fisher's final criticism concerns the use of the term "inductive behaviour"; this is Professor Neyman's field rather than mine.
(Ibid., 207)

Pearson Rejects the Long-Run Rationale

It seems clear that for Pearson the value of NP tests (in scientific or learning contexts) does *not* depend on the long-run error-rate rationale found in the decision model. Pearson raises the question as follows, the mention of "inference" already in contrast with Neyman:

> How far then, can one go in giving precision to a philosophy of statistical inference? (Pearson 1966a, 172)

He considers the rationale that might be given to NP tests in two types of cases, *A* and *B:*

> (A) At one extreme we have the case where repeated decisions must be made on results obtained from some routine procedure. . . . (B) At the other is the situation where statistical tools are applied to an isolated investigation of considerable importance. (Ibid., 170)

In cases of type *A*, long-run results are clearly of interest, while in cases of type *B*, repetition is impossible or irrelevant. For Pearson's treatment of the latter case (type *B*) the following passage is telling:

> In other and, no doubt, more numerous cases there is no repetition of the same type of trial or experiment, but all the same we can and many of us do use the same test rules to guide our decision, following the analysis of an isolated set of numerical data. Why do we do this? What are the springs of decision? Is it because *the formulation of the case in terms of hypothetical repetition helps to that clarity of view needed for sound judgment?* Or is it because we are content that the application of a rule, now in this investigation, now in that, should result in a long-run frequency of errors in judgement which we control at a low figure? (Ibid., 173; emphasis added)

Although Pearson leaves this tantalizing question unanswered, claiming, "On this I should not care to dogmatize," it is evident from his treatment of type *B* cases that, for Pearson, "the formulation of the case in terms of hypothetical repetition helps to that clarity of view needed for sound judgment." In addressing this issue, Pearson intends to preempt what he calls the "commonsense" objection to long-run justifications—precisely the objection lodged by contemporary critics of NP theory:

> Whereas when tackling problem A it is easy to convince the practical man of the value of a probability construct related to frequency of occurrence, in problem B the argument that "if we were to repeatedly do so and so, such and such result would follow in the long run" *is at once met by the commonsense answer that we never should carry out a precisely similar trial again.*

Nevertheless, it is clear that the scientist with a knowledge of statistical method behind him can make his contribution to a round-table discussion. (Ibid., 171)

Seeing how the scientist makes his contribution leads to substantiating my second claim, that Pearson rejects the routine use and interpretation of NP tests associated with the behavioral model. For the scientist's "contribution to a round-table discussion" turns on the thoughtful use of error probabilities to unearth causal knowledge—something not reducible to routine.

Nonroutine Uses of Tests: An Example of Type B

Weaving together strands found throughout Pearson's work, one can craft a picture of statistical tests much like the one I would promote, namely, as tools for learning about causal processes by enabling a piecemeal series of standard questions (about errors) to be posed and reliably answered. In the opening of a 1933 paper (jointly written with S. S. Wilks) Pearson writes:

> Statistical theory which is not purely descriptive is largely concerned with the development of tools which will assist in the determination from observed events of *the probable nature of the underlying cause system that controls them.* . . . We may trace the development through a chain of questionings: Is it likely, (*a*) that this sample has been drawn from a specified population, *P;* (*b*) that these two samples have come from a common but unspecified population; (*c*) that these *k* samples have come from a common but unspecified population? (Pearson and Wilks 1966, 81; emphasis added)

An example that Pearson often employs as a case of type *B*, where no repetition is intended, is the following:[9]

> *Example of type B.* Two types of heavy armour-piercing naval shell of the same calibre are under consideration; they may be of different design or made by different firms. . . . Twelve shells of one kind and eight of the other have been fired; two of the former and five of the latter failed to perforate the plate. (Pearson 1966a, 171)

Pearson's interest in this naval shell example stems from his own work on the statistical assessment of army weapons in World War II and after. The experimental variable observed (i.e., the statistic) is the difference, D, between the proportions that perforate the plate from the two types of shell. Its observed value, D_{obs}, equals $\frac{11}{24}$ (i.e., $\frac{10}{12} - \frac{3}{8}$). So

9. Pearson follows this naval shell example through a number of papers.

we have a standard case of a difference in proportions similar to our birth-control pill example in chapters 5 and 6. (In both cases, the null hypothesis predicts a zero difference.) Statistical tests aid the scientist's contribution here by answering a question under (*b*) about the causal origin of the two samples of naval shells:

> Starting from the basis that individual shells will never be identical in armour-piercing qualities, however good the control of production, he has to consider how much of the difference between (i) two failures out of twelve and (ii) five failures out of eight is likely to be due to this inevitable variability. (Ibid., 171)

Notably, Pearson does not simply report whether this observed difference falls in the rejection region (i.e., whether a test maps it to "reject *H*"), but calculates the probability "of getting as great or a greater positive difference" (p. 192) if hypothesis *H* were true and there was no difference in piercing qualities. This is, we know, the *significance level* of the observed difference—a measure that reflects the actual result observed.

Although testing the "no difference" hypothesis is standard, there is not just one plausible way to test it. More than one way has been proposed to describe the data and define a distance between data and hypotheses. This matter is the basis of a historical debate between Pearson and others, which I leave to one side. Although Pearson takes a position in this debate (arguing in favor of the test that he regards as more nearly describing the experimental situation), he does not feel that a single best test needs to be found. Pearson is not perturbed by the existence of this latitude in choosing tests, he does not see it as presenting a problem. It would only present a problem, he thinks, to one who regards tests as giving automatic routines; but, in striking contrast with the routine decision model, Pearson held that little turns on which of the various plausible tests one employs. Treating the (difference between two proportions) case in one way, Pearson obtains an observed significance level of .052; treating it differently (along Barnard's lines), he gets .025 as the (upper) significance level.[10] In an auto-

---

10. The first treatment falls under what Pearson calls Problem I (Barnard's "2x2 independence trial"). Here the question is restricted to just the 20 shells observed, the total number of failures being fixed at the observed one, 7. The test asks whether the observed difference is due to a random partition of the 20 individual shells, of whom 7 would fail to perforate in whichever group they are randomly included. The second way of treating this case views samples from the two processes as random samples from two populations, so the failure rates can vary from 0 to 12 and 0 to 8, respectively. The test asks whether the probability of failure is the same in both. This falls under what Pearson calls Problem II (Barnard's "2x2

matic routine use of tests this can make a substantial difference. Pearson rejects this use of tests.

> The result of either approach would raise considerable doubts as to whether the performance of the [second] type of shell was as good as that of the [first]. (Ibid., 192)[11]

In either case, the data indicate J: the first type of shell is better than the second, because in either case J passes a severe test (although one is more severe than the other). (Severity for passing J here is 1 minus the significance level.)

Pearson holds that in important cases the difference in error probabilities, depending upon which of these tests is chosen, makes no real difference to substantive judgments in interpreting the results:

> Were the action taken to be decided automatically by the side of the 5% level on which the observation point fell, it is clear that the method of analysis used would here be of vital importance. *But no responsible statistician, faced with an investigation of this character, would follow an automatic probability rule.* (Ibid., 192; emphasis added)

So, faced with this type of investigation, no responsible statistician would be a strict follower of the behavioristic model of NP tests.

Surprisingly, the same type of admonishment against an "automatic" use of tests, along with other remarks redolent of Pearson's "inferential" philosophy, occurs not only in Pearson's own papers, but also in one or two of the joint papers by Neyman and Pearson. In 1928, for example, "they" wrote:

> If then a statistician thoughtlessly decides, whatever be the test, to reject an hypothesis when $P \leq .01$, say, and accept it when $P > .01$, it will make a considerable difference to his conclusions whether he uses [one test statistic or another]. But as the ultimate value of statistical judgment depends upon a clear understanding of the meaning of the statistical tests applied, the difference between the values of the two P's should present no difficulty. (Neyman and Pearson 1967c, 18)

---

comparative trial"). For the naval shell example, Pearson regards the former treatment, though preferred by Barnard, as more artificial than the latter. Which of several ways to treat the 2x2 case had been much debated by Barnard and Fisher at that time. Pearson's position is that the appropriate sample space "is defined by the nature of the random process actually used in the collection of the data," which in turn directs the appropriate choice of test (Pearson 1966a, 190). But Pearson does not think there is a need for a rigid choice from among several plausible tests.

11. Pearson's conclusion inadvertently switches the observation to 2 of 12 and 5 of 8 *successful* perforations, where originally they had been failures. I have stated his conclusion to be consistent with the original results reported in this example.

(*P* here is equal to the significance level.) In other words, if the decision model of NP is taken literally, one accepts or rejects *H* according to whether or not the observed outcome falls in the preselected rejection region. Just missing the cutoff for rejection, say, because the observed significance level is .06 while the fixed level for rejection is .05, automatically makes the difference between an acceptance and a rejection of *H*. The "Pearsonian" view rejects such automation in scientific contexts because

> it is doubtful whether the knowledge that [the observed significance level] was really .03 (or .06) rather than .05 . . . would in fact ever modify our judgment when balancing the probabilities regarding the origin of a single sample. (Ibid., 27)

Most significant in this joint contribution is the declaration that

> if properly interpreted we should not describe one [test] as more *accurate* than another, but according to the problem in hand should recommend this one or that as providing information which is more *relevant* to the purpose. (Ibid., 56–57)

This introduces a criterion distinct from low error rates, namely, the *relevance* of the information. In addition, clues emerge for connecting tests (used nonroutinely) to learning about causes by probing key errors:

> The tests should only be regarded as tools which must be used with discretion and understanding. . . . We must not discard the original hypothesis until we have examined the alternative suggested, and have satisfied ourselves that it does involve a change in the real underlying factors in which we are interested; . . . that the alternative hypothesis is not error in observation, error in record, variation due to some outside factor that it was believed had been controlled, or to any one of many causes. (Ibid., 58)

This sentiment is clear enough: we should not infer some alternative to a hypothesis *H* until other alternative explanations for the discordancy with *H* have been ruled out. The surprise is only that such nonbehavioral talk should occur in a joint paper. Its very title—"On the Use and Interpretation of Certain Test Criteria for Purposes of Statistical Inference"—is at odds with Neyman's philosophy, which concerned behavior and not inference. A curious note by Neyman tucked at the end of this paper may explain its Pearsonian flavor.

> I feel it necessary to make a brief comment on the authorship of this paper. Its origin was a matter of close co-operation, both personal and

by letter. . . . Later I was much occupied with other work, and therefore unable to co-operate. The experimental work, the calculation of
tables and the developments of the theory of Chapters III and IV are
due solely to Dr Egon S. Pearson. (Neyman and Pearson 1967c, 66;
signed by J. Neyman)

This "joint" paper, it appears, was largely a contribution of Pearson's.

11.5 A PEARSONIAN PHILOSOPHY OF EXPERIMENTAL LEARNING

I want now to turn to Pearson's discussion of the steps involved in the
original construction of NP tests (of *H:* no difference). His discussion
underscores the key difference between the NP error statistical (or
"sampling") framework and approaches based on the likelihood principle. The previous chapters have amply illustrated the enormous consequences that this difference makes to an account of scientific testing.
This background should let us quickly get to the heart of why different
choices were made in the mathematical development of NP error statistics. The choices stem not only from a concern for controlling a test's
error probabilities, but also from a concern for ensuring that a test is
based on a plausible distance measure (between data and hypotheses).
By recognizing these twin concerns, we can answer a number of criticisms of NP tests.

Three Steps in the Original Construction of NP Tests

After setting up the *test* (or null) *hypothesis,* and the *alternative
hypotheses* against which "we wish the test to have maximum discriminating power" (Pearson 1966a, 173), Pearson defines three steps in
specifying tests:

Step 1. We must specify [the *sample space*[12]] the set of results which
could follow on repeated application of the random process used in
the collection of the data. . . .

Step 2. We then divide this set [of possible results] by a system of
ordered boundaries . . . such that as we pass across one boundary and
proceed to the next, we come to a class of results which makes us
more and more inclined, on the information available, to reject the hypothesis tested in favour of alternatives which differ from it by increasing amounts. (Pearson 1966a, 173)

Results make us "more and more inclined" to reject *H* as they get further away from the results expected under *H,* that is, as the results

12. Here Pearson calls it the "experimental probability set."

become more probable under the assumption that some alternative J is true than under the assumption that H is true. This suggests that one plausible measure of inclination is the likelihood of H—the probability of a result e given H. We are "more inclined" toward J as against H to the extent that J is more likely than H given e.

NP theory requires a third step—ascertaining the error probability associated with each measure of inclination (each "contour level"):

> *Step* 3. We then, if possible, associate with each contour level the chance that, if [H] is true, a result will occur in random sampling lying beyond that level. (Ibid.)[13]

For example, step 2 might give us the likelihood or the ratio of likelihoods of hypotheses given evidence, that is, the likelihood ratio. At step 3 the likelihood ratio is itself treated as a statistic, a function of the data with a probability distribution. This enables calculating, for instance, the probability of getting a high likelihood ratio in favor of H as against a specific alternative J', when in fact the alternative J' is true, that is, an error probability. We are already familiar with this kind of calculation from calculating severity.

Pearson explains that in the original test model step 2 (using likelihood ratios) did precede step 3, and that is why he numbers them this way. Only later did formulations of the NP model begin by first fixing the error value for step 3 and then determining the associated critical bounds for the rejection region. This change came about with advances in the mathematical streamlining of the tests. Pearson warns that

> although the mathematical procedure may put Step 3 before 2, we cannot put this into operation before we have decided, under Step 2, on the guiding principle to be used in choosing the contour system. That is why I have numbered the steps in this order. (Ibid., 173)

However, if the rationale is *solely* error probabilities in the long run, the need to *first* deliberate over an appropriate choice of measuring distance at step 2 drops out. That is why it is dropped in the standard behavioral model of NP tests. In the behavioral model, having set up the hypotheses and sample space (step 1), there is a jump to step 3, fixing the error probabilities, on the basis of which a good (or best) NP test determines the rejection region. In other words, the result of step 3 automatically accomplishes step 2. From step 3 we can calculate how the test, selected for its error probabilities, divides the possible out-

13. Where this is not achievable (e.g., certain tests with discrete probability distributions), the test can associate with each contour an upper limit to this error probability.

comes. Yet this is different from having first deliberated at step 2 about which outcomes are "further from" or "closer to" H in some sense, and thereby *should* incline us more or less to reject H. The resulting test, despite having adequate error probabilities, might have an inadequate distance measure. Such a test may fail to ensure that the test has an increasing chance of rejecting H the more the actual situation deviates from the one H hypothesizes. The test may even be irrelevant to the hypothesis of interest. The reason that critics can construct counterintuitive tests that appear to be licensed by NP methods, for example, certain mixed tests,[14] is that tests are couched in the behavioral framework from which the task Pearson intended for step 2 is absent.[15]

Likelihood Principle versus Error Probability Principles, Again

It might be asked, if Pearson is so concerned with step 2, why go on to include step 3 in the testing model at all? In other words, if Pearson is interested in how much a result "inclines us" to reject H, why not just stop after providing a measure of such inclination at step 2, instead of going on to consider error probabilities at step 3? This is precisely what many critics of NP have asked. It was essentially Hacking's (1965) point during his "likelihood" period. As briefly noted in previous chapters, Hacking's likelihood account held that the *likelihood ratio* (of H against alternative J) provides an appropriate measure of support for H against J.[16] In such a likelihood view, the tests *should* just report the measure of support or inclination (at step 2) given the data. For Bayesians also, the relevant evidence contributed by the data is fully contained in the likelihood ratio (or the Bayesian ratio of sup-

14. In a mixed test certain outcomes instruct one to apply a given chance mechanism and accept or reject H according to the result. Because long-run error rates may be improved using some mixed tests, it is hard to see how a strict follower of NP theory (where the lower the error probabilities the better the test) can inveigh against them. This is not the case for one who rejects the behavioral model of NP tests as Pearson does. A Pearsonian could rule out the problematic mixed tests as being at odds with the aim of using the data to learn about the causal mechanism operating in a given experiment. Ronald Giere presents a related argument against mixed tests, except that he feels it is necessary to appeal to propensity notions, whereas I appeal only to frequentist ones. See, for example, Giere 1976.

15. A notable exception is the exposition of tests in Kempthorne and Folks 1971 in which test statistics are explicitly framed in terms of distance measures. See also note 28.

16. Hacking later rejected this approach (e.g., Hacking 1972). Although he never clearly came out in favor of NP methods, in 1980 he reversed himself (Hacking 1980) on several of his earlier criticisms of NP methods.

port)—the thrust of the likelihood principle (LP).[17] We discussed the LP at length in chapter 10. To remind us, NP theory violates the LP because a hypothesis may receive the same likelihood on two pieces of data and yet "say different things" about what inference is warranted—at least to the error statistician's ears. To pick up on this difference requires considering not only the outcomes that did occur, but also the outcomes that might have occurred; and, as we saw, the Bayesian (or conditionalist) recoils from such considerations.

The debate in the philosophy of statistics literature often does little more than register the incompatibility between the NP approach on the one hand and the likelihood and Bayesian approaches on the other. Each side has a store of examples in which the other appears to endorse a counterintuitive inference. From the perspective of the aims of ampliative inquiry, I have been arguing, we can go further: control of error probabilities has a valid epistemological rationale—it is at the heart of experimental learning. The main lines of my argument may be found in Pearson. Here is where Pearson's rejection of the long-run rationale of error probabilities and his nonroutine use of tests come together with the Pearsonian logic of test construction.

Likelihoods Alone (Step 2) Are Insufficient for Pearsonian Reasoning

Pearson explains why he and Neyman held it essential to add the error probability calculations of step 3 to the "measures of inclination" at step 2. The concern was *not* pragmatic, with low error rates (in the long run of business), but with learning from experiments. Reflecting on this question (in "Some Thoughts on Statistical Inference"), Pearson tells of their "dissatisfaction with the logical basis—or lack of it—which seemed to underlie the choice and construction of statistical tests" at the time. He and Neyman, Pearson explains, "were seeking how to bring probability theory into gear with the way we think as rational human beings" (Pearson 1966e, 277).

17. The likelihood principle, we saw in chapter 10, falls out directly from Bayes's theorem. Birnbaum is responsible for showing, to the surprise of many, that it follows from two other principles, called sufficiency and conditionality (together, or conditionality by itself). For an excellent discussion of these principles see Birnbaum 1969. Birnbaum's result—while greeted with dismay by many non-Bayesians (including Birnbaum himself) who balked at the likelihood principle but thought sufficiency and conditionality intuitively plausible—was welcomed by Bayesians, who (correctly) saw in it a new corridor leading to a key Bayesian tenet. A third way would be to steer a path between the likelihood principle and advocating any principle that decreases error probabilities, thereby keeping certain aspects of sufficiency and conditionality *when and to the extent that they are warranted.*

> *But looking back I think it is clear why we regarded the integral of probability density within (or beyond) a contour as more meaningful than the likelihood ratio*—more readily brought into gear with the particular process of reasoning we followed.
>
> *The reason was this. We were regarding the ideal statistical procedure as one in which preliminary planning and subsequent interpretation were closely linked together—formed part of a single whole.* It was in this connexion that integrals over regions of the sample space were required. *Certainly, we were much less interested in dealing with situations where the data are thrown at the statistician and he is asked to draw a conclusion.* I have the impression that there is here a point which is often overlooked. (Ibid., 277–78; emphasis added)

I have the impression that Pearson is correct. The main focus of philosophical discussions is on what rival statistical accounts tell one to do once "data are thrown at the statistician and he is asked to draw a conclusion"; for example, to accept or reject for an NP test or compute a posterior probability for a Bayesian.

Why are error probabilities so important once the "preliminary planning and subsequent interpretation" are closely linked? First, if one of the roles of a theory of statistics is to teach how to carry out an inquiry, then some such before-trial rules are needed. By considering ahead of time a test's probabilities of detecting discrepancies of interest, one can avoid carrying out a study with little or no chance of teaching what one wants to learn; for example, one can determine ahead of time how large a sample would be needed for a given test to have a reasonably high chance (power) of rejecting H when in fact some alternative J is true. Few dispute this (before-trial) function of error probabilities.

But there is a second connection between error probabilities and preliminary planning, and this explains their relevance even after the data are in hand. It is based on the supposition that in order to correctly interpret the bearing of data on hypotheses one must know the procedure by which the data got there; and it is based on the idea that a procedure's error probabilities provide this information. The second role for error probabilities, then, is one of interpreting experimental results *after* the trial. It is on this "after-trial" function that I want to focus; for it is this that is denied by non-error-statistical approaches (those accepting the LP).[18] The Bayesians, paraphrasing LeCam's remark (chapter 10), have the magic that allows them to draw inferences

18. Some (e.g., Hacking 1965) have suggested that error probabilities, while acceptable for before-trial planning, should be replaced with other measures (e.g.,

from whatever aspects of data they happen to notice. NP statisticians do not.

Throughout this book I have identified several (after-trial) uses of error probabilities, but they may all be traced to the fact that *error probabilities are properties of the procedure that generated the experimental result.*[19] This permits error probability information to be used as a key by which available data open up answers to questions about the process that produced them. Error probability information informs about whether given claims are or are not mistaken descriptions of some aspect of the data generating procedure. It teaches us how typical given results would be under varying hypotheses about the experimental process.

We know how easy it is to be misled if we look only at how well data fit hypotheses, ignoring stopping rules, use-constructions, and other features that alter error probabilities. That is why fitting, even being the best-fitting hypothesis, is not enough. Step 2 assesses the fit, step 3 is needed to interpret its import. In a joint paper, Pearson and Neyman (1967) explain that

> if we accept the criterion suggested by the method of likelihood it is still necessary to determine its sampling distribution in order to control the error involved in rejecting a true hypothesis, because a knowledge of L [the likelihood ratio] alone is not adequate to insure control of this error. (P. 106; I substitute L for their λ)

Let L be the ratio of the likelihood of H and an alternative hypothesis J on given data x. That is,

$$L = \frac{P(x \mid H)}{P(x \mid J)}$$

(where in the case of composite hypotheses we take the maximum value of the likelihood). Suppose that L is small, say, .01, meaning H has a much smaller likelihood than J does. We cannot say that because

---

likelihoods) after the trial. Pearson took up and rejected this proposal, raised by Barnard in 1950, reasoning that

> if the planning is based on the consequences that will result from following a rule of statistical procedure, e.g., is based on a study of the power function of a test, and then, having obtained our results, we do not follow the first rule but another, based on likelihoods, what is the meaning of the planning? (Pearson 1966c, 228).

19. It may be objected that there are different ways of modeling the procedure. That is correct but causes no difficulty for the after-trial uses of error probabilities. Indeed, using different models is often a useful way of asking distinct but interrelated questions of the data.

L is a small value, "we should be justified in rejecting the hypothesis" *H*, because

> in order to fix a limit between "small" and "large" values of *L* we
> must know how often such values appear when we deal with a true
> hypothesis. That is to say we must have knowledge of . . . the chance
> of obtaining [*L* as small or smaller than .01] in the case where the
> hypothesis tested [*H*] is true. (Ibid., 106)

Accordingly, without step 3 one cannot determine the test's severity in passing *J*, and without this we cannot determine if there really is any warranted evidence against *H*.

The position I want to mark out even more strongly and more starkly than Pearson does is that the interest in a test's error probabilities (e.g., the probability of it passing hypotheses erroneously) lies not in the goal of ensuring a good track record over the long haul, but in the goal of learning from the experimental data in front of us. Comparisons of likelihoods or other magnitudes of fit can measure the observed difference between the data and some hypothesis, but I cannot tell if it *should* count as big or small without knowledge of error probabilities. In one particularly apt passage, Pearson explains that error probability considerations are valuable because

> *[they help] us to assess the extent of purely chance fluctuations that are pos-*
> *sible.* It may be assumed that in a matter of importance we should
> never be content with a single experiment applied to twenty individ-
> uals; but *the result of applying the statistical test with its answer in terms of*
> *the chance of a mistaken conclusion if a certain rule of inference were followed,*
> will help to determine the lines of further experimental work. (Pear-
> son 1966a, 176–77; emphasis added)

We saw how in certain cases of use-constructing hypotheses to fit data (e.g., gellerization cases), as well as in cases with optional stopping, the chance of mistaken conclusions may be very high. This error-probability information showed us how easily chance fluctuations could be responsible for a large extent of the results.

Let us go back to the case of Pearson's naval shell. The (after-trial) question asked was "how much of the difference between (i) two fail-ures out of twelve and (ii) five failures out of eight is likely to be due to this inevitable variability"? (Pearson 1966a, 171). It is asked by testing hypothesis *H*:

H: The observed difference is due to inevitable or "chance" variability.

(Alternative *J* would assert that it is due to a systematic discrepancy in the processes, with respect to successfully piercing the plate.) The difference statistic *D* is the difference between the proportions of successful perforations of the plate from the two types of shell. Using the experimental (or sampling) distribution of D he can calculate the statistical significance of a given observed difference D_{obs}:

The *statistical significance* of D_{obs} = P(a difference as great as D_{obs} | H).

He found D_{obs} to be improbably far from what would be expected were H correct. (The difference falls in the rejection region of a test of approximately .05 size.) Even if no repetitions are planned, this analysis is informative as to the origin of *this* difference. There are many ways of expressing this information.

One, paraphrasing Pearson, is that the observed difference (in piercing ability) is not the sort easily accounted for by inevitable variability in the shells and measurement procedures. The observed difference, rather, is indicative of the existence of a genuine (positive) difference in piercing ability. Were the two shells about as good, it is very probable that we would *not* have observed so large a difference—the severity in passing J is high. Finding the data indicative of hypothesis J, even with larger sample sizes than in this simple illustrative example, is just a first step. For simplicity, suppose that hypothesis J includes positive discrepancies in piercing rates between the first and second types of shell. One may also want to know which of the particular discrepancies are indicated by outcome D_{obs}. This further information may be obtained from the same experimental distribution, but the hypothesis to the right of the given bar would now be a member of J. We can thus learn how large a discrepancy in piercing rates would be needed to generate differences as large as D_{obs} fairly frequently. This calls for a custom-tailoring of the interpretation of test results to reflect the particular outcome reached. In the next section I shall consider two basic rules for interpreting test results that take into account the particular outcome observed. While they go beyond the usual NP test calculations, they fall out directly from the arguments based on severity calculations considered earlier.

11.6 Two Error Statistical Rules to Guide the Specification and Interpretation of NP Tests

Before proceeding with our next task, let me remind the reader that it pertains to just one piece, albeit a central one, of the series of tasks to which statistical considerations are put in the present account. In this piece, which is often regarded as statistical inference proper, statistical methods (tests and estimations) link data to experimental hypotheses, hypotheses framed in the experimental model of a given inquiry. Relating inferences about experimental hypotheses to primary scientific claims is, except in special cases, a distinct step. Yet an additional step is called for to generate data and check the assumptions of the statistical method. Restricting our focus to statistical tests, what I want to consider is how the nonbehavioral construal of tests that I favor supplies answers to two questions: how to specify tests, and how to interpret the results of tests. In so doing, the construal simultaneously answers the main criticisms of NP tests.

Because I think it is important to tie any proposed philosophy of experiment to the actual statistical procedures used, I am deliberately sticking to the usual kinds of test reports—either in terms of the statistical significance of a result, or "accept H" and "reject H"—although it might be felt that some other terms would be more apt. Rather than knock down the edifice of the familiar NP methods, I recommend effecting the nonbehavioral interpretation by setting out rules to be attached to the tests as they presently exist. They might be called "metastatistical" rules. To illustrate, it suffices to consider our by now familiar one-sided test with two hypotheses H and J:

H: μ equals μ_0
J: μ exceeds μ_0.

Parameter μ is the mean value of some quantity, and the experimental statistic for learning about μ is the sample mean, $\overline{X}$. As many of our examples showed, it is often of interest to learn whether observed differences, say in the positive direction, are due to actual discrepancies from some hypothesized value or are typical of chance deviations.

The difference statistic D is the positive difference between the observed mean and the mean hypothesized in H. That is,

$$D = \overline{X} - \mu_0.$$

The NP test, call it T^+, instructs H to be rejected whenever the value of variable $\overline{X}$ differs from H by more than some amount—that is, whenever it exceeds some cutoff point $\overline{X}^*$. One can work with $\overline{X}$ or with D to specify the cutoff point beyond which our test rejects H and accepts J. The cutoff is specified so that the probability of a type I error (rejecting H, given that H is true) is no more than α. Let us suppose that the test T^+ is a "best" NP test with small size α, say, for convenience,[20] that α is .03. Then

20. It is convenient because it corresponds to approximately a 2-standard-deviation cutoff point. If one were looking for discrepancies in both directions, that is, if this were a 2-sided rather than a 1-sided test, then the 2-standard-deviation cutoff would give, approximately, a test with size 0.05. See note 2, chapter 9.

Test T^+: Reject H at level .03 iff $\overline{X}$ exceeds μ_0 by 2 s.d.(x_n)

where s.d.(x_n) is the standard deviation of $\overline{X}$. For simplicity, let the sample size n be large enough to assume approximate normality of $\overline{X}$ according to the central limit theorem (say, n is greater than 30). Let us review the error probabilities of test T^+.

Error Probabilities of T^+

a. The type I error is rejecting H when H is true. The probability of this occurring, α, is no more than .02 because that is the preset size of the test. This holds because $\overline{X}$ exceeds its mean by 2 standard deviations less than 3 percent of the time.

b. The type II error is accepting (or failing to reject) H when H is false (and J is true). Hypothesis J is a "composite" alternative, since it contains all the μ values in excess of μ_0. The probability of the type II error varies depending on which value in J is the true one.

The probability of a type II error is usually written as β, but because it will depend for its value on which specific value in J is true, it is clearer to use $\beta(\mu')$ to refer to the probability of a type II error when μ' is true. One should read $\beta(\mu')$ as follows:

$\beta(\mu')$: the probability that test T^+ fails to reject H (and accept J) when alternative μ' is true.

The assertion that the mean equals μ' may be written as hypothesis J':

J': μ equals μ'.

J' is a particular "point" hypothesis within the composite alternative J. That is, $\beta(\mu')$ is the probability of committing a type II error when J' is true. So, $\beta(\mu')$ can also be written

$\beta(\mu')$: P(test T^+ fails to reject H | J' is true).

Notice that "failing to reject H" in test T^+ is equivalent to obtaining an $\overline{X}$ that is not so far from μ_0 as to reach the (2-standard-deviation) cutoff point $\overline{X}^*$. So $\beta(\mu')$ is the probability that $\overline{X}$ is less than $\overline{X}^*$, given that J' is true.

$$\beta(\mu') = P(\overline{X} < \overline{X}^* \mid J' \text{ is true}).$$

As is plausible, test T^+ has a decreasing probability of committing a type II error the "more false" H is—the further μ' is "to the right of" μ_0. One may wish to state this in terms of the complement to the probability of a type II error, namely, the *power* of the test to detect a specific simple alternative μ'. That is,

The power of T^+ *against* $J' = P(\overline{X} \geq \overline{X}* \mid J'$ is true$)$.

The power of the test to reject H when J' is true is $1 - \beta(\mu')$. As would be expected, the more discrepant μ' is from μ_0, the higher is test T^+'s power to detect this.

A test's error probabilities may be used to construct arguments from error, arguments based on severity. However, it is important to remember that severity is always calculated relative to a specific hypothesis that a given test passes on the basis of a given outcome. You cannot assess the severity of a test without considering the process of a test's passing a particular hypothesis with one or another outcome of a given experiment. So to relate error probabilities to arguments from error we need to consider specific kinds of inferences that test T^+ can license. We can begin with the simple dichotomy of standard NP tests: positive and negative results.

Positive Results: A Rejection of H

A positive result is the observation of a sample mean that exceeds the hypothesized mean μ_0 by a statistically significant amount. In the standard test T^+ with α set at .03, a statistically significant difference is one that exceeds μ_0 by 2 standard-deviation units. Within the NP model, this is taken as a rejection of H. That is, the cutoff point, $\overline{X}*$, is $\mu_0 + 2$ s.d.(x_n).

What is learned from an observed difference D_{obs} about the existence of a positive discrepancy from μ_0? For what value of μ' does J: μ exceeds μ' pass a severe test with T^+? From the pattern of arguing from error we get what might be called the rule of rejection (RR):

> RRi. A difference as large as D_{obs} is *a good indication* that μ exceeds μ' just to the extent that it is very probable that test T^+ would have resulted in a smaller difference if μ (the true mean) were as small as μ'.

Notice that this is the same as saying that D_{obs} is *a good indication* that μ exceeds μ' to the extent that J passes a severe test with D_{obs}. That is because "not-J" consists of μ values less than or equal to μ'.[21]

From RRi we get a companion rule for what an observed difference does *not* indicate. Let us set it out separately:

> RRii. D_{obs} is *a poor indication* that μ exceeds μ' if it is very probable that test T^+ yields so large a difference even if μ is no greater than μ'.

21. As discussed in chapter 6, to obtain severity for all of those values it is enough that P (a difference smaller than $D_{obs} \mid \mu$ equals μ') is high. See also the rule of acceptance (RA) below.

The if clause is the same as saying that the claim J: μ exceeds μ' fails to pass a severe test (i.e., the probability of not getting so large a difference even if J is false is low).

In addition to the rule of rejection (RR), I will be setting out a rule of acceptance (RA). These rules have many uses. They justify the standard prespecified small error probabilities, allow custom-tailoring of inferences after the trial, and serve to avoid common criticisms and misinterpretations of NP tests. Focusing first on rule RR, I will consider each of these uses in turn.

Rule RR Justifies Preset Significance Levels

The concern in the case of rejecting the null or test hypothesis H is that a rejection of H might be erroneous—that is, the concern is with the type I error. By stipulating that H be rejected only if the difference is statistically significant at some small level, it is assured that such a rejection—*at minimum*—warrants hypothesis J, that μ exceeds μ_0 by some amount or other. RRi makes this plain.

If H is rejected, then the hypothesis that passes the test is J, the assertion that μ exceeds the null value μ_0. To obtain severity, we have to consider one minus the probability of such a statistically significant result even if H is true (1 − the probability of a type I error). This will vary depending on how much the observed result exceeds the minimal boundary for declaring a result significant enough to reject H_0, namely, $\overline{X}^*$. Its lowest value, however, would be for a result that just makes it to the boundary $\overline{X}^*$. In this case, the observed mean, $\overline{X}_{obs}$, equals the cutoff value $\overline{X}^*$ for calling a result "positive." The severity for this "worst case" of rejecting H is one minus the probability of a type I error (i.e., $1 - \beta(\mu_0)$). So the assurance given by a test with a low type I error is that it tells me *ahead of time* that whenever T^+ rejects H_0, J has passed a severe test (at least to degree $1 - \alpha$).[22]

22. To review the argument with a bit more detail, remember that a test T^+ with low size or significance level α assures that the cutoff $\overline{X}^*$ beyond which point the sample mean is taken to reject H and accept J is one that occurs with no more than probability α when H is true. That is, it ensures that

1. P(test T^+ yields a sample mean that exceeds $\overline{X}^*$ | H is true) $\leq \alpha$.

But (1) ensures that whenever such a test passes J, the result is that J has passed a severe test, the severity being at least $1 - \alpha$. The reason is that (1) is equivalent to (2):

2. P(test T^+ passes J | H is true) $\leq \alpha$.

From which we get

3. P(test T^+ does *not* pass J | H is true) $> 1 - \alpha$ (or $\geq 1 - \alpha$ for continuous cases).

And so

4. if J passes test T^+, then J passes a severe test.

Custom-Tailoring

When outcomes deviate from H by even more than the α cutoff, rule RR justifies two kinds of custom-tailored results: (*a*) it warrants passing J at an even higher severity value than $1 - \alpha$, and (*b*) it warrants passing alternatives J': μ is greater than μ', where μ' is larger than μ_0, with severity $1 - \alpha$. With (*a*) we make the same inference—pass J—but with a higher severity than with a minimally positive result. With (*b*) we keep the same level of severity but make a more informative inference—that the mean exceeds some particular value μ' greater than the null value μ_0.

To illustrate (*b*), suppose that null hypothesis H: $\mu = .5$ in our lady tasting tea example is rejected with a result $\overline{X}_{obs}$ is equal to .7. (Mean μ, recall, is the same as the probability of success p.) That is, out of 100 trials, 70 percent are successes. The 2-standard-deviation cutoff was 60 percent successes, so this result indicates even more than that μ exceeds .5. The result is also a good indication that μ exceeds .6. That is because the observed difference exceeds .6 by 2 standard deviations. The probability of so large a difference from .6 if μ were no greater than .6 is small (about .03). Thus the assertion "μ exceeds .6" passes a severe test with result .7 (severity .97).

Avoiding Misinterpretations and Alleged Paradoxes for NP Tests

Where RRi shows that an α-significant difference from H (for small α) indicates *some* positive discrepancy from H_0, RRii makes it clear that it does *not* indicate any and all positive discrepancies. My failing exam score may indicate that I am ignorant of some of the material yet not indicate that I know none of it.

We can make the point by means of the usual error probabilities, even without customizing to the particular result. Consider the power of a test $(1 - $ the type II error$)$ regarding some alternative μ'. We know:

The power of T^+ against J': $\mu = \mu'$ equals $P(\overline{X} \geq \overline{X}^* \mid J'$ is true$)$, which equals $1 - \beta(\mu')$. The test's power may be seen as a measure of its sensitivity. The higher the test's probability of detecting a discrepancy μ', the more powerful or sensitive it is at doing so. If, however, a test has a good chance of rejecting H even if μ is no greater than some value μ', then such a rejection is a *poor* indication that μ is even greater than μ'. So—although this may seem odd at first—a statistically significant difference is indicative of a larger discrepancy the *less* sensitive or powerful the test is. If the test rings the alarm (i.e., rejects H_0) even for comparatively tiny discrepancies from the null value, then the ringing alarm is poor grounds for supposing that larger discrepancies exist. As obvious as this reasoning becomes using severity considerations, the exact opposite is assumed in a very common criticism of tests.

Before turning to this criticism, let us illustrate the reasoning in both parts of the RR by means of a medical instrument. Imagine an ultrasound probe to detect ovarian cysts. If the image observed is of the sort that very rarely arises when there is no cyst, but is common with cysts, then the image is a good indication a cyst exists. If, however, you learned that an image of the sort observed very frequently occurred with this probe even for cysts no greater than 2 inches, then you would, rightly, deny that it indicated a cyst as large as, say, 6 inches. The probe's result is a good indication of a cyst of some size, say ¼ inch, but a poor indication of a cyst of some other (much greater) size. And so it is with test results.

If a difference as large as the one observed is very common even if μ equals μ', then the difference does *not* warrant taking μ to exceed μ'. That is because the hypothesis that μ exceeds μ' would thereby have passed a test with poor severity. And in comparing outcomes from two different tests, the one that passes the hypothesis with higher severity gives it the better warrant. How do critics of NP tests get this backwards?

A Fallacy regarding Statistically Significant (Positive) Results

Criticisms of NP tests, we have seen, run to type, and one well-known type of criticism is based on cases of statistically significant (or positive) results with highly sensitive tests. The criticism begins from the fact that any observed difference from the null value, no matter how small, would be classified as statistically significant (at any chosen level of significance) provided the sample size is large enough. (While this fact bears a resemblance to what happens with optional stopping, here the sample size is fixed ahead of time.) There is nothing surprising about this if it is remembered that the standard deviation decreases as the sample size increases. (It is inversely proportional to n.) Indeed, my reason for abbreviating the standard deviation of the sample mean as s.d.(x_n) in this chapter was to emphasize this dependence on n. A 2-standard-deviation difference with a sample size of, say, 100 is larger than a 2-standard-deviation difference with a sample size of 10,000.

We can make out the criticism by reference to a Binomial experiment, such as in the lady tasting tea example. The null hypothesis H is that p, the probability of success on each trial, equals .5. Now, in a sample of 100 trials, the standard deviation of $\overline{X}$ is .05, while in 10,000 trials it is only .005. Accordingly, a result of 70 percent successes is a very significant result (it exceeds .5 by 4 standard deviations) in a sample of 100 trials. In a sample of 10,000 trials, an equally statistically significant result requires only 52 percent successes! An alleged paradox is that a significance test with large enough sample size rejects the null with outcomes that seem very close to, and by a Bayesian analysis are supportive of, the null hypothesis. This might be called the Jeffreys-Good-Lindley paradox, after those Bayesians who first raised it.

I discuss this paradox at length in Mayo 1985a and elsewhere, but here I just want to show how easy it is to get around a common criticism that is based on it. The criticism of NP tests results only by confusing the import of positive results. The fallacious interpretation results from taking a positive result as indicating a discrepancy beyond that licensed by RR. Howson and Urbach give a version of this criticism (along the lines of an argument in Lindley 1972). Their Binomial example is close enough to the one above to use it to make out their criticism (their p is equal to the proportion of flowering bulbs in a population). The criticism is that in a test with sample size 10,000, the null hypothesis $H: p = .5$ is rejected in favor of an alternative J, that p equals .6 even though .52 is much closer to .5 (the hypothesis being rejected) than it is to .6. And yet, the criticism continues, the large-scale test is presumably a better NP test than the smaller test, since it has a higher power (nearly 1) against the alternative that $p = .6$ than the smaller test (.5).[23]

The authors take this as a criticism of NP tests because "The thesis implicit in the [NP] approach, that a hypothesis may be rejected with increasing confidence or reasonableness as the power of the test increases, is not borne out in the example" (Howson and Urbach 1989, 168). Not only is this thesis not implicit in the NP approach, but it is the exact reverse of the appropriate way of evaluating a positive (i.e., statistically significant) result. The thesis that gives rise to the criticism comes down to thinking that if a test indicates the existence of some discrepancy then it is an even better indication of a very large discrepancy!

Looking at RRii makes this clear. Let us compare the import of the two 4-standard-deviation results, one from a test with sample size $n = 100$, the second from a test with sample size $n = 10,000$. In the experiment with 10,000 trials, the observation of 52 percent successes is an extremely *poor* indicator that p is as large as .6. For such a result is very probable even if the true value of p is actually less than .6,

---

23. I am calculating power here with the cutoff for rejection set at .6—the 2-standard-deviation cutoff.

say, if $p = .55$. Indeed, it is practically certain that such a large result would occur for p as small as .55. Were one to take such a result as warranting that p is .6, one would be wrong with probability very near one.

In contrast, the observation of 70 percent successes with $n = 100$ trials is a very good indication that p is as large as .6. The probability of getting so large a proportion of successes is very small (about .03) if μ is less than .6. The severity of a test that passes "p is as large as .6" with 70 percent successes out of 100 trials is high (.97).

Howson and Urbach's criticism, and a great many others with this same pattern, are based on an error to which researchers have very often fallen prey. The error lies in taking an α-significant difference (from H) with a large sample size as more impressive (better evidence of a discrepancy from H) than one with a smaller sample size.[24] That, in fact, it is the reverse is clearly seen with rule RR. The reasoning can be made out informally with an example such as our ultrasound probe. Take an even more homey example. Consider two smoke detectors. The first is not very sensitive, rarely going off unless the house is fully ablaze. The second is very sensitive: merely burning toast nearly always triggers it. That the first (less sensitive) alarm goes off is a *stronger* indication of the presence of a fire than the second alarm's going off. Likewise, an α-significant result with the *less* powerful test is *more* indicative of a discrepancy from H than with the more powerful test.[25] Interpreting the results accordingly, the authors' criticism disappears.

To be fair, the NP test, if regarded as an automatic "accept-reject" rule, only tells you to construct the best test for a small size α and then accept or reject. A naive use of the NP tools might seem to license the problematic inference. Rule RR is not an explicit part of the usual formulation of tests. Nevertheless, that rule, and the fallacious interpretation it guards against, is part of the error statistician's use of these tests.[26]

24. Rosenthal and Gaito (1963) explain the fallacy as the result of interpreting significance levels—*quite illicitly*—as E-R measures of the plausibility of the null hypothesis. In this view, the smaller the significance level, the less plausible is null hypothesis H, and so the more plausible is its rejection. Coupled with the greater weight typically accorded to experiments as the sample size increases, the fallacy emerges.

25. See Good 1980, 1982 for a Bayesian way of accommodating the diminishing significance of a rejection of H as the sample size increases.

26. The probabilities called for by RR would be obtained using the usual probability tables (e.g., for the Normal distribution). A good way to make use of rule RR without calculating exact severity values for each result is to substitute certain

Negative Results: Failures to Reject

Let us turn now to considering negative results, cases where the observed difference is *not* statistically significant at the specified small α level. Here the null hypothesis H ($\mu = \mu_0$) is not rejected. NP theory describes the result as "accept H," but one must be careful about how to interpret this. As we saw in section 6.5, it would not license the inference that μ is exactly μ_0—that μ does not exceed μ_0 at all. However, as we also saw, we may find a positive discrepancy that *can* be well ruled out. The pattern of reasoning again follows the pattern of arguing from error. We can capsulize this by the following rule of acceptance (RA):

> RAi. A difference as small as D_{obs} is a good indication that μ is less than μ' if and only if it is very probable that a larger difference would have resulted from test T^+ if the mean were as large as μ'.

That is, a statistically *insignificant* difference indicates that $J: \mu$ is less than μ' just in case J passes a severe test. As with the RR, we get a companion rule:

> RAii. A difference as small as D_{obs} is a *poor* indication that μ is less than μ' if it is very improbable that the test would have resulted in a larger difference even if the mean were as large as μ'.

Notice that when the result is negative, the error of interest is a false negative (a type II error)—that H will be accepted even though some alternative J is true.

Rule RA Directs Specifying Tests with High Power to Detect Alternatives of Interest

Now T^+ "accepts" H whenever $\overline{X}$ is less than[27] the .03 significance level cutoff. Before the test, one does not yet know what value of $\overline{X}$ will be observed. Ensuring ahead of time that test T^+ has a high power $1 - \beta$ against an alternative $J': \mu = \mu'$ ensures that a failure to reject

---

benchmarks for good and poor indications. Still focusing on test T^+, useful benchmarks for interpreting rejections of hypothesis H would be as follows:

1. A T^+ rejection of H is a *good* indication that μ exceeds $\overline{X}_{obs} - 2\text{s.d.}(x_n)$.

2. A T^+ rejection is a *poor* indication that μ exceeds $\overline{X}_{obs} + 1\text{s.d.}(x_n)$.

(1) corresponds to passing the claim that "μ exceeds $\overline{X}_{obs} - 2\text{s.d.}(x_n)$" with severity .97.

(2) corresponds to passing the claim that "μ exceeds $\overline{X}_{obs} + 1\text{s.d.}(x_n)$" with severity .16. For a more general discussion of benchmarks for both the RR and RA see Mayo 1983.

27. In continuous cases or discrete cases with fairly large n, it does not matter if we take it as $<$ or $\leq$.

H—a case where H passes—is a case that indicates that μ does not exceed μ'. That is, by assuring ahead of time that the power to detect μ' is high, the experimental tester is ensuring that accepting H constitutes passing severely the hypothesis H':

H': μ is no greater than μ'.

Tests should be specified according to the smallest discrepancy from μ_0 that is of interest.

Notice that this power calculation is a calculation of severity for the case where the result just misses the critical boundary for statistical significance. By custom-tailoring this calculation to the particular statistically insignificant result obtained, the after-trial analysis may warrant ruling out values of μ even closer to μ_0.

A variant on this after-trial question is to ask, with regard to a particular alternative μ'', whether the obtained negative result $\overline{X}_{obs}$ warrants ruling out a μ value as large as μ''. Severity tells you to calculate the probability that a mean larger than the one observed, $(\overline{X}_{obs})$, would have occurred, given that the true value of μ were equal to μ''. That is, you must calculate, still referring to test T^+,

$$P(\overline{X} > \overline{X}_{obs} \mid \mu \text{ equals } \mu'').$$

If this value is high, then $\overline{X}_{obs}$ indicates that μ is less than μ''. Equivalently, the claim that μ is less than μ'' passes a severe test with the obtained negative result $\overline{X}_{obs}$. For, were μ as large as μ'', the probability is high that a result greater than the one obtained would have occurred.

11.7 A NOTE ON OBJECTIVITY

The task of specifying the analytical tool for an experimental inquiry (e.g., tests) is a task we placed within the experimental model of our hierarchy. That it lies outside the formalism of standard NP tests has often led critics to charge that NP methods do not really get around the subjectivity that plagues the subjective Bayesian account. Deciding upon test statistics, sample sizes, significance levels, and so on, after all involves judgments—and these judgments, critics allege, are what the NP will "sweep under the carpet" (to use I. J. Good's phrase):

> You usually have to use subjective judgment in laying down your parametric model. Now the *hidebound* objectivist tends to hide that fact; he will not volunteer the information that he uses judgment at all. (Good 1976, 143)

A favorite line of subjective Bayesians is that by quantifying their subjective beliefs they are actually being more objective than users of non-
Bayesian, error probability methods. How do we respond to this
charge? First, the judgments the NP test requires are not assignments
of degrees of belief to hypotheses. Although subjective Bayesians seem
to think that all judgments come down to judgments of prior probabilities, I see no reason to accept this Bayesian dogma. Second, there is a
tremendous difference between the kinds of judgments in error statistical methods and subjective probability assignments.

The two main differences are these: First, the choice of statistical
test may be justified by specific epistemological goals. As rules RR and
RA helped us to see, the choice of NP test with low error probabilities
reflects a desire to substantiate certain standard types of arguments
from error. With increasing experience, experimenters learn which
types of tests are likely to provide informative results. There is leeway
in the specification, but it is of a rather restricted variety. Often, different studies will deliberately vary test specifications. Indeed, exploiting
different ways of analyzing results is often the basis for learning the
most. Second, and most important, the latitude that exists in the choice
of test does not prevent the determination of what a given result does
and does not say. The error probabilistic properties of a test procedure—*however that test was chosen*—allows for an objective interpretation of the results. Let us elaborate on these two points, making reference to the results we have already seen.

Severity and the Epistemological Grounds for Test Specifications

When tests are used in scientific inquiry, the basis for specifying
tests reflects the aims of learning from experiment. A low probability
of a type I error, for example, is of interest not because of a concern
about being wrong some small proportion of times in a long-run series
of applications. It is of interest because of what one wants to learn. If
you can split off a portion of what you wish to learn so that one of the
canonical experimental models can be used, then specifying the test's
error properties grows directly out of what one wants to know—what
kinds and extents of errors are of interest, what kinds of checks are
likely to be available, and so on.

What I am arguing, then, is that the grounds for specifying the
error probabilities of tests stem from the experimental argument one
wants to be able to sustain. By fixing the type I error at some low value
α the experimental tester ensures that any rejection of H, any passing
of J, is a good indication that J is the case. It should not be forgotten,
of course, that this depends on a suitable choice of distance measure

at step 2 in test construction. In the canonical tests, such as the one just described, the choice of distance measure is already accomplished for us.

But as Neyman and Pearson saw, this leaves too much latitude in the choice of a test. One must also consider the type II error—failing to reject H when H is false. The problem in cases where H is not rejected is that the test may have had little power (probability) of rejecting H even if a discrepancy from H exists. So severity considerations tell us that a failure to reject H cannot be taken as a good indication that H is precisely true, that no discrepancy from H exists. It is, however, possible to find some value of a discrepancy from H that the result "accept H" does warrant ruling out.

What I am proposing, I believe, is a way of drawing out the implications of Pearson's hints and suggestions. Before the trial, we are interested in how to ensure that the experiment is capable of telling us what we want to know, and we set these "worst case" values for the probabilities of type I and type II errors accordingly. After the trial, with the data in hand, Pearson says we should base our conclusions on the actual "tail area" found, which is tantamount to saying, "look at the severity values."

Telling the Truth with Error-Statistics

Of course there is no guarantee that an appropriate test will actually be run. Indeed, the existence of poorly specified and wrongly interpreted NP tests is at the heart of criticisms of that approach. We noted the problem of positive results with too-sensitive tests. An even more common problem arises when negative results arise from too-*insensitive* tests. As A. W. F. Edwards puts it:

> Repeated non-rejection of the null hypothesis is too easily interpreted as indicating its acceptance, so that on the basis of no prior information coupled with little observational data, the null hypothesis is accepted. . . . Far from being an exercise in scientific objectivity, such a procedure is open to the gravest misgivings. (Edwards 1971, 18)

Although such interpretations of negative results occur, it does not follow that they are licensed by the logic of error statistics. They are not. And because researchers must provide us with enough information to assess the error probabilities of their tests, we are able to check if what they want to accept is really warranted by the evidence.

To illustrate, we can pick up on the study on birth-control pills introduced in chapter 5 and scrutinized in section 6.5. The result, recall, was 9 cases of a blood-clotting disorder among women treated with the birth-control pill compared with 8 out of 5,000 in the control group. Suppose that the researchers reach the following interpretation of their result: "These results indicate that no more than 1 additional case of clotting disorders among 10,000 women on the pill would be expected." That is, using our abbreviation for the risk increase in the population, the researchers infer claim C:

C: the evidence indicates H_c: $\Delta < .0001$.

The rule of acceptance (RA) is the basis for denying that this is a warranted interpretation of the results.

The observed difference .0002 was not statistically significant; it reaches a significance level of .4. We can see right away that an observed difference of .0002 or one even more insignificant would occur 50 percent of the time even if the actual increased rate of the disorder was 2 in 10,000.[28] Hence RA tells us that the negative result from this study cannot be taken as ruling out increases as small as 2 in 10,000. The result is just the sort of thing that would occur half the time in studies of substances that cause 2 additional cases of the disorder per 10,000 women. Such an insignificant difference would therefore be even *more* probable if the pill caused only 1 additional case of the disorder in 10,000 women. Hence the result of this study is a poor indication of hypothesis H_c: $\Delta < .0001$. The inference in C is not warranted. Such an insignificant result would occur more than half the time even if H_c is false. Equivalently, the assertion H_c passes a test with severity of less than .5, on the basis of this result.

Utilizing a test's error probabilities in this manner, customizing even further to take account of the particular result, enables distinguishing warranted from unwarranted interpretations of the results, and it enables doing so objectively. The objectivity of the assessment is afforded by the objectivity of the error probability properties of the test. Even without calculating precise severity values, we can distinguish (reasonably) warranted and (flagrantly) unwarranted interpretations of results. Plenty of shortcut calculations are available for making this discrimination (see note 26), and more can be developed.[29]

28. This can be seen without any calculations. Label the supposition here as alternative hypothesis J': the increased risk is .0002. Now the observed outcome does not differ at all from what is hypothesized by J'. But even if J' is true, 50 percent of the time sample differences would be less than .0002, and 50 percent of the time they would be greater. (That is, half of the area under the normal curve would be "to the left of" J', and half "to the right.") See also the discussion of this example in section 6.5. A longer discussion occurs in Mayo 1985b.

29. Consider interpreting negative results, that is, acceptances of H, in test T^+. Rule RA directs us to find a value of μ, call it μ^+, such that the result indicates that $\mu < \mu^+$. Equivalently, we are to find the value μ^+ such that the claim "$\mu < \mu^+$"

The latitude in specifying tests is no different from that in the use of other kinds of reliable instruments in science. Understanding the properties of the instruments allows scrutinizing what a given reading does and does not say. The same holds for tests. It does not matter that test specifications might reflect the beliefs, biases, or hopes of the researcher. Perhaps the reason for selecting an insensitive test is your personal desire to find no increased risk, or perhaps it is due to economic or ethical factors. Those factors are entirely irrelevant to scrutinizing what the data do and do not say. They pose no obstacle to my scrutinizing any claims you might make based on the tests, nor to my criticizing your choice of test as inappropriate for given learning goals. There is no sort of comparable basis for criticizing your subjective degrees of belief.

Inferences without Numbers

There is one final objection that may be raised by Bayesians and others wedded to E-R accounts of inference. The present account of testing licenses claims about hypotheses that are and are not indicated by tests without assigning quantitative measures of support or probability to those hypotheses. But without such assignments of support or probability to hypotheses, the E-R theorist, I expect, will deny that the present account constitutes a genuine account of inductive or statistical inference. Yet this is just to assume that an E-R account is what is needed, and that is what those who embrace testing accounts of inference wish to deny. The Bayesian critic may persist that if I do not secretly really mean to assign some number to the inferences licensed by my tests, then what do I mean by evidence indicating hypotheses? My answer is the one I have been giving throughout this book. That data indicate hypothesis H means that the data indicate or signal that H is

---

passes a severe test. Say we take .97 as a benchmark for severity. Then μ^+ would equal $\overline{X}_{obs} + 2$s.d.(x_n). (See also section 6.5.)

Mathematically, the calculation of μ^+ (for the case of test T^+) is equivalent to formulating the *upper confidence bound* of a (one-sided) interval estimate at the corresponding level of confidence. However, unlike the report that "μ is somewhere between μ_0 and μ^+," RA instructs a distinct severity assessment for each value in the interval. More generally, RA directs us to understand what a specific negative statistical result indicates (more or less well) by calculating all or several of the upper bounds for different degrees of severity. This would yield what might be called *severity curves*. It most closely corresponds to forming a series of upper confidence intervals, one for each confidence level. I have recently come across an article by Poole (1987) using what are essentially severity curves in medical statistics. Similar curves are employed by Kempthorne and Folks (1971), but with a different interpretation. Clearly, more work is called for in studying statistical practice and in generalizing these ideas.

correct—much as I might say that a scale reading indicates my weight. Generally, several checks of a given indication of H (e.g., checks of the experimental assumptions) are required before reaching the inference that the data indicate the correctness of H. What does it mean to infer that H is indicated by the data? It means that the data provide good grounds for the correctness of H—good grounds that H correctly describes some aspect of an experimental process. What aspect, of course, depends on the particular hypothesis H in question. One can, if one likes, construe the correctness of H in terms of H being reliable, provided one is careful in the latter's interpretation. Learning that hypothesis H is reliable, I proposed (chapter 4), means learning that what H says about certain experimental results will often be close to the results that would actually be produced—that H will or would often succeed in specified experimental applications. What further substantive claims are warranted will depend on the case at hand.

What is learned receives a formal construal in terms of experimental distributions—assertions about what outcomes would be expected, and how often, if certain experiments were to be carried out. Informally and substantively, this corresponds to learning that data do or do not license ruling out certain errors and mistakes.

To those who insist that every uncertain inference must have a quantity attached, our position is that this insistence is seriously at odds with the kinds of inferences made every day, in science and in our daily lives. There is no assignment of probabilities to the claims themselves when we say things such as the evidence is a good (or a poor) indication that light passing near the sun is deflected, that treatment X prolongs the lives of AIDS patients, that certain dinosaurs were warm blooded, that my four-year-old can read, that metabolism slows down when one ingests fewer calories, or any of the other claims that we daily substantiate from evidence.

Concluding Remarks

To summarize, the key difference between standard NP methods and those based on the likelihood principle is that the former have an interest in and an ability to control error probabilities, whereas the latter do not. Criticisms of NP tests that are not merely misinterpretations arise from supposing that long-run error probabilities are all that matter in NP tests, and that the reason error probabilities matter in NP tests is their interest in ensuring a low probability of erroneous "acts" in the long run. A Pearsonian error statistician denies both of these suppositions. For a Pearsonian, the ability to control error probabilities matters (in a scientific context) because of the desire to correctly learn about underlying causes, distinguish genuine from spurious effects, and so on, to all that may be learned by arguing from error.

On the Pearsonian view of tests, the greater "seriousness" the behavioristic model attaches to the type I error goes over into the concern to be assured that a rejection of *H* is a good indication of the existence of a real departure from *H*, for example, a real effect. The particular balance chosen between the two types of errors is not an arbitrary matter reflecting pragmatic, decision-theoretic values, as Fisher had feared. In learning contexts, their specification is guided by the aims of inquiry, by what one wants to learn. After the results are in, utilizing these error probabilities is the key to scrutinizing objectively inferences based on test results.

In any substantive inquiry, NP methods would need to be used for a series of tests aimed at rejecting different types of alternatives and errors. Rejecting a "chance" hypothesis *H*, with its indication that some systematic factor is operating, is likely to be only a first step. Ruling out other substantive factors may be accomplished with subsequent statistical tests linking different experimental and data models. As Pearson stressed, there is no need to justify any single test as best; several tests may be used to learn the answers to different questions, as well as to check each other's assumptions. It is only by understanding how standard error statistical methods afford this type of *piecemeal* approach that one can capture the manner in which these tools are used in day-to-day experimental inquiries.

However, Pearson's advocacy of a piecemeal, inferential use of NP tests requires him to reject the basic tenets of the behavioral decision philosophy that has come to be associated with NP methods. There is no inconsistency in his rejection. While the interpretation of test results differs from the behavioral-decision one, still retained is what is central to error statistical theory: the focus on a procedure's error probabilities. The control of error probabilities has fundamental uses in learning contexts. The link between controlling error probabilities and experimental learning comes by way of the link between error probabilities and severity. The ability to provide methods whose actual error probabilities will be close to those specified by a formal statistical model, I believe, is the key to achieving experimental knowledge.

CHAPTER TWELVE

Error Statistics and Peircean Error Correction

> Induction (at least, in its typical forms) contributes nothing to our knowledge except to tell us approximately how often, in the course of such experience as our experiments go towards constituting, a given sort of event occurs. It thus simply evaluates an objective probability. Its validity does not depend upon the uniformity of nature, or anything of that kind. The uniformity of nature may tend to give the probability evaluated an extremely great or small value; but even if nature were not uniform, induction would be sure to find it out, *so long as inductive reasoning could be performed at all.* . . .
>
> But all the above is at variance with the doctrines of almost all logicians. . . . They commonly teach that the inductive conclusion approximates to the truth because of the uniformity of nature.
>
> —C. S. Peirce, *Collected Papers*, vol. 2, par. 775

I OPENED CHAPTER 1 with a quote from Popper: "The essays and lectures of which this book is composed are variations upon one very simple theme—the thesis that we can *learn from our mistakes.*" The theme of learning from error has a central place in the experimental program based on error statistics that I have been sketching. Nevertheless, from all we have said, is it apparent that Popper's account falls far short of showing how reliable knowledge is obtained from experiment or how that knowledge grows. The present account does not find its home in a Popperian framework. It is quite at home, however, within the experimental framework of another philosopher who also developed an account wherein scientific inference is based on learning from error and error correction, namely, C. S. Peirce. Nevertheless, the Peircean error correction thesis has been soundly criticized and found wanting. Indeed, as Nicholas Rescher (1978) remarks, "No part of Peirce's philosophy of science has been more severely criticized, even by his most sympathetic commentators, than this attempted validation of inductive methodology on the basis of its purported self-correctiveness" (p. 2).

Despite the hard times on which Peirce's validation of inductive methodology has fallen, I propose to revive the Peircean self-correcting doctrine. I want to do so not only to defend a great philosopher from whom I have gained many insights, but also, more selfishly, because by developing my view of Peirce's error-correcting justification of induction I will at the same time be developing the justification I need for error statistical methods in science. The justification for these methods lies in their ability to control error probabilities, hence sustain learning from error, hence provide for the growth of experimental knowledge. It seems to me that this is also the essence of Peirce's self-correcting rationale of inductive methods—when that thesis is properly understood. While on the one hand Neyman-Pearson and other contemporary methods increase the mathematical rigor and generality of Peirce's assertions about self-correcting methods, on the other, Peirce provides something the formal statistical tools lack (and Pearson only hinted at): an account of inductive inference and a philosophy of experiment ready-made for just such tools.

12.1 PEIRCEAN INDUCTION AND NEYMAN-PEARSON STATISTICS

Peirce's philosophy of experimental testing shares a number of key features with the Neyman and Pearson theory. For both, statistical methods provide not means for assigning degrees of probability, evidential support, or confirmation to hypotheses, but procedures for testing (and estimation) whose rationale is their predesignated high frequencies of leading to correct results in some hypothetical long run. The key similarities between Peirce and the methods later developed by Neyman and Pearson were first unearthed by Isaac Levi (1980b).[1]

> Peirce's inductions are inferences according to rules specified in advance of drawing the inferences where the properties of the rules which make the inferences good ones concern the probability of success in using the rules. These are features of the rules which followers of the Neyman-Pearson approach to confidence interval estimation would insist upon. (P. 138)

In describing his theory of inference, Peirce could be describing that of the error statistician:

1. Levi also relates Peirce's work to that of R. B. Braithwaite, who developed a theory of chance based on rules of testing akin to Neyman and Pearson tests. I regret being unable to discuss Braithwaite's work here, but good discussions exist in Hacking 1965 and Mellor 1980.

The theory here proposed does not assign any probability to the inductive or hypothetic conclusion, in the sense of undertaking to say how frequently *that conclusion* would be found true. It does not propose to look through all the possible universes, and say in what proportion of them a certain uniformity occurs; such a proceeding, were it possible, would be quite idle. The theory here presented only says how frequently, in this universe, the special form of induction or hypothesis would lead us right. The probability given by this theory is in every way different—in meaning, numerical value, and form—from that of those who would apply to ampliative inference the doctrine of inverse chances. (Peirce 2.748)[2]

One finds specific examples in Peirce that anticipate Neyman-Pearson hypothesis tests and confidence interval methods. A study of the statistical mathematics found in Peirce is of interest in its own right, but that is not my purpose here. My purpose is to explore how in Peirce's philosophy of experiment the formal NP tools become tools for scientific induction. (Neyman, remember, had denied them that function, and Pearson never fully worked out his "evidential" interpretation of NP tools.)

The place to begin is with the contrast that Peirce is at pains to draw between his view of induction and the more popular inductive accounts of his day. The most popular accounts of the time, Peirce tells us, are those of the "conceptualists"—the Bayesian theorists of Peirce's day—and the followers of Mill—essentially those who viewed induction as the straight rule, coupled with a premise as to the uniformity of nature. With admirable clarity Peirce compares these opposing views and forcefully argues against the popular ones. The main contrasts show up in the form of conclusion or inference (severe tests); the type of relevant information (preliminary planning, predesignation, random sampling); and the nature of its justification (self-correcting, growth of experimental knowledge). In each, Peirce takes the position of our error statistician.

What We Really Want to Know . . . Error Probabilities

The key disagreement was and is over the function of probability in statistical inference in science: whether probability provides a measure of evidential strength in a hypothesis, or whether it should be used only to characterize error probabilities of test procedures. Although the terminology has changed, it is clear that Peirce adopts the

2. All Peirce references are to C. S. Peirce, *Collected Papers*. References are cited by volume and paragraph number. For example, Peirce 2.777 refers to volume 2, paragraph 777.

second use of probability as the appropriate one for experimental inference. The supposition that the probability of the conclusion is needed, Peirce recognizes, stems from a faulty analogy with deductive inference. Since deductive inference tells us that if such and such premises are true, then a given conclusion is true, it might be thought that inductive inference should tell us that if such and such premises are true, then a given conclusion is probable. Peirce denies this:

> In the case of analytic inference we know the probability of our conclusion (if the premises are true), but in the case of synthetic inferences we only know the degree of trustworthiness of our proceeding. (2.693)

Those who modeled induction on the analogy with deduction were to Peirce what Bayesians or other E-R theorists are to NP or error statisticians of today, and the key themes of Peirce's work on induction mirror the key issues that divide E-R theorists from error statisticians: the importance of error probabilities, the rejection of prior probabilities, and the centrality of the mode of data and hypothesis generation to the analysis of test results.[3]

In Peirce's testing model, like that of Neyman and Pearson, the experimental conclusion concerns a hypothesis that either is or is not true about this one universe, and so the only probability that a frequentist could assign it is a trivial one, 1 or 0.[4] Assigning a probability to a particular conclusion, for Peirce (recall chapter 3), makes sense only "if universes were as plenty as blackberries" (2.684). If people had only been careful to keep to the relative frequency notion of probability, Peirce scolds, the mistake in analogizing induction to deduction would have been apparent. To view statistical inference as a matter of assigning a probability to a conclusion (an a posteriori probability), is, for a frequentist like Peirce, tantamount to seeing the problem as follows:

> Given a synthetic conclusion; required to know out of all possible states of things how many will accord, to any assigned extent with this conclusion. (2.685)

3. An interesting article of Hacking's is relevant in this connection. Hacking (1980), discussing Peirce and Braithwaite, admits to having promoted the rejection of Neyman-Pearson statistics on the grounds that, failing to provide an E-R account, it could not be seen as an account of inductive inference. With this article Hacking announces that he has changed his mind on this point, and allows that an error-statistical account does provide us with an account of inductive inference.

4. Peirce also has an account of probabilistic inference where that is appropriate. But this is not induction.

Here "all possible states of things" refers to all possible universes or possible ways in which this universe could be. This Peirce sees as "an absurd attempt to reduce synthetic to analytic reason, and that no definite solution is possible" (ibid.). Moreover, it "implies that we are interested in all possible worlds, and not merely the one in which we find ourselves placed" (2.686).

What we really want to know, according to Peirce, is this:

> Given a certain state of things, required to know what proportion of all synthetic inferences relating to it will be true within a given degree of approximation. (2.686)

In more modern terminology, what we want to know are the error probabilities associated with particular methods of reaching conclusions about this world. Peirce continues:

> Now, there is no difficulty about this problem (except for its mathematical complication); it has been much studied, and the answer is perfectly well known. And is not this, after all, what we want to know much rather than the other? (Peirce 2.686)

Peirce goes on to illustrate how, even in his day, "the answer is perfectly well known." His numerical illustration is important for us, and I will return to it in a later section.

Denying That Belief Has Anything to Do with It

A further reason that error statistical methods are congenial to Peirce's picture is that their error probability characteristics do not depend on subjective probabilities. Peirce held that

> subjective probabilities, or likelihoods, . . . express nothing but the conformity of a new suggestion to our prepossessions; and these are the source of most of the errors into which man falls, and of all the worst of them. (2.777)

An important part of Peirce's rejection of subjective probabilities is his insistence upon a distinction between the proper procedure for a scientific investigation and that for an individual seeking a practical basis for action. Peircean pragmatism (or pragmaticism) is not at all to be identified with practicalism!

While allowing that subjective beliefs and personal opinions may have to be appealed to in the area of practical conduct, where expediency is the rule, and where personal beliefs matter, Peirce thinks that "the word belief is out of place in the vocabulary of science" (7.185), except when considering actions based on science. In a scientific investigation Peirce declares, "I would endeavor to get to the bottom of the question, without reference to my preconceived notions" (7.177). The aim of science is to predict the future "or the means of conditionally predicting what would be perceived were anybody to be in a situation to perceive it" (7.186). Its aim is to predict what would be expected to occur with various relative frequencies were specified experiments carried out—in short, to obtain what I call experimental knowledge. Indeed, for Peirce "the essential character of induction is that it infers a *would-be* from actual singulars" (8.236).

Interestingly, Peirce's central arguments against the use of subjective probabilities have a naturalistic flavor: inferences based on subjective probabilities, he finds, make a poor showing when they themselves are put to the test of experiment. As an example Peirce considers their track record in archaeology. Finding the conclusions sanctioned by the practitioners of the subjective method "to be more or less fundamentally wrong in nearly every case," he declares the method "condemned by those tests" (7.182).[5]

This much has so far been brought to light about Peirce's theory of induction:

> In the case of analytic inference we know the probability of our conclusion (if the premises are true), but in the case of synthetic inferences we only know the degree of trustworthiness of our proceeding. As all knowledge comes from synthetic inference, we must equally infer that all human certainty consists merely in our knowing that the processes by which our knowledge has been derived are such as must generally have led to true conclusions. (2.693)

12.2 PEIRCEAN INDUCTION AS SEVERE TESTING

The scientific procedure in whose trustworthiness we are interested is, for Peirce, induction, but induction is to be understood as testing. The trustworthiness of inductive procedures, I maintain, is a matter of the test's severity, as measured formally (quantitative induction) or informally (qualitative induction). What is my evidence for this reading of Peirce?

First, there is the evidence that Peirce regards induction as severe testing. Induction, Peirce tells us, begins with a question or theory:

> The next business in order is to commence deducing from it whatever experiential predictions are extremest and most unlikely . . . in order

5. Peirce elsewhere gives astute criticisms of the use of the principle of indifference in assigning equal subjective probabilities, which I will not discuss.

to subject them to the *test of experiment*. (Peirce 7.182; emphasis added)

> The process of testing it will consist, not in examining the facts, in order to see how well they accord with the hypothesis, but on the contrary in examining such of the probable consequences of the hypothesis as would be capable of direct verification, especially those consequences which would be very unlikely or surprising in case the hypothesis were not true. (7.231)

> When the hypothesis has sustained a testing as severe as the present state of our knowledge . . . renders imperative, it will be admitted provisionally . . . subject of course to reconsideration. (Ibid.)[6]

Further passages to the same effect could easily be multiplied.

While these and other passages are redolent of Popper, Peirce differs from Popper in crucial ways—the same ways in which my own account differs. Peirce, unlike Popper, is primarily interested in the positive pieces of information provided by tests, that is, with the hypotheses, modified or not, that manage to pass severe tests. Indeed, Peirce often suggests that he equates the proper inductive part of a test of experiment with the inference that is reached when a hypothesis passes several stringent tests:

> When, however, we find that prediction after prediction, notwithstanding a preference for putting the most unlikely ones to the test, is verified by experiment, whether without modification or with a merely quantitative modification, we begin to accord to the hypothesis a standing among scientific results. This sort of inference it is, from experiments testing predictions based on a hypothesis, that is alone properly entitled to be called *induction*. (7.206)

A Peircean inductive inference, then, accords well with the thesis I have advocated: an inductive inference—that which is warranted to infer—is what passes a severe test. Whereas one could say nothing about the reliability of a Popperian corroboration procedure—the very reason I denied Popper supplies a genuine account of learning from error—the centerpiece of Peirce's experimental philosophy is his argument for the trustworthiness of proper inductive test procedures.

It is impossible to understand Peirce's argument, however, without understanding Peirce's doctrine of induction as self-correcting or as error-correcting. This requires us to open a door that many Peirce scholars already regard as closed, or at least to open it just far enough

6. Here Peirce is talking about historical hypotheses. See also (Peirce 1958, vol. 7, p. 89).

to give a different reading of the error-correcting doctrine. This new reading of the error-correcting doctrine, I believe, shows how the criticisms of the usual reading are avoided.

12.3 REVISITING PEIRCE'S ERROR-CORRECTING DOCTRINE

According to Peirce:

> The validity of induction is entirely different [from deduction]. . . . In the majority of cases, the method would lead to *some* conclusion that was true, and that in the individual case in hand, if there is any error in the conclusion, that error will get corrected by simply persisting in the employment of the same method. (2.781)

Throughout Peirce's work, a multitude of such passages can be found, each offering different clues to and different facets of his self-correcting doctrine. (Several are noted in Laudan 1981a.) What must be kept in mind, and often is not, is that induction for Peirce is testing, and testing of a certain sort (severe or reliable); it is testing (done severely) that he is claiming is self-corrective, and not other methods that philosophers often regard as inductive:

> Induction is the experimental testing of a theory. The justification of it is that, although the conclusion at any stage of the investigation may be more or less erroneous, yet the further application of the same method must correct the error. The only thing that induction accomplishes is to determine the value of a quantity. It sets out with a theory and it measures the degree of concordance of that theory with fact. (5.145)

Can Peirce sustain his self-correcting thesis as a way of giving a rationale for scientific induction? Critics and followers alike say no. The literature on this issue is too large to consider here, but fortunately, Rescher's excellent discussion (1978) lets me zero in on the key criticism, as waged by Larry Laudan and others. I will follow Laudan's abbreviation of the self-correcting thesis: (SCT). Let me begin with a brief summary of the main criticism and how I propose to deal with it.

The main criticism of the SCT is this: whereas Peirce claims to have substantiated the SCT for induction generally, he has at most done so regarding a certain species of induction, namely, quantitative or statistical induction. This criticism rests on two assumptions: the first concerns the nature of inductive testing for Peirce, of both the "quantitative" and "qualitative" varieties; the second concerns the question of what substantiating the SCT requires. As to the first, Peirce's critics

typically construe quantitative induction as classic enumerative induction or "the straight rule" (i.e., inference about a population proportion
from a sample proportion). By qualitative induction, critics understand
Peirce to mean hypothetico-deductive inference (Laudan 1981b, 238).
But from all we have already seen, it is clear that neither of these
modes of inference suffices for a test procedure that is trustworthy or,
in my terms, reliable or severe. So the first thing we need to do is to
revise the standard interpretation of Peirce's two types of induction.

I will be arguing that what distinguishes Peircean quantitative
from qualitative induction is not that the former is the straight rule
while the latter is a hypothetico-deductive inference. *Both* types of inference, in so far as they qualify as Peircean inductions, are inferences
based on tests with various degrees of severity. What distinguishes
them is the extent to which their severity or reliability can be quantitatively or only qualitatively determined. If the severity is quantitatively
specified, as in the case of the statistical significance test, then the inference is a quantitative induction. If severity is only qualitatively assessed, as for example in one of the informal arguments from coincidence we have considered, then it counts as a qualitative induction.
The difference is a matter of degree.

Turning to the second issue, critics are fairly clear on what they
suppose is required for an inductive method to be self-correcting: (*a*)
it must be capable of (eventually) rejecting false hypotheses, and (*b*) it
must provide a method of replacing rejected hypotheses with a better
(truer) one (Laudan 1981b, 229).[7] Their criticism of Peirce's SCT, in
the light of *their* understanding of Peircean quantitative and qualitative
induction, is this: although quantitative induction pretty well satisfies
both (*a*) and (*b*), qualitative induction only satisfies (*a*). Laudan puts
it plainly:

> Such qualitative inductions clearly satisfy the first condition for an
> SCM [self-correcting method], insofar as persistent application of the
> method of hypothesis will eventually reveal that a false hypothesis is,
> in fact, false. But the method . . . provides no machinery whatever
> for satisfying the second necessary condition. . . . Given that an hy-
> pothesis has been refuted, qualitative induction specifies no tech-
> nique for generating an alternative which is (or is likely to be) closer
> to the truth than the refuted hypothesis. (Laudan 1981b, 238–39)

Ilkka Niiniluoto (1984), in like fashion, assimilates Peircean self-
correcting to a view of scientific progress as replacing earlier theories

---

7. Laudan regards statement *b* as the strong thesis of self-correcting. A weaker
thesis would replace *b* with *b'*: science has techniques for unambiguously determin-

with those closer to the truth, leading him also to criticize Peirce for not having told us how induction affords such progress.[8] The technique for discovering a better alternative, moreover, is supposed to be mechanical or routine, and, not surprisingly, critics find that Peirce has not provided such a routine. Rescher (1978) objects to this requirement and defends the Peircean SCT as claiming only that it is the conglomeration of scientific methods that serves to find better alternatives. Rescher is right to object, but I think we can show that Peirce is saying something more specific about the error correcting role of inductive methodology in science. Inductive methods, properly construed, are very good at uncovering mistakes and this is what allows them to carry out effective tests to begin with. Their effectiveness consists in this: when they regard a hypothesis as having passed a test sufficiently well, that constitutes good grounds for that hypothesis.

These points lead to a reworking of the critics' two assumptions about the SCT. From the severity requirement we actually get a strengthened form of condition *a:* the inductive test procedure must have a *high*, not merely some, probability of rejecting false hypotheses. But we must not overlook, as critics seem to, the emphasis Peirce places on what is learned when such severe tests do not reject but instead pass their hypotheses. For Peirce, as I read him, the SCT is called upon to justify the *acceptance* of a hypothesis that has passed a severe test (e.g., 2.775). Inductive inference is the inference that is warranted when predictions hold up to severe testing. So the proper requirement for the SCT is not condition *b,* as the critics state it, but rather a condition that takes more literally what error-correction means.

A reworked condition *b* would have two parts: First, the method should be sufficiently good at detecting errors such that when no error is detected, when, try as we might, the effect will not go away, experimental knowledge (as we have defined it) is gained. Second, the method should be able to detect its own errors in the sense of checking its own assumptions or its "own premises" as Peirce puts it (i.e., assumptions of experimental tests and data), and it should be able to correct violations or "subtract them out" in the analysis. To show that scientific induction is self-correcting comes down to showing that severe testing methods exist and that they enable the growth of experi-

---

ing whether an alternative *T'* is closer to the truth than a refuted *T.* I reject both of these.

8. For some commentators, for example, Lenz (1964), what Peirce says about qualitative induction is so unclear that they restrict themselves to quantitative.

mental knowledge. The progress is not of the theory-dominated but of the experimentalist variety. My task now is to justify these claims.

The Path from Qualitative to Quantitative Induction

First I will argue my thesis about Peirce's notions of quantitative and qualitative induction. A major problem in understanding the self-correcting doctrine is that induction, for Peirce, takes several different forms corresponding to different types of test procedures. These different test procedures, in turn, are associated with different types of assessments of trustworthiness (i.e., of error probabilities) as well as different types of error-correcting tasks. What is more, throughout Peirce's work one finds a variety of attempts to delineate types of induction, and one may wonder which delineation to work with. In fact, Peirce does not think there is anything hard and fast about his classification attempts. Although critics are right to notice some shifts in Peirce's view on induction, his different schemes for classifying types of induction are almost entirely due to his directing himself to different kinds of experimental tests in different essays. Most important, if one looks at the big picture, a fairly clear-cut image emerges. Induction is testing, some qualitative, some quantitative or statistical—all agree on this. Where my reading of Peircean induction is new is that I view Peirce's delineation into quantitative and qualitative induction as a matter of classifying tests according to whether their trustworthiness (or severity) is quantitatively or only qualitatively ascertained. (This is the same construal, recall chapter 2, that I suggested for Kuhn's use of quantitative inference in normal science.)

In this reading of Peirce, the difference between qualitative and quantitative induction is really a matter of degree, and the degree is a function of how well developed its associated measures of trustworthiness are—in particular severity. This reading not only neatly organizes the long stories Peirce tells in classifying and subclassifying types of induction, it explains the way in which Peirce further subdivides types of inductions by their "strength" within a given classification.

First-order, rudimentary or crude induction. Take Peirce's delineation of types of induction in discussing scientific method. Here Peirce divides nonstatistical or qualitative induction into first and second orders. The first order is the lowest, most *rudimentary induction,* the so-called "pooh-pooh" argument. It is essentially an argument from ignorance: Lacking evidence for the falsity of *H,* provisionally adopt *H*—where *H* is some general claim or regularity. While Peirce holds this type of "crude induction" to be uneliminable in ordinary life, it has little place in scientific inquiry. (It corrects itself—but with a bang!) It is only in this very weakest sort of induction, crude induction, that one is limited to saying that a hypothesis would eventually be falsified if false. Crude induction, Peirce says, is "as weak an inference as any that I would not positively condemn" (8.237), and does not even make it into science. Once positive information is available, this most rudimentary induction is to go by the board. Hence, following Peirce, rudimentary induction is not to be included as scientific induction. It is, however, worthwhile to recognize why not: without some reason to think that evidence of *H*'s falsity would probably have been detected, failure to detect it is poor evidence for *H*. It is a highly unreliable error probe.

Second order (qualitative) induction. It is only with what Peirce calls the "Second Order" of induction that we arrive at a genuine test, and thereby scientific induction. Within second-order inductions, a stronger and a weaker type exist, and they correspond neatly to viewing the strength of a testing procedure as reflecting severity.

> The weaker of these is where the predictions that are fulfilled are merely of the continuance in future experience of the same phenomena which originally suggested and recommended the hypothesis. (7.116)

> The other variety of the argument from the fulfillment of predictions is where [they] . . . lead to new predictions being based upon the hypothesis of an entirely different kind from those originally contemplated and these new predictions are equally found to be verified. (7.117)

The weaker type, to put it in our terminology, occurs where violating use-novelty destroys the severity requirement. The stronger type is stronger because it generally yields a higher severity test. Peirce's divisions by strength within second-order inductions are also a function of severity, but the assessment of severity is qualitative, for example, very strong, weak, very weak.

> The strength of any argument of the Second Order depends upon how much the confirmation of the prediction runs counter to what our expectation would have been without the hypothesis. It is entirely a question of how much; and yet there is no measurable quantity. *For when such measure is possible the argument . . . becomes an induction of the Third Order* [statistical induction]. (7.115; emphasis added)

It is upon these and numerous like passages that I base my reading of Peirce. Furthermore, a qualitative induction, Peirce is quite clear, *becomes* a quantitative induction when the severity is quantitatively determined, when, as we might say, an objective error probability can be given.

Third order, statistical (quantitative) induction. This takes us to the third-order, statistical or quantitative induction. We enter the third order of induction when, to paraphrase Peirce, it is possible to quantify "how much" the prediction runs counter to what our expectation would have been without the hypothesis. Quantifying how much, as I hope is already clear from earlier discussions, permits quantifying trustworthiness by quantifying error probabilities.

To remind us, consider how a significance level measures how much a prediction runs counter to what is expected "without the hypothesis," where this refers to a simple null hypothesis H_0. As always, we see the following inversion: the lower the significance level, the more the prediction runs counter to the null hypothesis. Hence, the lower the significance level required before rejecting H_0 and accepting the nonnull hypothesis—call it H—the more improbable such an acceptance of H is, when in fact H_0 is true. And the more probable such an erroneous acceptance of H is, the higher the severity is of a result taken to pass H. This just rehearses what we already know. Other associated measures of "how much" are given by standard errors and probable errors, error probabilities all.

Notice that it is in order for the inductive *acceptance* of a hypothesis H to have strength that we meet the requirement that there be a high probability of rejecting hypothesis H, were H false. That is, Peircean induction refers to the positive inference—to what can be said to have passed a severe test:

> When we adopt a certain hypothesis, it is not alone because it will explain the observed facts, but also because the contrary hypothesis would probably lead to results contrary to those observed. So, when we make an induction, it is drawn not only because it explains the distribution of characters in the sample, but also because a different rule *would probably have led to the sample being other than it is.* (Peirce 2.628; emphasis added)[9]

This concern with the probability that the sample *would have been other than it is* in reasoning from the actual sample obtained puts Peirce squarely in the error statistics camp. And because one need not be able to point to some precise probability, the same self-correcting rationale is open to quantitative and qualitative tests.

As further evidence that Peirce understood the strength of an in-

---

9. By a "rule" here Peirce means a hypothesis such as most *As* are *Bs*.

duction in this way, Peirce often links the strength of induction—even in qualitative cases—to achieving what we would term a low standard deviation or low standard error (and, correspondingly, to a high severity):

> The results of non-quantitative researches also have an inexactitude or indeterminacy which is analogous to the probable error of quantitative determinations. To this inexactitude, although it be not numerically expressed, the term "probable error" may be conveniently extended. (7.139)

(A probable error is approximately .7 of a standard deviation.) It is convenient to extend the notion of a probable error for the same reason we found it convenient to use the term "severity" both when there was a numerical error probability that could be assigned to a test and when we could only argue that there was clearly a very high or a very low chance of error. They serve analogous roles in argument and, accordingly, qualitative and quantitative inductions are improved upon in analogous ways. The factors Peirce takes to increase or diminish the strength of procedure further illuminate the correspondence between the "strength of a proceeding" and our severity concept. Peirce explains that arguments are strengthened when certain invariabilities exist—in effect, factors that by diminishing a standard deviation would increase the chance of rejecting a false hypothesis. (See, for example, 7.125.)

Scientific induction, for Peirce, is inferring or accepting hypotheses that pass severe or trustworthy tests. The move from qualitative to quantitative induction is achieved by the acquisition of quantitative assessments of severity.

The SCT and Quantitative Induction

In inductive inference, unlike deductive or analytic inference, Peirce declared, what we really want to know is the trustworthiness of the proceeding or, in more modern terms, the error probabilities. Moreover, in the case of quantitative induction, Peirce said, the answer to the question we really want to know "is perfectly well known" (2.686). Let us now pick up Peirce where we left him in section 12.1.

In Peirce's example, the inductive inference estimates a Binomial parameter p on the basis of the number n of white balls observed in a sample of s balls. Referring to the difference between the observed proportion $\frac{n}{s}$ and the true proportion p as "the error", Peirce (2.686) explains that it is found that, if the true proportion of white balls is p, and s balls are drawn, then the error of the proportion obtained by the induction will be—

half the time within	0.477e
9 times out of 10 within	1.163e
99 times out of 100 within	1.821e
999 times out of 1,000 within	2.328e
9,999 times out of 10,000 within	2.751e
9,999,999,999 times out of 10,000,000,000 within	4.77e,

where I have substituted e for the square root of $[\dfrac{2\,p(1-p)}{s}]$.

Whereas simple enumerative induction, which is how critics construe quantitative induction, would merely estimate p to be the sample proportion, Peirce insists on a second step: attaching an error to this estimate. It may be in terms of the "probable error" concept of Peirce's day, or the more modern standard error, or, as in contemporary polls, a margin of error.

The SCT and Confidence Interval Estimation Procedures. The data from the above chart may be used to form confidence interval estimates (discussed in chapters 8 and 10). The inductive conclusion in the case of the interval estimation asserts that the observed proportion $\dfrac{n}{s}$ is within a certain distance from the true value of p, and attaches to that estimate a statement of the overall reliability of that method (as given by the confidence level). An example of such an estimate would assert that the observed proportion is within 1.821e of the true value p. Although the method does not assign a probability to this particular estimate being true, that probability being seen as either 0 or 1, the method can say that the inference comes from a procedure with .99 probability of covering the true value of p. The inferred estimate, that parameter p is within the interval formed, passes a severe test. So Neyman and Pearson confidence interval estimation satisfies Peirce's model of induction. In contrast to induction by simple enumeration (or the straight rule) as Peirce never tires of reminding us, the induction he espouses depends entirely on "the manner in which the instances have been collected" (2.765). But critics seem to overlook this contrast.

Isaac Levi puts his finger on how self-correcting works in the case of statistical estimation:

> Peirce is not claiming that induction is self correcting in the sense that
> following an inductive rule will, in the messianic long run, reveal the

true value of p. His thesis can be put this way: Either the conclusion reached *via* an inductive rule is correct or, if wrong, the revised estimate emerging from a new attempt at estimation based on a different sample will with probability at least equal to k be correct. (Levi 1980b, 138)

Suppose the induction is to a confidence interval with level k. The idea is that if one continues to sample (with replacement) and form a confidence interval with confidence level k, "he would be right with a relative frequency which would converge on k in the long run" (p. 136). The same type of argument is available for other cases of statistical estimation.

In the case of hypothesis testing, a claim parallel to Levi's on estimation can be made: if a particular conclusion is wrong, subsequent severe (or highly powerful) tests will with high probability detect this. For example, in a good test hypothesis H_0 is rejected by results improbably far from what is expected were H_0 true. Then, if we are wrong to reject H_0 (and H_0 is actually true), we would find we were rarely able to get so statistically significant a result to recur, and in this way we would discover our original error. If, on the other hand, we find that it is easy to keep getting results statistically significantly far from H_0, then we have grounds for saying that a real departure from H_0 exists. To say we have experimental knowledge of a real or systematic departure from H_0 is to say that H_0 would be rejected about as often as expected if such a departure exists. (The expectation comes from the laws of large numbers, discussed in chapter 5.)

Peirce discussed the Gaussian or Normal case as well as the Binomial. Modern statistical theory greatly extends the cases for which "the answer is well-known," but the rationale for the inferences it licenses is essentially the one that Peirce had already articulated:

> While the induction is probable in this sense, that though it may happen to give a false conclusion, yet in most cases in which the same precept of inference was followed, a different and approximately true inference (with the right value of p) would be drawn. (2.703)

More needs to be said about how formal statistical arguments supply tools for substantive error-correcting and learning. Here is where Peirce's stress on the *intended use* of these methods comes in.

Quantitative Induction and Canonical Models of Error. In developing the error statistical approach to testing, I have urged that the role of quantitative models such as the Binomial goes far beyond the case in which the primary aim is to infer the proportion of Bs in a population of As—

even though Binomial inference is formally couched in those terms. This formal statistical case serves largely as a canonical model for imaginatively asking questions about errors, about experimental assumptions, about the reality of a given effect, about quantities in laws and theories, about causes. I find evidence of this idea in Peirce, if not explicitly, then by considering how he applies statistical models in his examples.

Immediately after listing the different error ranges for the Binomial case above, Peirce remarks that "the use of this may be illustrated by an example" (2.687) that sounds very much like running an NP statistical significance test. As in many other cases, Peirce applies it to testing if a difference is real or systematic as opposed to due to chance. Peirce reports that an observed proportion of white males under one year (according to the census of 1870) is .5, while that of nonwhite children is .498, the difference being about .01. Peirce asks, "Can this be attributed to chance," or is it systematic? The largeness of the observed difference excludes it even from the largest interval formed; it falls beyond $4.77e$, "and such a result would happen, according to our table, only once out of 10,000,000,000 censuses, in the long run" (2.687). In short, the observed difference is indicative of a real rather than a chance difference. Were it due to chance it would, with high probability, have been included in the interval. The procedure was a reliable probe of the error of ruling out chance; so we can argue that this error is absent.

In the above illustration, the hypothesis concerned the ordinary kind of Binomial population. But Peirce extends this analysis to assess hypotheses that are not themselves statistical, but where *introducing* statistical considerations enables a question of interest to be modeled as inquiring about a Binomial parameter p. In particular, a question that can often be framed by means of parameter p is to let p be the probability with which a given agreement or fit between the experiment and a given hypothesis H would occur. Such a question may be probed statistically, even where hypothesis H itself is not statistical. Let us see how the SCT enters:

> It is true that the observed conformity of the facts to the requirements of the hypothesis may have been fortuitous. But if so, we have only to persist in this same method of research and we shall gradually be brought around to the truth. (Peirce 7.115)

But the correction is not a matter of getting estimates closer to p. It is a matter of finding out whether the agreement is fortuitous; whether it is generated *about as often as would be expected* were the agreement of the chance variety. The measure of severity reflects how fast the correction is likely to be.

The SCT and the Importance of Hypothesis and Data Generation

This error-correcting capacity, Peirce stresses, depends upon the *predesignation* of the Binomial property p (or, at least on an argument that its violation does not vitiate the induction).[10] I limit myself to one of Peirce's many instructive examples: that of Dr. Lyon Playfair. It illustrates both a mistake resulting from violating predesignation, as well as how, arguing from cases where error probabilities are sustained, the original mistake is corrected. Error correction is not a hope for tomorrow, it *is* the inductive conclusion of tests we run today.

The Example of Dr. Playfair. Peirce describes how "so accomplished a reasoner" as Dr. Playfair violates predesignation in testing a hypothesis about a regularity between the specific gravity of a metal and its atomic weight (2.738). Looking at the specific gravities of 3 forms of carbon, Peirce tells us, Playfair seeks and discovers a formula connecting them: each is a root of the atomic weight of carbon, which is 12. Peirce describes the test Playfair carries out to judge whether this regularity can be expected to hold generally for metals, showing that several alleged instances of the formula really involve modifications not specified in advance. If one limits the instances to ones for which the formula is predesignated, only half satisfy Playfair's formula. Peirce reasons:

> Having thus determined [the] ratio, we proceed to inquire whether an agreement half the time with the formula constitutes any special connection between the specific gravity and the atomic weight of a metalloid. (2.738)

Of particular interest here is the creative use of a canonical test of a proportion. The proportion refers to the *proportion or probability of agreements* with the formula. There is hardly a limit to the kinds of cases where a question about this proportion could be posed.

Peirce then subjects the hypothesis that there *is* a special connection (between the specific gravity and the atomic weight of a metal) to a test of experiment. The falsity of this hypothesis is that the observed agreement is "due to chance" (2.738)—a variant of the standard null hypothesis. Peirce asks, How often would such an agreement be found even if it were due to chance? To answer this question, Peirce *introduces* statistical considerations into an otherwise nonstatistical case.

Peirce introduces a hypothetical chance distribution by matching the specific gravity of a set of elements not with its own atomic weight but with the atomic weight of some other element with which it is

---

10. Peirce qualifies this. It is sufficient that the Binomial property to be estimated or tested be prespecified; the value of the proportion need not be.

arbitrarily paired. For example, the specific gravity of carbon is compared with the atomic weight of iodine. Note that Peirce is not running more trials of Playfair's experiment, but considering "on paper" how often agreements with Playfair's formula would occur in a case designed so that such agreements could only be due to chance, and using this information about *what would occur* to argue about the cause of the agreements actually found. This strategy is analogous to the other introductions of statistics we have seen, whether they are by random pairings of treatments and subjects, by manipulations done on paper (e.g., Perrin), or by simulation (neutral currents). The logic applied to the results is the same as well.

Peirce finds about the same number of cases satisfying Playfair's formula in this chance pairing of elements as Playfair found in comparing the specific gravities and atomic weights of a given element. Peirce concludes,

> It thus appears that there is no more frequent agreement with Playfair's proposed law than what is due to chance. (2.738)

So Playfair was mistaken in thinking that the evidence showed a special or systematic connection. This example, which merits more attention than I can give it here, is used by Peirce to make a point about predesignation. His point is that the popular inductive accounts are insensitive to the effects of violating predesignation, and as a result they allow one to persist in Playfair's error.

While it would be going too far to see in Peirce the anticipation of Armitage (chapter 10), it is no stretch to see that error probability considerations play identical roles for Peirce and for the error statistician: before the trial their role is to ensure the severity of the test, after the trial it is to assess what induction is warranted. Peirce recognized that violations of predesignation need not preclude severity (see chapter 9). By introducing the hypothetical chance element, Peirce is ascertaining whether Playfair's inference is warranted *despite* the violation of predesignation. He shows that it is not. Whether it is Peirce's handwritten pairings on paper or twentieth-century Monte Carlo simulations in high energy physics, the basic strategy is the same. We find a way of modeling *what it would be like* (in this case, in terms of proportions of agreements) if the agreement is accidental or "due to chance." In Playfair's case, the actual situation is much like what we would expect were the observed agreements accidental.

The SCT and the Relevance of Preliminary Planning

The concern with "the trustworthiness of the proceeding" for Peirce, like the concern with error probabilities for Pearson and error statisticians generally, is directly tied to their view that statistical method should closely link experimental design and data collection with subsequent inferences. Pearson, remember, railed against the tendency to see statistical inference as beginning once "data are thrown at the statistician and he is asked to draw a conclusion" (Pearson 1966e, 278). Peirce had the same problem with the popular inductive accounts of his day. Peirce regarded as a conclusive refutation of Mill that "an induction, unlike a demonstration, does not rest solely upon the facts observed, but upon the manner in which those facts have been collected" (2.766). Peirce even introduces a term, "quasi-experimentation," to include the entire process of generating and analyzing the data *and* using them to test a hypothesis. And "this whole proceeding," Peirce declares, "I term Induction" (7.115, editor's note). Accordingly, for Peirce, the "true and worthy" task of logic is to "tell you how to proceed to form a plan of experimentation" (7.59).

It is this emphasis on the manner in which the data and hypotheses to test are generated, Peirce stresses, that really distinguishes his view of scientific induction from the two far more popular views of his day (Mill and the conceptualists). That is why the rationale for Peircean induction cannot be divorced from experimental rules for controlling error probabilities.

> This account of the rationale of induction is distinguished from others in that it has as its consequences two rules of inductive inference which are very frequently violated. . . . The first . . . is that the sample must be a random one. . . . The other rule is that the character [about which claims are to be tested] must not be determined by the character of the particular sample taken. (Peirce 1.95)

Hence induction, Peirce says, "must by the rule of predesignation, be a deliberate experiment" (5.579). One wishes that Peirce's critics had made more of the importance he attaches to these rules of data and hypothesis generation. Recognizing their importance is the key to understanding Peirce's SCT: they show that this self-correcting rationale has to do with the control of error probabilities.

Peirce's arguments for these rules are strikingly similar to those arising from the contemporary controversy between Neyman-Pearson "sampling" and nonsampling philosophies, that is, between error probability principles and the likelihood principle. As we saw in previous chapters (e.g., chapter 10), for those who accept the likelihood principle (e.g., Bayesians), once the data are obtained, it is irrelevant for assessing their evidential import how they were selected, or whether the hypothesis was predesignated (the so-called irrelevance of the sampling rule). For these do not alter likelihoods. But they do alter error

probabilities. Just as NP theorists insist on the relevance of predesignation—along the lines detailed in chapter 9—Peirce is highly critical of
predesignation being "singularly overlooked by those who have
treated of the logic of [induction]" (2.738).

> It is of the essence of induction that the consequence of the theory
> should be drawn first in regard to the unknown . . . result of experi-
> ment. . . . For if we look over the phenomena to find agreements with
> the theory, it is a mere question of ingenuity and industry how many
> we shall find. (2.775)

Just as it is only by planning ahead of time that a test can be regarded
as a reliable error probe, for Peirce "reasoning tends to correct itself,
and the more so, the more wisely its plan is laid" (5.575).

Learning from Qualitative Induction

Now critics claim that for qualitative inductive testing to be self-
corrective, it would have to provide a method of replacing substantive
hypotheses with better ones, for example, condition *b*. But Peirce calls
inferences from data to substantive hypotheses abduction or presumption, not induction. As abductive inference is free to violate predesignation, Peirce holds, it enjoys no such general error-correcting guarantee. Since induction is said only to have the power to correct any errors
into which *it* may lead, it is no part of the SCT to show that abduction
is trustworthy. However, the self-correcting rationale is all-important
when it comes to putting an abductively arrived at hypothesis to the
test of experiment.

Inductive testing of the qualitative variety has to do not with replacing falsified hypotheses with brand new ones, but with learning
from rejected hypotheses. A central aim is to learn what modifications
are called for by the experiments. The rationale for subjecting such
an abduction to a severe test of experiment is to learn about these
modifications. Among types of qualitative induction, Peirce places a
case that "tests a hypothesis by sampling the possible predictions that
may be based upon it. . . . We cannot say that a collection of predictions drawn from a hypothesis constitutes a strictly random sample of
all that can be drawn. Sometimes we can say that it appears to constitute a very fair, or even a severe sample of the possible prediction"
(7.216). Here the correction of hypotheses is expected to come about
through gradual modification. Peirce illustrates with the case of the
kinetic theory of gases.

> It began with a number of spheres almost infinitesimally small occa-
> sionally colliding. It was afterward so far modified that the forces be-

tween the spheres, instead of merely separating them, were mainly attractive, that the molecules were not spheres, but systems. (7.216)

These modifications "were partly merely quantitative, and partly such as to make the formal hypothesis represent better what was really supposed to be the case, but which had been simplified for mathematical simplicity" (7.216). Peirce grants that there is "no new hypothetical element in these modifications," but it is precisely with these kinds of modifications that induction is concerned. One poses a question, say, "Suppose I tried to model molecules as having uniform radius?" and then learns from the given experimental data how similar or divergent that model would be from the experimental phenomena.

The quantity of interest is not how much the evidence confirms the hypothesis tested—in any of the senses of confirmation—but *how discordant* evidence shows a given model to be in a specified respect. The problem of assessing the approximate accordance of a model is quite different from that of assigning it some E-R measure. There are a handful of methods for putting forward deliberately oversimplified or canonical hypotheses, because, with the appropriate methodology of testing, they serve for learning about these modifications (from rejected hypotheses). Experimental learning requires not some update of the probability assignment that I start out with, but tools to build, correct, and fill out a model. What justifies Peirce's SCT is that induction—understood as severe testing—supplies such tools.

Economy and the Piecemeal Breakdown of Inquiries

Having identified the aim of inductive testing, it is easy to understand Peirce's advice as to the type of hypotheses that are useful to test. Peirce considers "what principles should guide us in abduction, or the process of choosing a hypothesis" (7.219). He lists three: First, the hypothesis selected for tests "must be capable of being subjected to experimental testing"; second, the hypothesis must explain surprising facts. "In the third place," Peirce continues, "is the consideration of economy" (7.220).

The first two are familiar, but the third is rather unique to Peirce. While a concern for economy sounds as if pragmatic or practical considerations are being appealed to, Peirce's concern is in fact wholly epistemological. Considering "economy" in choosing a hypothesis to test means we should consider strategically what questions can be put to a reasonably severe test with the data that are likely to be actually obtainable. This aim, I have argued, leads to "getting small" and to a piecemeal approach to inquiry. It is likewise for Peirce. Under economy

Peirce cites the kind of strategy that makes for a shrewd playing of 20 questions:

> Twenty skillful hypotheses will ascertain what two hundred thousand stupid ones might fail to do. The secret of the business lies in the caution *which breaks a hypothesis up* into its smallest logical components, and only risks one of them at a time. (7.220; emphasis added)

These are questions amenable to the yes/no types of answers typical of standard statistical tests.

Considerations of economy, Peirce says, also direct one to try the same kind of model to account for the same kinds of phenomena, but in different areas. Peirce considers how the model used in the kinetic theory (chapter 7) "accounts for those phenomena . . . by representing that they are results of chance; or . . . of the law of high numbers" (7.221).

Giving Good Leave. A third important consideration under economy is "that it may give a good 'leave,' as the billiard-players say. If it does not suit the facts, still the comparison with the facts may be instructive with reference to the next hypothesis" (7.221). Even if we primarily want to know whether a quadratic equation holds between quantities, we would do well to test a linear model first "because the residuals will be more readily interpretable." The residuals, or errors—the differences between the observed and predicted value—may teach more about the next hypothesis to try. Hence, "even although we imagine that by complicating the hypothesis it could be brought nearer the truth" (ibid.), testing a simpler one may be justified because it will teach us more.

An adequate philosophy of experiment, I agree with Peirce, should include methodological rules directed at asking fruitful questions and arriving at local hypotheses to test, as well as rules for data generation and modeling. The former type has generally been left out of discussions of philosophy of statistics, and yet an important asset of standard statistical methods is that they can offer canonical models and rules for both of these types of rules. The nature and aims of the rules are very much in the spirit of Peirce's considerations. They are not mechanical or algorithmic, but neither are they mere guesswork. The logic of science, for Peirce, is not formal but a systematic methodology for experiment. In a favorite passage, Peirce describes the aim of a theory of experiment thus:

> It changes a fortuitous event which may take weeks or may take many decennia into an operation governed by intelligence, which will be finished within a month. (7.78)

The idea that a central aim of statistical method is to speed things up in this way, while overlooked in philosophical discussions, is at the heart of the rationale of error statistical methods. The concern is not with the kind of speeding up of production that Fisher so disliked (chapter 11), but rather, we might say, with making good on the "long-run" claims in the short long run, if not "within a month," then within a year or the usual amount of the time for a given scientific research project.

That we have a workable theory of experiment, that we make progress with this theory is what the SCT is all about. However, we are not quite finished with justifying this thesis; we have to go back down to the models of data, experimental design, and data generation.

12.4 Induction Corrects Its Premises

Justifying experimental inferences depends on being able to justify the assumptions of the experimental and data models required. Self-correcting, or error-correcting, enters here too, and precisely in the way that Peirce recognized. This leads me to consider something apparently missed by critics of the SCT, namely, Peirce's insistence that induction "not only corrects its conclusions, *it even corrects its premises*" (3.575; emphasis added).

Induction corrects its premises by checking, correcting, or validating its own assumptions. One way that induction corrects its premises is by correcting and improving upon the accuracy of its data. The idea is a fundamental part of what allows induction—understood as severe testing—to be genuinely ampliative. It is why, in an important sense, statistical considerations allow one to come out with more than is put in. At times, even "garbage in" need not mean "garbage out."

Peirce comes to his philosophical stances from his experiences with astronomical observations:

> Every astronomer, however, is familiar with the fact that the catalogue place of a fundamental star, which is the result of elaborate reasoning, is far more accurate than any of the observations from which it was deduced. (5.575)

Daily use of the method of least squares taught Peirce how knowledge of errors of observation can be used to infer an accurate observation from highly shaky data.[11] Peirce proceeds to apply the same strategy

11. The method of least squares is a method of finding the best estimate of a parameter value. Given a set of observations made independently, the differences of the observed values from the best estimate are the residuals or errors. The theory of least squares directs one to find the value for which the sum of the squares of

from astronomy to an informal, qualitative example to illustrate how "a properly conducted Inductive research corrects its own premisses":

> That Induction tends to correct itself, is obvious enough. When a man undertakes to construct a table of mortality upon the basis of the Census, he is engaged in an inductive inquiry. And lo, the very first thing that he will discover from the figures . . . is that those figures are very seriously vitiated by their falsity. (5.576)

The premises here are reports on age, and it is discovered that there are systematic errors in these reports. How? By noticing, Peirce explains, that the number of men reporting their age as 21 far exceeds those who are 20, while in all other cases ages are much more likely to be expressed in round numbers. How is it that induction helps to uncover that there is this subject bias, that those under 21 tend to put down that they are 21? It does so by means of formal models of age distributions along with informal background knowledge of the root causes of such bias. "The young find it to their advantage to be thought older than they are, and the old to be thought younger than they are" (5.576). Moreover, statistical considerations often allow one to correct for bias, that is, by estimating the number of "21" reports that are likely to be attributable to 20-year-olds. As with the star catalogue in astronomy, the data thus corrected are *more accurate* than the original data. That is Peirce's main point. The thrust of the thesis that induction corrects its own premises is easy to put in terms of our error statistical framework: by means of an informal tool kit of key errors and their causes, coupled with systematic tools to model them, experimental inquiry checks and corrects its own assumptions for the purpose of carrying out some other (primary) inquiry.

These cases of correcting premises underscore what I have maintained for Peircean self-correction generally. It is not a matter of saying that with enough data we will get better and better estimates of the star positions or the distribution of ages in a population. It is a matter of being able to employ methods right now to detect and correct mistakes in a given inquiry. The methods stem from canonical models of error, here for errors in observations of different types (e.g., from instruments, subjects, etc.). To get such methods off the ground, we need not build a careful tower where evidence rests on a pile of inferences, each as shaky as the ones before (like piles driven into a swamp). Properly collected and cleverly used, inaccurate observations give way to far more accurate data.

---

residuals is minimum. That is the best estimate of the value. This canonical method was used in the eclipse experiments discussed in chapter 8.

Induction Fares Better than Deduction at Correcting Its Errors

Consider how this reading of Peirce makes sense of his holding inductive science as better at self-correcting than deductive science.

> Deductive inquiry . . . has its errors; and it corrects them, too. But it is by no means so sure, or at least so swift to do this as is Inductive science. (5.577)

An example he gives is that the error in Euclid's elements was undiscovered until non-Euclidean geometry was developed. Other everyday examples arise in checking and rechecking calculations to uncover arithmetical errors. "It is evident that when we run a column of figures down as well as up, as a check," or look out for possible flaws in a demonstration, "we are acting precisely as when in an induction we enlarge our sample for the sake of the self-correcting effect of induction" (5.580). In both cases we are appealing to various methods we have devised because we find they increase our ability to correct our mistakes.

What is distinctive about the methodology of inductive testing is that it deliberately directs itself to devising tools for reliable error probes. This is not so for mathematics. Granted, "once an error is suspected, the whole world is speedily in accord about it" (5.577) in the case of deductive reasoning. But for the most part mathematics itself does not supply tools for uncovering flaws. (Consider, in this connection, the recent dispute about the correctness of an alleged proof of Fermat's last theorem.)

> So it appears that this marvelous self-correcting property of Reason . . . belongs to every sort of science, although it appears as essential, intrinsic, and inevitable only in the highest type of reasoning, which is induction. (5.579)

In one's inductive or experimental tool kit, one finds explicit models and methods whose single purpose is the business of detecting patterns of irregularity, checking assumptions, assessing departures from canonical models, and so on. Where experimental tests are unable to do this—where methods are unable to mount severe tests—then they fail to count as scientific induction.

Random Sampling and the Uniformity of Nature

In addition to the rule of predesignation, Peirce's SCT requires that the selection of the experimental sample be random or approximately so.[12] In fact Peirce is generally credited with defining a random sample. Yet the assumption of random sampling is often thought to be an obstacle to justifying statistical inference. In an interesting footnote, Hans Reichenbach (1971) makes this remark about Peirce:

> The self-corrective nature of induction was emphasized by C. S. Peirce. . . . I have not been able . . . to find a passage in Peirce's work where he clearly states a reason for his contention. The fact that he constantly connects the problem of induction with that of a fair sample . . . seems to indicate that he bases the self-corrective nature of induction on Bernoulli's theorem. . . . Such an argument is invalid, of course, since the justification of induction must be given before the use of probability considerations. (P. 446, n. 1)

There are many intriguing similarities between Peirce and Reichenbach that merit attention, but here I want to dwell on a key point of contrast that this passage points up. For it is this classical view of what is required to justify induction that Peirce is anxious to deny.

Peirce views the problem of justifying induction as explaining why inductive testing is so successful when it is. He contrasts his explanation with those favored by followers of Mill and "almost all logicians" of his day, who "commonly teach that the inductive conclusion approximates to the truth because of the uniformity of nature" (2.775). Inductive inference, as Peirce conceives it (i.e., severe testing) does not use the uniformity of nature as a premise. Rather, the justification is sought in the manner of obtaining data and specifying hypotheses to test. It is a matter of showing that methods exist with good error probabilities. For this it suffices that randomness be met only approximately, that inductive methods check their own assumptions, and that inductive methods can often detect and correct departures from randomness. Says Peirce:

> A sample is a *random* one, provided it is drawn by such machinery . . . that in the long run any one individual of the whole lot would get taken as often as any other. Therefore, judging of the statistical composition of a whole lot from a sample is judging by a method which will be right on the average in the long run, and, by the reasoning of the doctrine of chances, will be nearly right oftener than it will be far from right.

> It has been objected that the sampling cannot be random in this sense. But this is an idea which flies far away from the plain facts. Thirty

12. All that is really required is that a statistical relationship between the sampling and the population of interest be known approximately.

throws of a die constitute an approximately random sample of all the throws of that die; and that the randomness should be approximate is all that is required. (Peirce 1.94)

This again shows that Peirce was in the know about mathematical results (the central limit theorem). (Thirty is the magic number for which the distribution of the sample mean is nearly normal, regardless of the underlying distributions.)

Peirce backs up his defense with robustness arguments. For example, in an (attempted) Binomial induction Peirce asks, "What will be the effect upon inductive inference of an imperfection in the strictly random character of the sampling?" (2.728). What if, for example, a certain proportion of the population had twice the probability of being selected? Peirce shows that "an imperfection of that kind in the random character of the sampling will only weaken the inductive conclusion, and render the concluded ratio less determinate, but will not necessarily destroy the force of the argument completely" (2.728). This is particularly so if the sample mean is near 0 or 1. Yet a further safeguard is at hand, Peirce reminds us:

> Nor must we lose sight of the constant tendency of the inductive process to correct itself. This is of its essence. This is the marvel of it. . . . Even though doubts may be entertained whether one selection of instances is a random one, yet a different selection, made by a different method, will be likely to vary from the normal in a different way, and if the ratios derived from such different selections are nearly equal, they may be presumed to be near the truth. (2.729)

Here the "marvel" is its ability to correct the attempt at random sampling. Numerous, even more marvelous methods exist today to check randomness and other assumptions. Still, Peirce cautions, we should not depend so much on the self-correcting virtue that we relax our efforts to get a random and independent sample. But if our effort is not successful, and our method not robust, we will probably discover it. "This consideration makes it extremely advantageous in all ampliative reasoning to fortify one method of investigation by another" (ibid.).

"The Supernal Powers Withhold Their Hands and Let Me Alone"

Peirce turns the tables on those skeptical about satisfying random sampling—or, more generally, about satisfying the assumptions of a statistical model. He declares himself "willing to concede, in order to concede as much as possible, that when a man draws instances at random, all that he knows is that he *tries* to follow a certain precept" (2.749). There might be a "mysterious and malign connection between the mind and the universe" that deliberately thwarts such efforts. Peirce considers betting on the game of *rouge et noire*. "Could some devil look at each card before it was turned, and then influence me mentally" to bet or not, the ratio of successful bets might differ greatly from .5 (ibid.). But this would equally vitiate *deductive* inferences about the expected ratio of successful bets. We would find systematic departures from the Binomial model with $p = .5$, even where the card game did have an equal chance of a red or black card.

Peirce's argument can be seen as the counterpart to Neyman's justification for the use of mathematical models of random experiments (from chapter 5). Neyman (1952), recall, had explained how probabilistic models adequately represent certain real experimental procedures "whenever we succeed in arranging the technique of a random experiment, such that the relative frequencies of its different results in long series approach" sufficiently the mathematical probabilities in the sense of the law of large numbers (Neyman 1952, 19). We can check whether we have succeeded in satisfying the statistical model sufficiently. But the experimental procedure whose assumptions are found to be satisfied where p is known should work as well when p is unknown. To suppose otherwise, Peirce is saying, would be akin to supposing a mysterious power can read my mind and deliberately thwart my efforts to satisfy assumptions just when p is unknown.

Peirce therefore grants that the validity of induction is based on assuming "that the supernal powers withhold their hands and let me alone, and that no mysterious uniformity . . . interferes with the action of chance" (2.749). But this is very different from the uniformity of nature assumption.

> The negative fact supposed by me is merely the denial of any major premiss from which the falsity of the inductive or hypothetic conclusion could . . . be deduced. Nor is it necessary to deny altogether the existence of mysterious influences adverse to the validity of the inductive . . . processes. So long as their influence were not too overwhelming, the wonderful self-correcting nature of the ampliative inference would enable us, even so, to detect and make allowance for them. (2.749)

This is the reason for having standard mechanisms, for example, a coin-tossing mechanism such as our canonical Binomial experiment with $p = .5$. Finding systematic departures from the deductively derived statistical distribution would be one way of detecting that we had failed in a particular case to get the experiment to accord with the standard Binomial. We could then subtract out its influence.

Not only do we not need the uniformity of nature assumption, but also, Peirce declares, "That there is a general tendency toward uniformity in nature is not merely an unfounded, it is an absolutely absurd, idea in any other sense than that man is adapted to his surroundings" (2.750). But the validity of inductive inference does not depend on this.

The ability to make successful inductions, our success in obtaining experimental knowledge, is explained by the properties of our methods. The properties of the methods are error probabilities. Because we can frame questions of interest in term of hypotheses amenable to severe testing, we are able to learn from error and in so doing obtain experimental knowledge. That is what Peirce's SCT requires and what Peirce means by saying that "the true guarantee of the validity of induction" is that it is a method of reaching a conclusion that is able to detect errors:

> This it will do . . . because it is manifestly adequate . . . to discovering any regularity there may be among experiences, while utter irregularity is not surpassed in regularity by any other relation of parts to whole, and is thus readily discovered by induction to exist where it does exist, and the amount of departure therefrom to be mathematically determinable from observation. . . . The doctrine of chances . . . is nothing but the science of the laws of irregularities. . . . There is no possibility of a series of experiences so wanting in uniformity as to be beyond the reach of induction, provided there be sufficiently numerous instances of them, and provided the march of scientific intelligence be unchecked. (2.769)[13]

In the final chapter, I shall have more to say about how the error statistical program explains the success of scientific induction.

13. In aligning myself with Peirce, it should not be thought that I agree with a position popularly attributed to him, namely, that truth is the final opinion to which inquiry would eventually lead. One gloss is unproblematic, however: the true but *fixed* value of a population mean is the average of all the possible sample means. Hence, the average of sample means would eventually equal it.

CHAPTER THIRTEEN

Toward an Error-Statistical Philosophy
of Science

IN THE PRECEDING CHAPTERS I have attempted to set out the main ingredients for a non-Bayesian philosophy of science that may be called the error-statistical account. The account utilizes and builds upon several methods and models from classical and Neyman-Pearson statistics, but in ways that depart from what is typically associated with these approaches enough to warrant some new label. Because the chief feature that my approach retains from Neyman-Pearson methods is the centrality of error probabilities, the label "error statistics" seems about right. Moreover, what fundamentally distinguishes this approach from others is that in order to determine what inferences are licensed by data it is necessary to take into account the error probabilities of the experimental procedure. In referring to an error-statistical philosophy of science, I have in mind the various ways in which statistical methods based on error probabilities may be used in philosophy of science. At present, when it comes to appealing to statistical ideas in philosophy of scientific inference, the Bayesian Way is sometimes thought to be the only game in town. What I wish to impress upon the reader is that an error-statistical philosophy of science presents a viable alternative to the Bayesian Way.

The application of Bayesian statistics to philosophy of science—the Bayesian Way—for all of its flaws, enjoys a simplicity and unity of statement: Evidence is linked to hypotheses by way of Bayes's theorem and scientific inference is modeled as the application of Bayesian confirmation (e.g., evidence confirms hypothesis H when the posterior probability of H exceeds the prior probability of H). The rationality of a scientific episode is assessed according to how well it admits of a Bayesian reconstruction, and methodological rules are appraised according to whether they can be justified by Bayesian principles of support and confirmation. Although there is serious disagreement among Bayesians about the result of applying Bayesian principles to each of these tasks, at least there is a kind of overarching framework that prospective Bayesian philosophers can look to. Can anything as succinct and unified be said of an error-statistical philosophy of science? Indeed it can.

In the error theorist's approach, experimental inquiry is viewed in terms of a series of models: primary models, experimental models, and data models. In an experimental inference, primary hypotheses are linked to models of data by means of experimental tests, and hypotheses are inferred according to whether they pass severe tests. Methodological rules are regarded as claims about strategies for coping with, and learning from, errors in furthering the overarching goal of severe testing, and rules are assessed according to their role in promoting that end.

I do not wish to downplay the complexities of the error-statistical approach. It is necessarily more complex to state than an evidential-relationship view because, rather than starting its work with evidence or data, it includes the task of arriving at data—a task that it recognizes as calling for its own inferences. A second feature of the error-statistical approach that introduces a level of complexity is that it does not equate the scientific inference with a direct application of some statistical inference scheme. This is in contrast with other attempts that model scientific inference on statistical inference, whether Bayesian or non-Bayesian.

For example, to apply Neyman-Pearson statistics in philosophy of science, it is typically thought, requires viewing scientific inference as a matter of accepting or rejecting hypotheses according to whether outcomes fall in rejection regions of Neyman-Pearson tests. Finding that this distorts scientific inference, it is concluded that it is inappropriate to appeal to Neyman-Pearson statistics in erecting an account of inference in science. This conclusion, I have argued, is quite unwarranted because it overlooks the ways in which Neyman-Pearson methods, and standard statistics in general, are actually used in science. What I am calling the error-statistical account, I believe, reflects these actual uses.

In the error-statistical account, formal statistical methods relate to experimental hypotheses, hypotheses framed in the experimental model of a given inquiry. Relating inferences about experimental hypotheses to primary scientific claims is, except in special cases, a distinct step. Yet a third step is called for to link raw data to data models—the real material of experimental inference. The indirect and piecemeal nature of our use of statistical methods, far from introducing an undesirable complexity into our approach, is what enables it to serve as an account of inference that is truly *ampliative*. The complexities notwithstanding, I wish to impress upon the reader that there is a structure and a logic to the error-statistical approach, that its parts hang together to provide a full-bodied experimental philosophy.

To this end, I will address in this closing chapter the main lines taken by the error-statistical approach to three chief tasks to which statistical accounts are put in philosophy of science: modeling scientific inference, solving problems about evidence and inference, and performing a critique of methodological rules. In the final section I shall consider yet another task for which one might rightly look to an account of statistics: that of explaining the success of science—the fact that we are so good at predicting, controlling, and learning about experimental phenomena. The present philosophy of experiment locates scientific progress in the growth of experimental knowledge, and explains that growth in terms of the properties of the methods making up error statistics, as broadly conceived. These methods work because (1) we are, at least some of the time, able to carry out real random experiments (in the sense of chapter 5), and can test if we have done so in particular cases, and (2) we are often able to put questions about errors in terms of questions that can be answered by real—or simulated—random experiments. Thanks to the limit theorems of statistics (and their empirical correlates), (1) and (2) give us reliable experimental knowledge.

My aim in this final chapter is to identify for the reader the main threads with which the error-statistical approach ties together and performs the tasks expected of an account of experimental inference. These sketches do not substitute for the fuller arguments and examples given throughout the book. At several points I will deliberately identify gaps that still remain to be filled. My hope is to organize the various ways in which the error-statistical program might be further pursued by others.

13.1 Modeling Experimental Inquiry

In the error-statistical account, experimental inference must be understood within a framework of inquiry. You cannot just throw some "evidence" at the error statistician and expect an informative answer to the question of how well it warrants a hypothesis. A framework of inquiry incorporates methods of experimental design, data generation, modeling, and testing, all of which can be organized around the hierarchy of models set out in chapter 5. The framework of models does double duty; it also allows addressing systematically the key questions of an epistemology of experiment: questions about what data to collect, how to model them, how to check their assumptions, how to use them to learn about experimental processes, and how to relate experimental knowledge to scientific hypotheses.

Experimental Knowledge and Arguing from Error

Experimental knowledge is obtained by learning about the (actual or hypothetical) future performance of experimental processes, about the outcomes that would be expected (with specified frequencies) if certain experiments were carried out. Hypotheses about experimental processes may be inferred regardless of whether they are part of a substantive scientific theory. They have their own homes within various experimental models.

Although a single inquiry involves a network of models, an overall logic of experimental inference emerges: data e indicate the correctness of hypothesis H, to the extent that H passes a severe test with e. All the tasks of the interconnected models are directed toward substantiating this piece of reasoning. To remind us, hypothesis H passes a severe test with e if e fits H, and the test procedure had a high probability of producing a result that accords *less well* with H than e does, if H were false or incorrect.

The severe testing inference corresponds to an informal pattern of *arguing from error* or *learning from error.* The underlying thesis is this:

> It is learned that an error is absent to the extent that a procedure of inquiry with a high probability of detecting the error if and only if it is present nevertheless detects no error.[1]

Its failing to detect the error means it produces a result (or set of results) that is in accordance with the absence of the error. Such a procedure of inquiry may be called a *reliable (or highly severe) error probe.* We argue that an error is absent if it fails to be detected by a highly reliable error probe. Correspondingly, an assertion that the error is absent has passed a severe test. We identified an analogous argument for inferring the presence of an error.

The experimental inference that is licensed is whatever can be regarded as having passed a severe test by the given result and the given test procedure. The hypothesis that is indicated may not be the full hypothesis of interest, but only that part or aspect that has passed a severe test. One has to look, in other words, at the argument from

---

1. In terms of a hypothesis H, the argument from error may be construed as follows: Data in accordance with hypothesis H indicate the correctness of H to the extent that the data result from a procedure that with high probability would have produced a result more discordant with H, were H incorrect.

error that is substantiated (if any) in order to infer which hypothesis or which aspect of a hypothesis is experimentally warranted.

To infer that H is indicated by the data does not mean that a high degree of probability is assigned to H—no such probabilities are wanted or needed in the error-statistical account. The entire attempt to find a quantitative measure of evidential relationship, an E-R measure, between evidence and hypotheses is rejected. That H is indicated by the data means that the data provide good grounds for the correctness of H. One can, if one likes, construe the correctness of H in terms of H being reliable, provided care is taken in its interpretation. Learning that hypothesis H is reliable, I proposed (in chapter 4), means learning that what H says about certain experimental results will often be close to the results actually produced—that H will or would often succeed in specified experimental applications. By means of statistical tests, we check whether in fact this has been learned.

Take one kind of hypothesis we discussed several times, one asserting that a given effect is real or systematic. Perhaps hypothesis H asserts that a real correlation exists between a specific gene and a type of cancer in a given population of individuals. Suppose experimental test results indicate hypothesis H. One thing that the correctness of H may be taken to assert is that the incidence of this gene among cancer patients will not vary in the manner expected for a chance correlation. This leads to pinpointing a corresponding notion of success. We may regard an application of H as successful when it is statistically significantly different from what would be expected if the correlation between the gene and cancer were of the chance variety. More formally, a successful outcome is one that a specified experimental test would take as failing the null hypothesis H_0 that the observed correlation is merely chance. That H is frequently successful—that H is reliable—asserts that the null hypothesis would frequently be rejected by a given statistical test (about as frequently as the test specifies).

A parallel interpretation emerges if instead the results indicate that H is false and the correlation is spurious. This tells us that things would be as if the null hypothesis H_0 were true: differences (from what the null hypothesis asserts) that are statistically significant at level α would be expected about as rarely as the test indicates (i.e., $\alpha[100$ percent] of the time).

I am stating all this in a general way. Information about the actual experimental test of interest would be required to interpret, in a more specific way, a given assertion that the test result indicates a hypothesis. I discussed the question of interpreting tests in chapter 11 (especially section 11.7).

Three Tasks for Error Statistics

Experimental inquiry is a matter of building up, correcting, and filling out the models needed for substantiating severe tests in a step-by-step manner. Individual steps are split off and tackled according to the same pattern of argument. Standard statistical ideas and tools enter into this picture of experimental inference in a number of ways, all of which are organized around the three chief models of inquiry. They serve three main roles by providing (1) techniques of data generation and modeling along with tests for checking if the resulting data satisfy the experimental assumptions; (2) tests and estimation methods that allow control of error probabilities; and (3) canonical models of low-level questions with associated tests and data modeling techniques. I readily admit that the ways in which error-statistical tools serve these functions do not fall out directly from the mathematical framework found in statistical texts. There are important gaps that need to be filled in by the methodologist and philosopher of experiment as well as by statistical practitioners. I have only scratched the surface here.

The three tasks just listed relate to the models of data, experiment, and primary hypotheses, respectively. I shall consider them briefly in turn:

1. The first task involves issues of pretrial planning to generate data likely to justify assumptions of the analysis of interest, and after-trial checking to test whether the assumptions are satisfactorily met. The conglomeration of methods and models from standard error statistics, error analysis, experimental design, and cognate methods is the place to look for forward-looking procedures in order to obtain data in the first place. The work for experimentalists building on error statistics is to identify those procedures and the roles they play in substantive inquiries. The scientific episode that I have treated most fully is the case of Brownian motion (chapter 7). A number of other examples discussed by philosophers of science cry out for an analysis of their modes of data modeling. Certainly, the whole area of testing assumptions (e.g., by distribution-free methods and by special experimental designs) calls for a much more detailed and sophisticated discussion than I had time for in this book. Newer techniques from the field of exploratory data analysis offer yet another treasure chest of canonical error discernment strategies based upon visual manipulations (on paper or on computers).

2. The second task centers on what is typically regarded as statistical inference proper, namely, specifying and carrying out statistical tests (and the associated estimation procedures) or informal analogs to

these tests. The error-statistical program brings with it reinterpretations of the standard methods as well as extensions of their logic into informal arguments from error (e.g., as discussed in chapters 5, 9, and especially 11). The criteria for selecting tests depart from those found in classic behavioristic models of testing. One seeks not the "best" test according to the desiderata of low-error probability alone, but rather sufficiently informative tests.

Accordingly, what directs the choice of a test statistic, and its associated reference set and error probabilities, is the goal of ensuring that something relevant is likely to be learned. Tests are not used as automatic accept or reject rules—accepting or rejecting hypotheses according to whether outcomes fall in the rejection region of a test. Rather, one infers those hypotheses that pass severe tests in the manner just described. However, assessing severity typically calls for considerations that go beyond the standard test rule, and for custom-tailored interpretations of results after the trial. For one thing, because the severity calculation must be sensitive to the actual outcome reached, it is not enough to know whether the result fell in the rejection region. Second, the most informative assertion warranted is rarely one of the preset statistical hypotheses themselves, but more commonly a claim about the discrepancies from those hypotheses that are or are not indicated by the data (e.g., chapters 6 and 11). The value of the standard tests of preset hypotheses is that they provide the basis for learning about such discrepancies. For example, it useful to test a null hypothesis asserting a zero difference in means—even knowing that the null is strictly false—because it teaches the extent to which an effect differs from 0.

Systematic ("metastatistical") rules may be specified to guide the appropriate interpretation of statistical results after the trial, such as the two I set out in chapter 11: the rule of acceptance (RA) and the rule of rejection (RR). While directing the interpretation of test results, these rules also prevent vintage misinterpretations of tests and let us make short work of what have been mistaken as damaging criticisms of Neyman and Pearson tests. For example, the rules RR and RA let one see that the choice of an NP test with prespecified low error probabilities reflects the goal of substantiating certain standard types of arguments from error. There is leeway in the specification, but it is of a rather restricted variety. Second, and most important, the latitude that exists in the choice of test does not prevent the determination of what a given result does and does not say. The error probabilistic properties of a test procedure—no matter how that test was chosen—allow for an objective interpretation of the results. This is the basis for criticizing a given test (e.g., as too sensitive or not sensitive enough) and for finding certain inferences unwarranted.

3. Experimental inquiries are broken down into piecemeal questions such that they can be reliably probed by statistical tests or analogs to those tests. But how does this piecemeal breakdown work? Even more than the first two, the third role I assign to statistics takes us beyond statistical methodology, as it is usually conceived, and into tasks that call for a full-blown (error-statistical) philosophy of experiment. Here again I have only taken a few preliminary steps.

I proposed that inquiries are broken down into local questions referring to standard types of errors or mistakes, construed broadly. Strategies for investigating these errors often run to type. I delineate four such standard or canonical types of errors: mistaking chance effects or spurious correlations for genuine correlations or regularities; mistakes about a quantity or value of a parameter; mistakes about a causal factor; mistakes about experimental assumptions. Statistical models are relevant because they model patterns of irregularity that are useful for studying these errors.

In the preceding chapters I have tried to illustrate how reliable inferences are made by learning how to tap into one of the known patterns of variability. This is not just a matter of splitting things up; generally it is necessary to introduce or inject statistical considerations into inquiries. Statistical considerations are introduced in two main ways: (1) by means of the collection of data, and (2) by means of the modeling of the data (manipulations on paper). I discuss these in turn:

Injecting Statistical Considerations

1. Suppose that one is interested in a quantity μ, say the mean radius of one of Perrin's populations of gamboge grains. One way of introducing statistical considerations is to randomly sample n members from the population and average up their values for this quantity. In the experimental model this can be described as: observe the value of the statistic $\overline{X}$—the sample mean. The value of the mean that would be observed may be viewed as a random sample from a *hypothetical* population consisting of all the possible n-fold samples of grains that could have been taken, with the average radius in each n-fold sample recorded. The distribution of these values is the experimental (or sampling) distribution of $\overline{X}$. Why should we be interested in this hypothetical population of sample means? Because the probability distribution of $\overline{X}$ gives us the error probabilities (both of tests and estimates). By collecting data in a certain way (e.g., so that random sampling is ap-

proximately satisfied), Perrin gets his sample mean to be *related statisti-cally* to the population mean. The mean of $\overline{X}$ is itself equal to the popu-lation mean μ and the variability of $\overline{X}$ in the hypothetical population is related in a known way to the variability in the real population of grains. Thus the observed sample mean can be used to give an interval estimate of μ and Perrin can attach error probabilities to this estimate. Through this trick, what Perrin learns about his hypothetical popula-tion teaches him about the real population.

2. The second type of deliberate introduction of statistics is by way of data modeling. This was illustrated in Perrin's introduction of statis-tical manipulations "on paper" to learn about the distribution of his Brownian particles. We saw, for example, how Perrin took the ob-served displacements of a group of Brownian particles and condensed them into 9 pigeonholes to see if they would distribute themselves as shots fired would distribute themselves around a bull's-eye. By means of this and other models of data, Perrin was able to use types of known variable phenomena, for example, from random walks, to answer a question about the unknown variability of the displacement of Brownian particles. (See section 7.4.)

What justifies these manipulations on paper is not that they repre-sent actual phenomena, but that by manipulating the data (e.g., into bull's-eye rings), Perrin could deduce the probability that a displace-ment would fall in each of the 9 rings under the hypothesis being tested. With this, statistical tests could be run and their error probabili-ties calculated. And these error probabilities (e.g., severity) *do* refer to actual experimental phenomena.

Future Projects

Statistical ideas and tools are clearly taking on a heavy load in the present approach, performing many tasks that go beyond extant texts on statistical methodology. This is particularly so of the third task of breaking down a substantive primary inquiry into piecemeal inquiries into errors. The philosopher of experiment seeking to pursue this task will be taking up questions that statistics texts barely ask, such as how do scientists appeal to a handful of canonical exemplars of learning from error?[2] An open question, of course, is how well this picture of

2. This question, as I understand it, enjoins one to identify such exemplars in strategies and reasonings in scientific practice, as well as to explain how they work in facilitating learning from error. There are descriptive and normative compo-nents. Work from several areas will be important to draw upon, several which have been noted throughout this book (e.g., case studies from the New Experimental-ists). Some of the empirical work from philosophy of psychology and cognitive sci-

scientific inference is borne out in general. But several gaps first need to be filled.

Several additional examples of particular canonical models of error and of error avoidance strategies need to be articulated. I have largely dealt with questions that may be articulated within statistical models, and yet experimental inquiry relies on numerous informal exemplars and models from outside statistics.

In addition, I have for the most part restricted my focus to experimental inquiries that could be broken down into relatively low-level primary hypotheses. Many, I suspect, would press me to go higher up in the hierarchy to more full-blown scientific theories. Even granting my main point, that experimental knowledge is highly robust through changes in higher-level theories, much more can and should be said about the relationships between experimental knowledge and more global theories. A task that would need to be tackled by those with an understanding of particular fields is how to partition theories so that their hypotheses can be severely tested. A place to begin, I think, is with the experimental tests of specific laws and key hypotheses, as we did, for example, with Perrin's experiments for the Brownian motion hypothesis within the kinetic theory, and, in less detail, with the eclipse tests for the gravitation law in Einstein's theory.

As for what might be done next, recall the program for partitioning and eliminating whole chunks of theories of gravity (e.g., "nonmetric" theories of gravity), noted in section 6.3, from a discussion by John Earman (1992). Looking at historical and perhaps also current theories, methodologists of experiment might be able to do something analogous. They may be able to show how sets of theories were or were not distinguished by given experiments, and perhaps identify some general strategies for making progress in discriminating types of hypotheses. Something like this kind of effort at partitioning and discriminating theories is what the present program would call for, if one were to pursue it at the level of large-scale theories.

13.2 ARTICULATING AND APPRAISING METHODOLOGICAL RULES

In the latter half of this book, I have paid a lot of attention to scrutinizing particular methodological rules: about varying data, about novelty, about predesignation and about stopping rules. I will not review that

---

ence could also be valuably appropriated to our project; and in so doing, I submit, those areas might be enhanced with a normative dimension that they generally do not address.

material here, but rather will sketch in a general way what the error-statistical program calls for when it comes to the task of articulating and scrutinizing methodological rules. This task is shaped by the following three theses:

First, methodological rules can be seen as aiming to avoid key errors that get in the way of good experimental arguments. Second, they reflect the requirement that before a hypothesis is warranted it must have passed a severe experimental test. Third, one may get at how rules work, if they do, by asking, If any and all violations of the rule were permitted, how might (flagrantly) unreliable arguments be licensed thereby?[3] I showed that these three theses hold up with respect to some of the more controversial methodological rules. I put them forward to be tested by others who seek to pursue the task of meta-methodology. Let me flesh out and draw some consequences from these theses.

A Naturalistic Metamethodology

In the present account, methodological rules for experimental learning are regarded as strategies that enable learning from common types of experimental mistakes. They may be seen to systematize the day-to-day learning from mistakes discussed in chapter 1. From the history of mistakes made in reaching a type of inference arises a repertoire of errors; methodological rules are techniques for circumventing and uncovering these errors. Some refer to before-trial experimental planning, others to after-trial analysis of the data. The former include rules about how specific errors are likely to be avoided or circumvented, the latter, rules about checking the extent to which given errors are committed or avoided in specific contexts. They work together to form reliable arguments from error.

Methodological rules are empirical claims about how to find certain things out by arguing from experiments. Accordingly, these hypotheses are open to an empirical appraisal: their truth depends upon what is actually the case in experimental inquiries. At the same time, the present account is normative, in that the strategies are claims about how to actually proceed in given contexts to learn from experiments.

Since the rules are claims about strategies for avoiding mistakes and learning from errors, their appraisal turns on understanding how

---

3. Even many that are not explicitly about experimental learning would turn out, on closer inspection, to be rooted in the aim of arguing from error. I am thinking of a rule such as: prefer simple hypotheses.

methods enable avoidance of specific errors. One has to examine the methods themselves, their roles, and their functions in experimental inquiry. To know which rule is called for in a particular context of inquiry requires knowing something about what the rule does and does not do—how it helps and when it does not help. This will also inform us of other ways to get around the same mistake. There is never just one way to skin a cat, nor to avoid an error.

Canonical Models of Error

To further flesh out these ideas, consider an uncontroversial method of experimental control. Given the empirical realities of inquiring about many causal factors of interest—namely, that extraneous factors may also be responsible for an observed effect E—we can reason out why perfect experimental control is desirable: it enables avoiding erroneously attributing causes in arguing from observed effect E. If the method of control is violated and factors are left uncontrolled, then, as a matter of empirical fact, the experimental results are also showing the influences of these other factors. If, however, a way can be found to avoid confounding the effects of extraneous factors with the effect of the factor of interest F, then one can still arrive at an experimental argument as or nearly as reliable as the one provided by literal control.

An experimental argument based on an ideal case of literal experimental control illustrates the notion of a canonical model. It is an exemplary case of how a causal inference may be warranted with high (if not maximal) reliability. A canonical model for a certain type of claim is the basis for two kinds of spin-off strategies: first, arriving at further canonical models, and second, discovering and scrutinizing methods for satisfying the assumptions of these models. As an example of the first strategy, the ideal of perfect control gives rise to exemplary arguments that approximate or simulate the ideal case, such as arguments based on control groups (e.g., randomized treatment-control studies). Having justified the approximation to the ideal case, the stage is set for the second type of strategy. As an example of the second strategy, consider how rules about keeping subjects and experimenters blind, using placebos, and so on, stem directly from the goals of a valid comparison of treated and untreated groups.

Although specific methods and rules about their use are numerous, I suggest that they are all tied to detecting and avoiding a handful of error types. More correctly, they are all tied to the handful of canonical arguments for inquiring about the key errors delineated in chapter 1: mistakes about a causal factor, an experimental artifact or chance effect, a quantity or value of a parameter, and an assumption of the experimental data. There is nothing firm about this list; perhaps it should be expanded or contracted. It is simply one convenient grouping, reflecting, it seems to me, the manner in which canonical arguments now in use tend to run to type.

Consider the example of statistical significance tests of a null hypothesis H_0: the effect is "due to chance" within a standard Binomial model. This corresponds to a formal statement of a reliable argument for affirming an effect is "real" and not due to chance. It is an example of a formal canonical model. But as we have seen throughout this book, the same pattern of argument is followed in informal arguments, such as Hacking's argument from coincidence to rule out the artifact explanation (of dense bodies), and Galison's to rule out escaping muons (e.g., chapter 3).

Canonical experimental models—whether formally or informally given—exemplify cases in which the kinds of errors known to be possible or problematic in the given type of investigation are handled well—that is, ruled out. Also included would be examples of infamous mistaken cases, especially cases initially thought to have surmounted key problems. It is to be expected, and is wholly unproblematic, that different practitioners will worry more or less about certain errors or will appeal to different canonical models. The results are communicable, and are capable of being checked with different (intersubjective) standards.

Uncovering a Rule's Rationale

To get at the underlying rationale of a methodological rule we ask: if experiments were allowed to violate freely the methodological requirement in question, would some flagrant error be countenanced? some clearly unreliable argument allowed? The answer may stem from actual episodes or our past repertoire of errors. This leads to the second kind of canonical model: an example of an infamous mistake, a classic, demonstrably unreliable argument. Such a chamber of horrors need only be populated by a handful of examples to serve its function.

Having extracted a rule's rationale, it does not follow that satisfying the rule is necessary to avoid the given threat to reliability. What follows is that if the method is violated, then the onus is on the experimenter to show that reliability is nevertheless achieved. If it proves too impractical or too difficult to circumvent the error while violating the rule, then we begin to understand why, in certain contexts, a given rule makes good sense. For such experimental contexts, epistemological as well as pragmatic and economic, considerations may justify the rule as a kind of *cautionary rule,* a way to play it safe. Several examples come to mind relating to statistical significance tests:

- Always require predesignation of the hypothesis to be tested.
- Routinely preset tests with a .05 level of significance.
- Never decide when to stop sampling *after* looking at the data.

Two consequences for the methodological enterprise emerge: First, rather than keeping score—either between instances where a rule is satisfied and where it is violated, or between proponents and opponents of a rule—we should seek to find a specific rationale for such rules. Second, researchers need not be hamstrung by cautionary rules when they can defend the reliability of particular experimental arguments despite their violation. (See, for example, chapter 9.)

A Role for History

A naturalistic experimental methodology along these lines requires (*a*) articulating canonical models or paradigm cases of experimental arguments and errors, and (*b*) appraising and arriving at methodological rules by reference to these models. Historical cases, if handled correctly, provide a unique laboratory for these tasks. The example of Brownian motion focused on task *a* while the Eddington eclipse experiment focused on task *b*.

In promoting historical cases as providing real data to analyze for our methodology of experiment, I do not mean what many historically minded philosophers have meant. The data we need do not consist of the full scientific episode, all finished and tidied up. The data we need are the experimental data that scientists have actually analyzed, debated, used, or discarded. Whether the episode concerns the bearing of evidence on large scale or local appraisal, whether the case is recent or long past, the focus for these tasks must be on the ways specific data were used in arguing for the evidence in the first place. The particular data, and discussions and debates over their generation and interpretation, would need to take center stage, rather than debates about high-level theorizing. Often the best sources are the scientific reports and published debates themselves. This is what made the case of Perrin and Brownian motion so illuminating (chapter 7).

Especially revealing are the processes and debates that take place before the case is settled and most of the learning is going on, particularly where there is a reliable record of the data analysis involved. Here one can often find how a method's violation may lead to specific problems in reaching experimental conclusions. This type of scrutiny also helps us to better understand the "rationality" of the episode; what

stood in the way of obtaining certain information, and why it took a
certain amount of time for a good argument to be reached.

I do not think it farfetched to suppose that an adequate methodology of experiment would be relevant to the improvement of experimental methods.[4] It should help to make explicit the ways in which
inquirers add to their repertoire of errors, classifying by subject matter
(e.g., rats, microscopic vegetable particles, star positions) as well as by
type of error. Looking at historical cases with an eye for failed experimental arguments may prevent their being committed again. A related
area of research, of interest in its own right, would be to investigate
how certain mistakes had to be relearned several times until they became "canonized" in a systematic way.

13.3 PHILOSOPHICAL PROBLEMS OF EVIDENCE AND INFERENCE:
DUHEM'S PROBLEM

I regard the error-statistical philosophy of experiment as providing the
general framework and tools for carrying out the main goals of the
New Experimentalist program (see chapter 3). I put it forward as a
fresh perspective from which to reask a number of philosophical challenges regarding scientific inference based on empirical data. I have
argued that its requirement of a severe test provides the key to answering the "alternative hypothesis" challenge (by distinguishing between
hypotheses that "fit" the data equally well), and that error-statistical
reasoning is the proper way to solve Duhem's problem. The former
issue is treated at length in chapter 6. This may be a good place to
sketch the error-statistical treatment of Duhem's problem (chapters 4,
5, and 6), and why I regard it as superior to the Bayesian treatment.
Hopefully others will pursue this treatment further.

The task Duhem's problem poses for philosophers of science is to
provide a way to determine which of the hypotheses used to derive a
predicted consequence should be rejected or disconfirmed when experiment disagrees with that prediction. In actual scientific episodes H
is sometimes taken to blame; at other times H is retained while auxiliary assumptions are said to be responsible for the anomalous result.
An adequate model of testing should account for this.

When Bayesians say they can solve Duhem's problem, as they do,
what they mean is that they can show how certain subjective probability assignments can justify a given apportioning of blame. Solving Du-

---

4. It might be possible to actually construct, particularly with the use of computers, a tool kit of errors and error simulations.

hem comes down to a homework assignment of how various assumptions and priors allow the scientific inference reached to be in accord with that reached via Bayes's theorem. Nothing is said about whether the assignments are warranted or, more important, how a scientist should go about determining where the error really lies. Assigning the probabilities differently puts blame elsewhere, and the Bayesian "solution" is not a solution for adjudicating such assignments. (See chapter 4.)

But scientists regularly tackle and often enough solve Duhemian problems, and when they do they employ (formally or informally) the logic and a methodology of error statistics. In the error-statistics approach, the task of finding out whether auxiliary hypotheses are satisfied is split off from that of appraising the primary hypothesis H. In fact, the chief reason for separating out the models relating data and hypotheses is to achieve the aim of correctly apportioning blame (as well as praise). Before experimental results can speak for or against a hypothesis under test, it is necessary to check and estimate the extent of any errors along the way—regarding the data and the auxiliaries. This calls for methods to discern whether the experiment was well run, to distinguish real effects from artifacts, estimate backgrounds, and "subtract out" influences of factors other than some intended one. The methods and principles from standard-error statistics are regularly appealed to in carrying out and giving structure to these tasks. A scientist may claim that an extraneous factor, and not H, is to be blamed for an anomalous result, but to *warrant* that claim requires it to have passed a reliable test. Some examples may be found by going back to the eclipse experiments discussed in chapter 8.

Several staunch defenders of Newton could well be seen as satisfying the subjective probability assignments that would have warranted taking the anomaly as only very slightly decreasing belief in Newton and greatly decreasing their belief in auxiliary A that no factor other than gravity was operating to produce the observed effect. But no one (not even the staunchest Newtonian defenders) thought that their strong degree of belief in Newton was evidence of the correctness of the various hypotheses they put forward with which to save Newton and account for the anomalous deflection. Proposed factors by which to save Newton were evaluated according to whether they stood up to severe scrutiny—when they did not they were shot down. The concern was with a classic error—that the evidence failed to constitute a reliable test in favor of the auxiliary factor hypothesized to accommodate the anomaly. The concern turned out to be one that the Newtonian defenders could not put to rest. It turned out that in order to construe the anomalous deflection as passing Newton it was necessary to employ precisely the kinds of strategies that would make it *easy* for hypotheses to pass, even if they are false (i.e., high error probabilities).

The eclipse episode, we saw, also included instances where an apparent anomaly was explained away successfully. I am thinking of how Eddington was able to explain away the anomaly stemming from one of the sets of eclipse results (from Sobral). The evidential appraisal turned on the question of whether there were grounds for an error—either in the data or in the background factors assumed not to be responsible. The debate was engaged in by scientists with very different opinions about Einstein's theory versus Newton, but the appraisal did not turn on these opinions, and there was no need to imagine having a prior in all the other possible hypotheses (i.e., the so-called catchall). There was a need, however, for tools to discriminate signals from noise, rule out artifacts, distinguish backgrounds, and so on. The relevant argument turned on a rather esoteric piece of data analysis showing (holdouts notwithstanding) that the mirror distortion was implicated. The conglomeration of methods and models from standard error statistics—even at that early stage of their development—provided the needed tools along with the repertoire of errors gleaned from other astronomical experiments.

The above examples instantiate two distinct strategies by which the present approach grapples with Duhemian problems: The first is to criticize and bar attempts to explain away anomalies (e.g., as due to the Newton-saving factors) on the grounds that they fail to pass severe tests (or, even more strongly, that their denials pass severe tests). The second is to show that an anomaly may be legitimately blamed on an auxiliary factor F (e.g., a mirror distortion) by showing that "F is responsible" passes a severe test. Clearly, we do not always have a warranted way to attribute blame—we can not always satisfy the requirement of the second strategy. But this requirement directs progress with Duhemian problems—and it explains the lengths to which scientists work to test auxiliaries.

Allocating blame was possible in the eclipse experiments because enough was known to distinguish the patterns of different auxiliaries. Just as in our day-to-day repertoire of errors (chapter 1), so it is in science. A "too much salt" error in rice is distinguishable from a "too much water" error, and a mirror distortion is distinguishable from a deflection effect. Duhem's problem is built on the supposition that an error or anomaly is silent about its source—and indeed it is silent when approached by the white gloves of logical analysis. But in the hands of shrewd inquisitors of error it may be made to speak volumes, and often a whisper is enough to distinguish its source from the others.

13.4 The Growth of Experimental Knowledge

The error-statistical philosophy of experiment locates scientific progress in the growth of experimental knowledge, and holds that to understand the growth of experimental knowledge, one should look to tests of local hypotheses (normal experimental testing). The aim of these tests is not to update the probability assignment in hypotheses, but to build, correct, and fill out a model by means of severe tests. What justifies error statistics is that its methods have properties that enable it to provide such tools. These methods work because (1) we are often able to put questions about errors—regularities and irregularities—in terms of questions that can be answered by real, or simulated, random experiments, and (2) we are, at least some of the time, able to carry out real random experiments (in the sense of chapter 5), check if we have done so in particular cases, and make reliable inferences even where assumptions are violated. (1) and (2) are the bases for severe tests and, by means of these, the growth of experimental knowledge. To substantiate these points requires weaving together the pieces of the preceding twelve chapters. Here are a few threads.

Learning What It Would Be Like

In a very real sense, applying statistical methods may be seen as continuing experimentation by other means.[5] Where data are inexact, noisy, and incomplete, where extraneous factors are uncontrolled or physically uncontrollable, and where these facts pose the most serious threats to our ability to find our way around, we often enough have learned from our mistakes. From our first successes in our day-to-day learning from mistakes, the challenge, the fun, of outwitting and outsmarting drives us to find ways to learn what it would be like to control, manipulate, and change in situations where we cannot literally control, manipulate, or change. *The conglomeration of systematic tools designed for these ends I call statistics.*

Statistical models inform us about what *would be expected* under various assumptions about aspects of the underlying experimental process. Whether it is by pointing to a statistical calculation, a canonical

---

5. I am mimicking a claim by van Fraassen to the effect that an important aim of *"experimentation is the continuation of theory construction by other means"* (van Fraassen 1980, 77).

exemplar, a pictorial display, or a computer simulation, the "what would it be like" question is answered by means of an experimental distribution: a statement of the relative frequency with which certain results would occur in an actual or hypothetical sequence of experiments.

Such answers are not enough by themselves: they are informative because of their links to the statistical tools of analysis (tests and estimation methods) that make use of them. This information is used in a variety of ways, as we have seen. I will consider two types:

1. We can use this information to ask what it would be like if various different hypotheses about the underlying experimental process were incorrect or misdescriptions of a specific experimental process. It teaches us what it would be like were it a mistake to suppose that a given effect were nonsystematic or due to chance, what it would be like were it a mistake to attribute the effect to a given factor, what it would be like were it a mistake to hold that a given quantity or parameter had a certain value, and what it would be like were it a mistake to suppose that experimental assumptions are satisfactorily met. Statistical tests can then be designed to magnify the differences between what it would be like under various hypotheses. These "formal" mistakes, as it were, are used to probe real mistakes.

A central role of statistical tests is to demonstrate how results that *appear* to count against the presence of an error can actually be produced fairly often where the error is committed. A test can be designed so that with high probability it would yield a result that it could show to be "reasonably typical of a process where the error is committed," if in fact it is being committed—but not otherwise. Accordingly, if the result is one that the test finds practically incapable of arising under the assumption of error, the result is taken as indicating that the error is absent. This "indication" generally needs to be bolstered by other tests (with the same or different data) before a strong argument from error is sustained.

After learning enough about certain types of mistakes and the ways of making them show up, it can be argued that finding no indication of error despite the battery of deliberate probing with several, well-understood methods, is excellent grounds for taking the error to be absent. To suppose otherwise is itself to adopt a highly unreliable method. In the language of chapter 2, this would violate the canons of good normal science.

Appealing to experimental distributions also allows inverting the question of what would it be like (under various hypotheses about the experimental process). Starting with an experimental result one may ask: What hypotheses would be indicated as having passed a severe test with this result? Which hypotheses would pass with only very low degrees of severity? (This is essentially what goes on in an estimation problem.) Alternately, this inverted question asks: What arguments from error (if any) can be substantiated with this experimental result? And which are very poorly substantiated?

2. Information on "what it would be like" serves a second set of functions. These pertain to checking experimental assumptions. Where we cannot literally control extraneous factors we are often able to estimate the likely extent to which such factors could influence results in the experiment at hand. We can then "subtract them out" in order to discern the effects of nonextraneous factors of interest. The standard statistical significance test, for example, lets us subtract out the effects of "sampling error" to discern systematic effects. We can utilize the same canonical argument to model lots of effects we would like to subtract out. Recall how in discussing Galison's example of neutral currents (chapter 3), researchers could subtract out approximately the effect of escaping muons. Where manipulation and control are present, we can say "this is what it is like when we vary this factor." Where they are absent, we may still be able to ascertain what it would be like *statistically* if we were to vary this factor.

Discriminating backgrounds, signal from noise, real effects from artifact, is the cornerstone of experimental knowledge. Beginning with canonical examples—given either in a statistical model (e.g., the Binomial with $p = .5$) or by way of an exemplary case (e.g., the lady tasting tea)—scientists have been able to arrive at spin-off strategies for accomplishing these ends in diverse fields. These tools work because they employ assumptions that can be checked independently and need only to be approximately satisfied. The methods are robust enough that violating assumptions may weaken but do not destroy the validity of inferences, and the trouble such violations cause is itself often detectable and correctable. Finally, even if we fail to satisfy assumptions and even if our methods are not sufficiently robust, there is a high probability that this will be discovered in subsequent tests.

In sum, standard statistical models afford very effective tools for approximating the experimental distributions needed to convey "what it would be like" under varying hypotheses about the process generating the experimental data, and the error probabilistic properties of these tools enable this information to substantiate arguments from error. One need not look any deeper to justify their use. Adherence to misconceptions about what a theory of statistics should do to provide a philosophically adequate account of inference, supposing, in particular, that it should provide a quantitative measure of the relationship between evidence and hypotheses, has let this interesting and powerful role of statistical ideas go unappreciated.

Performing Real Random Experiments

If a hypothesized error is not detected by a procedure with an overwhelmingly high chance of detecting it, then there are grounds for the claim that the error is absent. We can affirm something positive, that the particular error is absent (or is no greater than a certain amount). What justifies affirming something positive, it is important to see, is not only that the past tests had certain properties—that they were highly probative and yet failed to detect a given error. It is also the fact that the claim that is inferred is an experimental claim—it asserts what would be expected to occur in other trials where given experimental assumptions are approximately satisfied. It matters not whether any other experiments will actually be carried out; it matters only that the inferred claim be understood as an experimental claim.

Of course I may be wrong about a hypothesis even if it has passed severe tests, but if I am I have several effective procedures for finding this out: (*a*) I will find that I am unable to substantiate the assumptions of a subsequent experiment, or (*b*) with high probability I will get discordant results in a subsequent experiment. This hinges on my ability to discern whether experimental assumptions are met sufficiently to distinguish (*a*) from (*b*).

This discernment is a product of being able to run experiments such that the results may be approximated by certain statistical models. These "real random experiments" are found in nature and in games of chance (see the discussion of Neyman in chapter 5). We canonize these, as it were, in standard mechanisms: mechanisms known to generate results that are as close to what the statistical model predicts about as often as it stipulates. We extend our ability to generate or simulate such "real random experiments" by exploiting a series of standard tests that would with high probability tell us if we are failing. Because there are only a few assumptions that need to be checked on any given application of a statistical procedure, there are only a handful of ways that we can fail to satisfy their assumptions.

The goal of arriving at reliable means of checking and, often enough, satisfying experimental assumptions gives rise to distinct methodological strategies at the level of experimental design and data generation. This goal is one more reason for having standard mechanisms, for example, a coin-tossing mechanism as in the canonical Binomial experiment. By varying a known Binomial process so as to violate one of the assumptions deliberately (e.g., cause it to violate independence) we can arrive at tests that are very good at detecting the violation just when it should. Observing the frequencies of outcomes to be in conflict with the deductively derived statistical distribution would be one way of detecting that for some reason we were unable to get the experiment to accord with the standard Binomial. We could then subtract out its influence.

It should be kept in mind that for a statistical model to approximate an actual experimental procedure, the procedure need not exemplify all the properties of the statistical model. There only needs to be closeness with respect to a rather coarse property: the relative frequencies of certain outcomes calculated from the model need to be close to the actual relative frequencies—in the sense defined by statistical tests. Appealing to the central limit theorem allows going even further. It assures us that, regardless of the underlying distribution, the sample mean is approximately normally distributed, even where randomness is violated. Using the experimental procedure and deliberately altering a known value of mean μ, it can be checked that the procedure detects these discrepancies in just the way that is expected. This shows us that our instrument is working, that the experimental method is good enough to distinguish between parameter values where these are unknown. Once we have a handful of such canonical procedures, we can rely on them to build an increasingly probative and varied arsenal to simulate and detect departures from those procedures.

None of this would satisfactorily explain experimental knowledge were it not also the case that we are rather good at asking questions of interest in terms of questions about experiments that we can actually run or simulate. We may start with a quantity or parameter—it may be part of a theory or not—and reign it in, as it were, so that it is a parameter of an experimental model. Our statistical inferences relate to the experimental model, true, but at the same time they teach about the substantive quantity of interest.

Typically, we introduce statistical features into experiments (through data generation, or manipulations "on paper") because we are aware that this will create a desired link between an experimental parameter and a substantive quantity or question. But we need not suppose that this introduction of statistics results from a conscious effort to appeal to this explanation of successful learning from sample data. It is sufficient that certain data generation procedures *in fact* establish a relationship between sample data and populations, and that this permits reliable inferences from the former to the latter. Triggering this relationship, in some cases, is an accidental byproduct of day-to-day research strategies. What the statistical methodologist can do is make these triggering mechanisms explicit. In so doing we would understand at last how by carrying out certain experimental procedures, statistical models are approximated sufficiently so that—whether one knows it or not—the limit theorems of statistics are working their magic.

By splitting up the inquiry appropriately, the possible irregularities are constrained to be related to known patterns of variability. By learning about these, we learn about experimental phenomena—about what would be expected to happen with certain frequencies whenever certain experimental conditions are fulfilled. This is the basis of experimental knowledge. Such experimental knowledge resembles what Peirce meant by the "experimental purport" of hypotheses. If there is experimental knowledge to be had of a phenomenon, then it will be detectable by means of these methods. The ability to make successful inductions, our success in obtaining experimental knowledge, is explained by the error-statistical properties of our methods. We make progress in experimental knowledge—experimental knowledge grows—because we have methods that are manifestly adequate for learning from errors.

References

Achinstein, P. 1994. Jean Perrin and molecular reality. *Perspectives on Science* 2:396–427.

Achinstein, P., and O. Hannaway, eds. 1985. *Observation, experiment, and hypothesis in modern physical science.* Cambridge, Mass.: MIT Press.

Ackermann, R. 1985. *Data, instruments, and theory: A dialectical approach to understanding science.* Princeton, N.J.: Princeton University Press.

————. 1989. The new experimentalism. *British Journal for the Philosophy of Science* 40:185–90.

Andersen, B., and P. Holm. 1984. *Problems with p: Significance testing in medical research.* Basel, Switzerland: F. Hoffmann-La Roche & Co.

Anderson, A. 1919. The displacement of light rays passing near the sun. *Nature* 104 (4 December): 354.

————. 1920. Deflection of light during a solar eclipse. *Nature* 104 (1 January): 436.

Anscombe, F. J. 1954. Fixed sample-size analysis of sequential observations. *Biometrics* 10:89–100.

Armitage, P. 1961. Contribution to discussion in Consistency in statistical inference and decision, by C. A. B. Smith. *Journal of the Royal Statistical Society (B)* 23:1–37.

————. 1962. Contribution to discussion in *The foundations of statistical inference,* edited by L. Savage. London: Methuen.

————. 1975. *Sequential medical trials.* 2d ed. New York: John Wiley & Sons.

Baccus, F., H. E. Kyburg Jr., and M. Thalos. 1990. Against conditionalization. *Synthese* 85:475–506.

Bailey, D. 1971. *Probability and statistics: Models for research.* New York: John Wiley & Sons.

Barnard, G. A. 1947. The meaning of significance level. *Biometrika* 34:179–82.

————. 1949. Statistical inference (with discussion). *Journal of the Royal Statistical Society (B)* 11:115–39.

————. 1962. Contribution to discussion in *The foundations of statistical inference,* edited by L. Savage. London: Methuen.

————. 1971. Scientific inferences and day to day decisions. In *Foundations of statistical inference,* edited by V. Godambe and D. Sprott, 289–300. Toronto: Holt, Rinehart and Winston of Canada.

————. 1972. The logic of statistical inference. Review of *The logic of statistical inference,* by I. Hacking. *British Journal for the Philosophy of Science* 23:123–32.

465

————. 1985. *A coherent view of statistical inference.* Technical Report Series. Waterloo, Ontario: University of Waterloo.

Barnett, V. 1982. *Comparative statistical inference.* 2d ed. New York: John Wiley & Sons.

Bartlett, M. S. 1971. Comments on The estimation of many parameters, by D. V. Lindley. In *Foundations of statistical inference,* edited by V. Godambe and D. Sprott, 447. Toronto: Holt, Rinehart and Winston of Canada.

Beatty, J. 1980. What's wrong with the received view of evolutionary theory? In *PSA 1980,* vol. 2, edited by P. Asquith and R. Giere, 397–426. East Lansing, Mich.: Philosophy of Science Association.

Berger, J. O., and R. L. Wolpert. 1988. *The likelihood principle.* 2d ed. Hayward, Calif.: Institute of Mathematical Statistics.

Birnbaum, A. 1962. On the foundations of statistical inference (with discussion). *Journal of the American Statistical Association* 57:269–326.

————. 1969. Concepts of statistical evidence. In *Philosophy, science, and method: Essays in honor of Ernest Nagel,* edited by S. Morgenbesser, P. Suppes, and M. White, 112–43. New York: St. Martin's Press.

————. 1977. The Neyman-Pearson theory as decision theory, and as inference theory; with a criticism of the Lindley-Savage argument for Bayesian theory. *Synthese* 36:19–49.

Born, M. 1964. *Natural philosophy of cause and chance.* New York: Dover Publications.

Braithwaite, R. 1953. *Scientific explanation: A study of the function of theory, probability and law in science.* New York: Cambridge University Press.

Bross, I. 1971. Critical levels, statistical language, and scientific inference. In *Foundations of statistical inference,* edited by V. Godambe and D. Sprott, 500–513. Toronto: Holt, Rinehart and Winston of Canada.

Brown, R. 1828. A brief account of microscopical observations made in the months of June, July and August, 1827, on the particles contained in the pollen of plants; and on the general existence of active molecules in organic bodies. *Philosophical Magazine* 4:161–73.

Brush, S. 1968. Mach and atomism. *Synthese* 18:192–215.

————. 1977. A history of random processes, I: Brownian movement from Brown to Perrin. In *Studies in the history of statistics and probability,* vol. 2, edited by Sir M. Kendall and R. L. Plackett, 347–82. (New York: Macmillan). First published in *Archive for the History of the Exact Sciences* 5(1968): 1–36.

————. 1989. Prediction and theory evaluation: The case of light bending. *Science* 246:1124–29.

Campbell, R., and T. Vinci. 1983. Novel Confirmation. *British Journal for the Philosophy of Science* 34:315–41.

Carnap, R. 1962. *Logical foundations of probability.* Chicago: University of Chicago Press.

Carnap, R., and R. Jeffrey. 1971. *Studies in inductive logic and probability.* Vol. 1. Berkeley: University of California Press.

Cartwright, N. 1983. *How the laws of physics lie.* Oxford: Clarendon Press.

―――. 1989. *Nature's capacities and their measurement.* Oxford: Clarendon Press.

Chandrasekhar, S. 1954. Stochastic problems in physics and astronomy. In *Selected papers on noise and stochastic processes,* edited by N. Wax, 3–91 (New York: Dover). First published in *Reviews of Modern Physics* 15(1943):1–89.

Clark, P. 1976. Atomism versus thermodynamics. In *Method and appraisal in the physical sciences,* edited by C. Howson, 41–105. Cambridge: Cambridge University Press.

Cramér, H. 1974. *Mathematical methods of statistics.* Princeton: Princeton University Press.

Crommelin, A. C. D. 1919. Contribution to discussion in Joint eclipse meeting of the Royal Astronomical Society. *The Observatory* 42:389–98.

Darden, L. 1991. *Theory change in science: Strategies from Mendelian genetics.* New York: Oxford University Press.

de Finetti, B. 1972. *Probability, induction and statistics: The art of guessing.* New York: John Wiley & Sons.

DeGroot, M. H. 1973. Doing what comes naturally: Interpreting a tail area as a posterior probability or as a likelihood ratio. *Journal of the American Statistical Association* 68:966–69.

Delsaulx, J. 1877. Thermodynamic origin of the Brownian motion. *Monthly Microscopical Journal* 18:1–7.

Descartes, R. [1644] 1984. *Principles of Philosophy.* Pt. 3. Translated by V. Miller and R. Miller, 104. Reprint, Dordrecht, The Netherlands: D. Reidel.

Donovan, A., L. Laudan, and R. Laudan, eds. 1988. *Scrutinizing science: Empirical studies of scientific change.* Dordrecht, The Netherlands: Kluwer.

Dorling, J. 1979. Bayesian personalism, the methodology of scientific research programmes, and Duhem's problem. *Studies in History and Philosophy of Science* 10:177–87.

Duhem, P. 1954. *The aim and structure of physical theory.* Translated by P. Wiener. New York: Atheneum.

Dyson, E. W., A. S. Eddington, and C. Davidson. 1920. A determination of the deflection of light by the sun's gravitational field, from observations made at the total eclipse of May 29, 1919. *Philosophical Transactions* 220:291–333.

Earman, J. 1992. *Bayes or bust? A critical examination of Bayesian confirmation theory.* Cambridge, Mass.: MIT Press.

―――. 1993. Underdetermination, realism, and reason. In *Midwest Studies in Philosophy,* vol. 18, edited by P. French, T. Uehling Jr., and H. Wettstein. Notre Dame: University of Notre Dame Press.

―――. ed. 1983. *Testing scientific theories.* Minnesota Studies in the Philosophy of Science, vol. 10. Minneapolis: University of Minnesota Press.

Earman, J., and C. Glymour. 1980. Relativity and eclipses: The British eclipse expeditions of 1919 and their predecessors. *Historical Studies in the Physical Sciences* 11:49–85.

Eddington, A. S. 1918. Gravitation and the principle of relativity. *Nature* 101 (14 March): 34–36.

———. 1919. Joint eclipse meeting of the Royal Astronomical Society. *Observatory* 42:389–98.

———. [1920] 1987. *Space, time and gravitation: An outline of the general relativity theory.* Reprint, Cambridge Science Classics series, Cambridge: Cambridge University Press.

Edwards, A. W. F. 1971. Science, statistics and society. *Nature* 233:17–19.

———. 1972. *Likelihood: An account of the statistical concept of likelihood and its application to scientific inference.* Cambridge: Cambridge University Press.

Edwards, W., H. Lindman, and L. Savage. 1963. Bayesian statistical inference for psychological research. *Psychological Review* 70:193–242.

Efron, B. 1986. Why isn't everyone a Bayesian? *The American Statistician* 40, no. 1:1–4.

Einstein, A. [1926] 1956. *Investigations on the theory of the Brownian movement.* Edited by R. Fürth and translated by A. D. Cowper. Reprint, New York: Dover Publications.

Feller, W. 1940. Statistical aspects of ESP. *Journal of Parapsychology* 4:271–98.

———. 1971. *An introduction to probability theory and its applications.* Vol. 2. New York: John Wiley & Sons.

Fetzer, J. H. 1981. *Scientific knowledge: Causation, explanation, and corroboration.* Dordrecht, The Netherlands: D. Reidel.

Finch, P. D. 1976. The poverty of statisticism. In *Foundations of probability theory, statistical inference and statistical theories of science,* vol. 2, edited by W. L. Harper and C. A. Hooker, 1–44. Dordrecht, The Netherlands: D. Reidel.

Fisher, R. A. [1925] 1970. *Statistical methods for research workers.* Reprint, New York: Macmillan.

———. 1947. *The design of experiments.* 4th ed. Edinburgh: Oliver and Boyd.

———. 1955. Statistical methods and scientific induction. *Journal of the Royal Statistical Society (B)* 17:69–78.

———. 1956. *Statistical methods and scientific inference.* Edinburgh: Oliver and Boyd.

Franklin, A. 1986. *The neglect of experiment.* Cambridge: Cambridge University Press.

———. 1990. *Experiment, right or wrong.* Cambridge: Cambridge University Press.

Fuertes-De La Haba, A., J. Curet, I. Pelegrina, and I. Bangdiwala. 1971. Thrombophlebitis among oral and nonoral contraceptive users. *Obstetrics and Gynecology* 38, no. 2 (August): 259–63.

Galison, P. 1987. *How experiments end.* Chicago: University of Chicago Press.

Garber, D. 1983. Old evidence and logical omniscience in Bayesian confirmation theory. In *Testing scientific theories,* edited by J. Earman, 99–131. Minnesota Studies in the Philosophy of Science, vol. 10. Minneapolis: University of Minnesota Press.

Gardner, M. 1979. Realism and instrumentalism in 19th century atomism. *Philosophy of Science* 46:1–34.

———. 1982. Predicting novel facts. *British Journal for the Philosophy of Science* 33:1–15.

Giere, R. N. 1969. Bayesian statistics and biased procedures. *Synthese* 20:371–87.

———. 1976. Empirical probability, objective statistical methods, and scientific inquiry. In *Foundations of probability theory, statistical inference and statistical theories of science*, vol. 2, edited by W. L. Harper and C. A. Hooker, 63–101. Dordrecht, The Netherlands: D. Reidel.

———. 1983. Testing theoretical hypotheses. In *Testing scientific theories*, edited by J. Earman, 269–98. Minnesota Studies in the Philosophy of Science, vol. 10. Minneapolis: University of Minnesota Press.

———. 1984a. *Understanding scientific reasoning.* 2d ed. New York: Holt, Rinehart and Winston.

———. 1984b. Toward a unified theory of science. In *Science and reality*, edited by J. T. Cushing, C. F. Delaney, and G. M. Gutting, 5–31. Notre Dame: University of Notre Dame Press.

———. 1988. *Explaining science.* Chicago: University of Chicago Press.

Glymour, C. 1980. *Theory and evidence.* Princeton: Princeton University Press.

Glymour, C., R. Scheines, P. Spirtes, and K. Kelly. 1987. *Discovering causal structure: Artificial intelligence, philosophy of science, and statistical modeling.* Orlando: Academic Press.

Godambe, V., and D. Sprott, eds. 1971. *Foundations of statistical inference.* Toronto: Holt, Rinehart and Winston of Canada.

Good, I. J. 1956. Contribution to discussion of a paper by G. S. Brown in *Information theory: Third London symposium 1955*, edited by C. Cherry, 13–14. London: Butterworths.

———. 1976. The Bayesian influence, or how to sweep subjectivism under the carpet. In *Foundations of probability theory, statistical inference and statistical theories of science*, vol. 2, edited by W. L. Harper and C. A. Hooker, 125–74. Dordrecht, The Netherlands: D. Reidel.

———. 1980. The diminishing significance of a P-value as the sample size increases. *Journal of Statistical Computation and Simulation* 11:307–9.

———. 1982. Standardized tail-area probabilities. *Journal of Statistical Computation and Simulation* 13:65–66.

———. 1983a. *Good thinking.* Minneapolis: University of Minnesota Press.

———. 1983b. Some logic and history of hypothesis testing. In *Good thinking* (Minneapolis: University of Minnesota Press). Originally published in J. C. Pitt, ed., *Philosophy in economics* (Dordrecht, The Netherlands: D. Reidel, 1981).

Gooding, D., T. Pinch, and S. Schaffer, eds. 1989. *The uses of experiment: Studies in the natural sciences.* Cambridge: Cambridge University Press.

Goodman, N. 1955. *Fact, fiction, and forecast.* Cambridge, Mass.: Harvard University Press.

Gouy, L. 1888. Note sur le mouvement brownien. *Journal de Physique* 7:561–64.

Greenspan, E., ed. 1982. *Clinical interpretation and practice of cancer chemotherapy.* New York: Raven Press.

Grünbaum, A. 1978. Popper vs. inductivism. In *Progress and rationality in sci-*

ence, edited by G. Radnitzky and G. Andersson, 117–42. Boston Studies in the Philosophy of Science, vol. 58. Dordrecht, The Netherlands: D. Reidel.

―――. 1979. Is Freudian psychoanalytic theory pseudo-scientific by Karl Popper's criterion of demarcation? *American Philosophical Quarterly* 16:131–41.

―――. 1989. The degeneration of Popper's theory of demarcation. In *Freedom and rationality: Essays in honor of John Watkins,* edited by F. D'Agostino and I. Jarvie, 141–61. Dordrecht, The Netherlands: Kluwer.

Hacking, I. 1965. *Logic of statistical inference.* Cambridge: Cambridge University Press.

―――. 1972. Likelihood. *British Journal for the Philosophy of Science* 23:132–37.

―――. 1980. The theory of probable inference: Neyman, Peirce and Braithwaite. In *Science, belief and behavior: Essays in honour of R. B. Braithwaite,* edited by D. H. Mellor, 141–60. Cambridge: Cambridge University Press.

―――. 1983. *Representing and intervening: Introductory topics in the philosophy of natural science.* Cambridge: Cambridge University Press.

―――. 1992a. The self-vindication of the laboratory sciences. In *Science as practice and culture,* edited by A. Pickering, 29–64. Chicago: University of Chicago Press.

―――. 1992b. Statistical language, statistical truth, and statistical reason: The self-authentification of a style of scientific reasoning. In *The social dimensions of science,* edited by E. McMullin, 130–57. Notre Dame: University of Notre Dame Press.

Harman, G. 1965. Inference to the best explanation. *Philosophical Review* 74:88–95.

Harper, W. L., and C. A. Hooker, eds. 1976. *Foundations of probability theory, statistical inference and statistical theories of science.* Vol. 2. Dordrecht, The Netherlands: D. Reidel.

Hempel, C. 1965. *Aspects of scientific explanation: And other essays in the philosophy of science.* New York: Free Press.

Hesse, M. 1980. *Revolutions and reconstructions in the philosophy of science.* Bloomington: Indiana University Press.

Hodges, J., and E. Lehmann. 1970. *Basic concepts of probability and statistics.* 2d ed. San Francisco: Holden-Day.

Horwich, P. 1982. *Probability and evidence.* Cambridge: Cambridge University Press.

Howson, C. 1984. Bayesianism and support by novel facts. *British Journal for the Philosophy of Science* 35:245–51.

―――. 1989. Accommodation, prediction and Bayesian confirmation theory. In *PSA 1988,* vol. 2, edited by A. Fine and J. Leplin, 381–92. East Lansing, Mich.: Philosophy of Science Association.

―――. 1990. Fitting your theory to the facts: Probably not such a bad thing after all. In *Scientific theories,* edited by C. W. Savage, 224–44. Minnesota Studies in the Philosophy of Science, vol. 14. Minneapolis: University of Minnesota Press.

Howson, C., and A. Franklin. 1991. Maher, Mendeleev and Bayesianism. *Philosophy of Science* 58:574–85.

Howson, C., and P. Urbach. 1989. *Scientific reasoning: The Bayesian approach.* La Salle, Ill.: Open Court.

Hoyningen-Huene, P. 1993. *Reconstructing scientific revolutions: Thomas S. Kuhn's philosophy of science.* Chicago: University of Chicago Press.

Hughes, A., and D. Grawoig. 1971. *Statistics: A foundation for analysis.* Reading, Mass.: Addison Wesley.

Hull, D. 1988. *Science as a process: An evolutionary account of the social and conceptual development of science.* Chicago: University of Chicago Press.

Jaynes, E. T. 1976. Common sense as an interface. In *Foundations of probability theory, statistical inference and statistical theories of science,* vol. 2, edited by W. L. Harper and C. A. Hooker, 218–57. Dordrecht, The Netherlands: D. Reidel.

Jeffrey, R. 1965. *The logic of decision.* Chicago: University of Chicago Press.

———. 1983. Bayesianism with a human face. In *Testing scientific theories,* edited by J. Earman, 133–56. Minnesota Studies in the Philosophy of Science, vol. 10. Minneapolis: University of Minnesota Press.

———. ed. 1980. *Studies in inductive logic and probability.* Vol. 2. Berkeley: University of California Press.

Jeffreys, H. 1919a. Contribution to Discussion on the theory of relativity. *Monthly Notices of the Royal Astronomical Society* 80 (December):96–118.

———. 1919b. On the crucial test of Einstein's theory of gravitation. *Monthly Notices of the Royal Astronomical Society* 80 (December): 138–54.

———. 1957. *Scientific inference.* 2d ed. Cambridge: Cambridge University Press.

Jones, B. 1870. *The life and letters of Faraday.* Vol. 1. London: Longmans, Green & Co.

Kadane, J., M. Schervish, and T. Seidenfeld. (forthcoming). Reasoning to a foregone conclusion. *Journal of the American Statistical Association.*

Kahneman, D., P. Slovic, and A. Tversky, eds. 1982. *Judgment under uncertainty: Heuristics and biases.* Cambridge: Cambridge University Press.

Kalbfleisch, J. G. 1979. *Probability and statistical inference, I.* New York: Springer-Verlag.

Kempthorne, O. 1972. Theories of inference and data analysis. In *Statistical papers in honor of George W. Snedecor,* edited by T. A. Bancroft, 167–91. Ames, Iowa: Iowa State University Press.

Kempthorne, O., and L. Folks. 1971. *Probability, statistics, and data analysis.* Ames, Iowa: Iowa State University Press.

Keynes, J. M. [1921] 1952. *A treatise on probability.* Reprint, New York: St. Martin's Press.

Kish, L. 1970. Some statistical problems in research design. In *The significance test controversy,* edited by D. Morrison and R. Henkel, 127–41. Chicago: Aldine.

Kuhn, T. 1962. *The structure of scientific revolutions.* Chicago: University of Chicago Press.

———. 1970. Logic of discovery or psychology of research? and Reflections on my critics. In *Criticism and the growth of knowledge,* edited by I. Lakatos and A. Musgrave, 1–23, 231–77. Cambridge: Cambridge University Press.

———. 1977. *The essential tension: Selected studies in scientific tradition and change.* Chicago: University of Chicago Press.

Kyburg, H. E., Jr. 1971. Probability and informative inference. In *Foundations of statistical inference*, edited by V. Godambe and D. Sprott, 82–103. Toronto: Holt, Rinehart and Winston of Canada.

———. 1974. *The logical foundations of statistical inference*. Dordrecht, The Netherlands: D. Reidel.

———. 1983. *Epistemology and inference*. Minneapolis: University of Minnesota Press.

———. 1984. *Theory and measurement*. Cambridge: Cambridge University Press.

———. 1993. The scope of Bayesian reasoning. In *PSA 1992*, vol. 2, edited by D. Hull, M. Forbes, and K. Okruhlik, 139–52. East Lansing: Philosophy of Science Association.

Kyburg, H. E., Jr., and H. Smokler, eds. 1964. *Studies in subjective probability*. New York: John Wiley & Sons.

Lakatos, I. 1978. *The methodology of scientific research programmes*. Edited by J. Worrall and G. Currie. Vol. 1 of *Philosophical papers*. Cambridge: Cambridge University Press.

Lakatos, I., and A. Musgrave, eds. 1970. *Criticism and the growth of knowledge*. Cambridge: Cambridge University Press.

Laudan, L. 1977. *Progress and its problems*. Berkeley: University of California Press.

———. 1981a. Peirce and the trivialization of the self-correcting thesis. In *Science and hypothesis: Historical essays in scientific methodology*, 226–51 (Dordrecht, The Netherlands: D. Reidel). Originally published in R. Giere and R. Westfall, eds., *Foundations of scientific method: The 19th century*, 275–306 (Bloomington: Indiana University Press, 1973).

———. 1981b. *Science and hypothesis: Historical essays on scientific methodology*. Dordrecht, The Netherlands: D. Reidel.

———. 1984a. Explaining the success of science: Beyond epistemic realism and relativism. In *Science and reality: Recent work in the philosophy of science*, edited by J. T. Cushing, C. F. Delaney, and G. M. Gutting, 83–105. Notre Dame: University of Notre Dame Press.

———. 1984b. *Science and values: The aims of science and their role in scientific debate*. Berkeley: University of California Press.

———. 1987. Progress or rationality? The prospects for normative naturalism. *American Philosophical Quarterly* 24:19–31.

———. 1990a. Demystifying underdetermination. In *Scientific theories*, edited by C. W. Savage, 267–97. Minnesota Studies in the Philosophy of Science, vol. 14. Minneapolis: University of Minnesota Press.

———. 1990b. Normative naturalism. *Philosophy of Science* 57:44–59.

———. 1990c. *Science and relativism: Some key controversies in the philosophy of science*. Chicago: University of Chicago Press.

———. 1996. Beyond positivism and relativism. Boulder, Colo.: Westview Press.

Laudan, L., A. Donovan, R. Laudan, P. Barker, H. Brown, J. Leplin, P. Thagard, and S. Wykstra. 1986. Testing theories of scientific change: Philosophical models and historical research. *Synthese* 69:141–223.

LeCam, L. 1977. A note on metastatistics or "An essay toward stating a problem in the doctrine of chances." *Synthese* 36:133–60.

Lenz, J. 1964. Induction as self-corrective. In *Studies in the philosophy of Charles Sanders Peirce*, 2d ser., edited by E. Moore and R. Robin, 151–62. Amherst: University of Massachusetts Press.

Leplin, J. 1990. Renormalizing naturalism. *Philosophy of Science* 57:20–33.

Levi, I. 1967. *Gambling with truth: An essay on induction and the aims of science*. New York: Knopf.

———. 1980a. *The enterprise of knowledge: An essay on knowledge, credal probability, and chance*. Cambridge, Mass.: MIT Press.

———. 1980b. Induction as self correcting according to Peirce. In *Science, belief and behavior: Essays in honor of R. B. Braithwaite*, edited by D. H. Mellor, 127–40. Cambridge: Cambridge University Press.

———. 1982. Ignorance, probability and rational choice. *Synthese* 53:387–417.

———. 1984. On the seriousness of mistakes. In *Decisions and revisions: Philosophical essays on knowledge and value*, 14–33 (Cambridge: Cambridge University Press). First published in *Philosophy of Science* 29 (1962): 47–65.

Lindemann, F. A. 1919. Contribution to Discussion on the theory of relativity. *Monthly Notices of the Royal Astronomical Society* 80 (December):96–118.

Lindley, D. V. 1957. A statistical paradox. *Biometrika* 44:187–92.

———. 1971. The estimation of many parameters. In *Foundations of statistical inference*, edited by V. Godambe and D. Sprott, 435–47. Toronto: Holt, Rinehart and Winston of Canada.

———. 1972. *Bayesian statistics, a review.* Philadelphia: Society for Industrial and Applied Mathematics.

———. 1976. Bayesian statistics. In *Foundations of probability theory, statistical inference and statistical theories of science*, vol. 2, edited by W. L. Harper and C. A. Hooker, 353–62. Dordrecht, The Netherlands: D. Reidel.

Lloyd, E. 1988. *The structure and confirmation of evolutionary theory.* Westport, Conn.: Greenwood Press.

Lodge, O. 1919. Contribution to Discussion on the theory of relativity. *Monthly Notices of the Royal Astronomical Society* 80 (December):96–118.

Maher, P. 1988. Prediction, accommodation, and the logic of discovery. In *PSA 1988*, vol. 1, edited by A. Fine and J. Leplin, 273–85. East Lansing, Mich.: Philosophy of Science Association.

———. 1993a. Acceptance in Bayesian philosophy of science. In *PSA 1992*, vol. 2, edited by D. Hull, M. Forbes, and K. Okruhlik, 153–60. East Lansing, Mich.: Philosophy of Science Association.

———. 1993b. *Betting on theories.* Cambridge: Cambridge University Press.

———. 1993c. Discussion: Howson and Franklin on prediction. *Philosophy of Science* 60: 329–40.

Masterman, M. 1970. The nature of a paradigm. In *Criticism and the growth of Knowledge*, edited by I. Lakatos and A. Musgrave, 59–89. Cambridge: Cambridge University Press.

Mayo, D. 1980. Testing statistical testing. In *Philosophy in Economics*, edited by J. C. Pitt, 175–203. Dordrecht, The Netherlands: D. Reidel.

———. 1981. In defense of the Neyman-Pearson theory of confidence intervals. *Philosophy of Science* 48:269–80.

———. 1982. On after-trial criticisms of Neyman-Pearson theory of statistics. In *PSA 1982*, vol. 1, edited by P. Asquith and T. Nickles, 145–58. East Lansing, Mich.: Philosophy of Science Association.

———. 1983. An objective theory of statistical testing. *Synthese* 57:297–340.

———. 1985a. Behavioristic, evidentialist, and learning models of statistical testing. *Philosophy of Science* 52:493–516.

———. 1985b. Increasing public participation in controversies involving hazards: The value of metastatistical rules. *Science, Technology, and Human Values* 10:55–68.

———. 1986a. Cartwright, causality, and coincidence. In *PSA 1986*, vol. 1, edited by A. Fine and P. Machamer, 42–58. East Lansing, Mich.: Philosophy of Science Association.

———. 1986b. Understanding frequency-dependent causation. *Philosophical Studies* 49:109–24.

———. 1988. Brownian motion and the appraisal of theories. In *Scrutinizing science: Empirical studies of scientific change*, edited by A. Donovan, L. Laudan, and R. Laudan, 219–43. Dordrecht, The Netherlands: Kluwer.

———. 1989. Toward a more objective understanding of the evidence of carcinogenic risk. In *PSA 1988*, vol. 2, edited by A. Fine and J. Leplin, 489–503. East Lansing, Mich.: Philosophy of Science Association.

———. 1991a. Novel evidence and severe tests. *Philosophy of Science* 58:523–52.

———. 1991b. Sociological versus metascientific views of risk assessment. In *Acceptable evidence: Science and values in risk management*, edited by D. Mayo and R. Hollander, 249–79. New York: Oxford University Press.

———. 1992. Did Pearson reject the Neyman-Pearson philosophy of statistics? *Synthese* 90: 233–62.

———. 1993. The test of experiment: C. S. Peirce and E. S. Pearson. In *Charles S. Peirce and the philosophy of science: Papers from the Harvard sesquicentennial congress*, edited by E. C. Moore, 161–74. Tuscaloosa: University of Alabama Press.

———. 1994. The new experimentalism, topical hypotheses, and learning from error. In *PSA 1994*, vol. 1, edited by D. Hull, M. Forbes and R. Burian, 270–79. East Lansing, Mich.: Philosophy of Science Association.

Meehl, P. 1990. *Psychological Reports* 66:195–244.

Mellor, D. H., ed. 1980. *Science, belief and behaviour: Essays in honour of R. B. Braithwaite*. Cambridge: Cambridge University Press.

Mill, J. S. 1888. *A System of Logic*. 8th ed. New York: Harper and Brothers.

Miller, R. 1987. *Fact and method: Explanation, confirmation and reality in the natural and the social sciences*. Princeton: Princeton University Press.

Morrison, D., and R. Henkel, eds. 1970. *The significance test controversy*. Chicago: Aldine.

Moyer, D. 1979. Revolution in science: The 1919 eclipse test of general relativity. In *On the path of Albert Einstein*, edited by A. Perlmutter and L. Scott, 55–102. New York: Plenum Press.

Musgrave, A. 1974. Logical versus historical theories of confirmation. *British Journal for the Philosophy of Science* 25:1–23.

———. 1978. Evidential support, falsification, heuristics, and anarchism. In *Progress and rationality in science,* edited by G. Radnitzky and G. Andersson, 181–201. Boston Studies in the Philosophy of Science, vol. 58. Dordrecht, The Netherlands: D. Reidel.

———. 1980. Kuhn's second thoughts. In *Paradigms and revolutions: Appraisals and applications of Thomas Kuhn's philosophy of science,* edited by G. Gutting, 39–53. Notre Dame: University of Notre Dame Press.

———. 1989. Deductive heuristics. In *Imre Lakatos and theories of scientific change,* edited by K. Gavroglu, Y. Goudaroulis, and P. Nicolacopoulos, 15–32. Dordrecht, The Netherlands: Kluwer.

Newall, H. F. 1919. Contribution to Joint eclipse meeting of the Royal Society and the Royal Astronomical Society. *The Observatory* 42 (November):389–98.

———. 1920. Note on the physical aspect of the Einstein prediction. *Monthly notices of the Royal Astronomical Society* 80:22–25.

Neyman, J. 1950. *First course in probability and statistics.* New York: Henry Holt.

———. 1952. *Lectures and conferences on mathematical statistics and probability.* 2d ed. Washington, D.C.: U.S. Department of Agriculture.

———. 1967a. Outline of a theory of statistical estimation based on the classical theory of probability. In *A selection of early statistical papers of J. Neyman,* 250–90 (Berkeley: University of California Press). First published in *Philosophical Transactions of the Royal Society (A),* no. 236 (1937):333–80.

———. 1967b. *A selection of early statistical papers of J. Neyman.* Berkeley: University of California Press.

———. 1971. Foundations of behavioristic statistics. In *Foundations of statistical inference,* edited by V. Godambe and D. Sprott, 1–19. Toronto: Holt, Rinehart and Winston of Canada.

———. 1977. Frequentist probability and frequentist statistics. *Synthese* 36:97–131.

Neyman, J., and Pearson, E. S. 1967a. *Joint statistical papers.* Berkeley: University of California Press.

———. 1967b. On the problem of the most efficient tests of statistical hypotheses. In *Joint statistical papers,* 140–85 (Berkeley: University of California Press). First published in *Philosophical Transactions of the Royal Society (A)* (1933):231, 289–337.

———. 1967c. On the use and interpretation of certain test criteria for purposes of statistical inference. Part I. In *Joint statistical papers,* 1–66 (Berkeley: University of California Press). First published in *Biometrika* 20(A) (1928): 175–240.

Nickles, T. 1987. Lakatosian heuristics and epistemic support. *British Journal for the Philosophy of Science* 38:181–205.

Niiniluoto, I. 1984. *Is science progressive?* Dordrecht, The Netherlands: D. Reidel.

Nye, M. J. 1972. *Molecular reality.* London: Macdonald.

Ostwald, W. 1907. The modern theory of energetics. *Monist* 17:481–515.

———. 1909. *Gundriss der allgemeinen Chemie,* 4. Leipzig: Verlag von Wilhelm Engelmann.

Parzen, E. 1960. *Modern probability theory and its applications.* New York: John Wiley & Sons.

Pearson, E. S. 1955. Statistical concepts in their relation to reality. *Journal of the Royal Statistical Society (B)* 17:204–7.

———. 1966a. The choice of statistical tests illustrated on the interpretation of data classed in a 2×2 table. In *The selected papers of E. S. Pearson,* 169–97 (Berkeley: University of California Press). First published in *Biometrika* 34(1947): 139–67.

———. 1966b. The efficiency of statistical tools and a criterion for the rejection of outlying observations. In *The selected papers of E. S. Pearson,* 118–30 (Berkeley: University of California Press). First published in *Biometrika* 28(1936):308–20.

———. 1966c. On questions raised by the combination of tests based on discontinuous distributions. In *The selected papers of E. S. Pearson,* 217–32 (Berkeley: University of California Press). First published in *Biometrika* 37(1950):383–98.

———. 1966d. *The selected papers of E. S. Pearson.* Berkeley: University of California Press.

———. 1966e. Some thoughts on statistical inference. In *The selected papers of E. S. Pearson,* 276–83 (Berkeley: University of California Press). First published in *Annals of Mathematical Statistics* 33(1962):394–403.

Pearson, E. S., and J. Neyman. 1967. On the problem of two samples. In *Joint statistical papers,* by J. Neyman and E. S. Pearson, 99–115 (Berkeley: University of California Press). First published in *Bull. Acad. Pol. Sci.* (1930):73–96.

Pearson, E. S., and S. Wilks. 1966. Methods of statistical analysis appropriate for k samples of two variables. In *The selected papers of E. S. Pearson,* by E. S. Pearson, 81–106 (Berkeley: University of California Press). First published in *Biometrika* 25(1933):353–78.

Peirce, C. S. 1931–35. *Collected papers.* Vols. 1–6. Edited by C. Hartshorne and P. Weiss. Cambridge: Harvard University Press.

———. 1958. *Collected papers.* Vols. 7–8. Edited by A. Burks. Cambridge: Harvard University Press.

Perrin, J. [1913] 1990. *Atoms.* Translated by D. L. Hammick. Reprint, Woodbridge, Conn.: Ox Bow Press.

———. 1950. *Oeuvres scientifiques de Jean Perrin.* Paris: Centre National de la Recherche Scientifique.

Pickering, A. ed. 1992. *Science as practice and culture.* Chicago: University of Chicago Press.

Poincaré, J. 1905. The principles of mathematical physics. In *Congress of Arts and Science, Universal Exposition, St. Louis, 1904,* vol. 1, 604–22. New York: Houghton Mifflin.

Poole, C. 1987. Beyond the confidence interval. *American Journal of Public Health* 77:195–99.

Popper, K. 1959. *The logic of scientific discovery.* New York: Basic Books.

———. 1962. *Conjectures and refutations: The growth of scientific knowledge.* New York: Basic Books.

————. 1970. Normal science and its dangers. In *Criticism and the growth of knowledge*, edited by I. Lakatos and A. Musgrave, 51–58. Cambridge: Cambridge University Press.

————. 1974. Replies to my critics. In *The philosophy of Karl Popper*. Bk. 2. Edited by P. A. Schilpp, 961–1197. La Salle, Ill.: Open Court.

————. 1979. *Objective knowledge: An evolutionary approach*. Oxford: Oxford University Press.

————. 1983. *Realism and the aim of science*. Totowa, N.J.: Rowman and Littlefield.

Pratt, J. W. 1977. "Decisions" as statistical evidence and Birnbaum's "confidence concept." *Synthese* 36:59–69.

Putnam, H. 1981. The "corroboration" of theories. In *Scientific revolutions*, edited by I. Hacking, 60–79. Oxford: Oxford University Press.

Quine, W. V. 1969. *Ontological relativity and other essays*. New York: Columbia University Press.

Radnitzky, G., and G. Andersson, eds. 1978. *Progress and rationality in science*. Boston Studies in the Philosophy of Science, vol. 58. Dordrecht, The Netherlands: D. Reidel.

Ramsay, W. 1882. On Brownian or pedetic motion. *Proceedings of the Bristol Naturalists' Society* 3:299–302.

Redhead, M. 1986. Novelty and confirmation. *British Journal for the Philosophy of Science* 37:115–18.

Reichenbach, H. 1971. *The theory of probability: An inquiry into the logical and mathematical foundations of the calculus of probability*. 2d ed. Translated by E. Hutten and M. Reichenbach. Berkeley: University of California Press.

Rescher, N. 1978. *Peirce's philosophy of science: Critical studies in his theory of induction and scientific method*. Notre Dame: University of Notre Dame Press.

Robbins, H. E. 1952. Some aspects of the sequential design of experiments. *Bulletin of the American Mathematical Society* 58:527–35.

Rosenkrantz, R. 1977. *Inference, method and decision: Towards a Bayesian philosophy of science*. Dordrecht, The Netherlands: D. Reidel.

————. 1983. Why Glymour *Is* a Bayesian. In *Testing scientific theories*, edited by J. Earman, 69–97. Minnesota Studies in the Philosophy of Science, vol. 10. Minneapolis: University of Minnesota Press.

Rosenthal, R. 1987. *Judgment studies: Design, analysis, and meta-analysis*. Cambridge: Cambridge University Press.

Rosenthal, R., and J. Gaito. 1963. The interpretation of levels of significance by psychological researchers. *Journal of Psychology* 55:33–38.

Salmon, W. 1966. *The foundations of scientific inference*. Pittsburgh: University of Pittsburgh Press.

————. 1967. Carnap's inductive logic. *Journal of Philosophy* 64, no. 21:725–39.

————. 1983. Carl G. Hempel on the rationality of science. *Journal of Philosophy* 80 (no. 10):555–62.

————. 1984. *Scientific explanation and the causal structure of the world*. Princeton: Princeton University Press.

———. 1988. Dynamic rationality: Propensity, probability, and credence. In *Probability and causality: Essays in honor of Wesley C. Salmon*, edited by J. H. Fetzer, 3–40. Dordrecht, The Netherlands: D. Reidel.

———. 1990. Rationality and objectivity in science, *or* Tom Kuhn meets Tom Bayes. In *Scientific theories*, edited by C. W. Savage, 175–204. Minnesota Studies in the Philosophy of Science, vol. 14. Minneapolis: University of Minnesota Press.

———. 1991. The appraisal of theories: Kuhn meets Bayes. In *PSA 1990*, vol. 2, edited by A. Fine, M. Forbes, and L. Wessels, 325–32. East Lansing, Mich.: Philosophy of Science Association.

Savage, C. W. ed. 1990. *Scientific theories*. Minnesota Studies in the Philosophy of Science, vol. 14. Minneapolis: University of Minnesota Press.

Savage, I. R. 1968. *Statistics: Uncertainty and behavior*. New York: Houghton Mifflin.

Savage, L. ed. 1962. *The foundations of statistical inference: A discussion*. London: Methuen.

———. 1964. The foundations of statistics reconsidered. In *Studies in subjective probability*, edited by H. Kyburg and H. Smokler, 173–88. New York: John Wiley & Sons.

———. 1972. *The foundations of statistics*. New York: Dover Publications.

Scheffler, I. 1974. *Four pragmatists*. London: Routledge & Kegan Paul.

———. 1982. *Science and subjectivity*. Indianapolis: Hackett.

Schilpp, P. A. ed. 1963. *The philosophy of Rudolph Carnap*. La Salle, Ill.: Open Court.

———. ed. 1974. *The philosophy of Karl Popper*. Bk. 2. La Salle, Ill.: Open Court.

Schuster, A. 1920. The influence of small changes of temperature on atmospheric refraction. *Proceedings, Physical Society of London* 32:135–40.

Seidenfeld, T. 1979a. *Philosophical problems of statistical inference: Learning from R. A. Fisher*. Dordrecht, The Netherlands: D. Reidel.

———. 1979b. Why I am not an objective Bayesian; Some reflections prompted by Rosenkrantz. *Theory and Decision* 11:413–40.

Seidenfeld, T., M. Schervish, and J. Kadane. 1990. When fair betting odds are not degrees of belief. In *PSA 1990*, vol. 1, edited by A. Fine, M. Forbes, and L. Wessels, 517–24. East Lansing, Mich.: Philosophy of Science Association.

Selvin, H. 1970. A critique of tests of significance in survey research. In *The significance test controversy*, edited by D. Morrison and R. Henkel, 94–106. Chicago: Aldine.

Sewell, W. 1952. Infant training and the personality of the child. *American Journal of Sociology* 58 (September): 150–59.

Shapere, D. 1984. *Reason and the search for knowledge: Investigations in the philosophy of science*. Boston Studies in the Philosophy of Science. Dordrecht, The Netherlands: D. Reidel.

Shimony, A. 1970. Scientific inference. In *The nature and function of scientific theories: Essays in contemporary science and philosophy*, edited by R. Colodny. Pittsburgh: University of Pittsburgh Press.

Snedecor, G., and W. Cochran. 1980. *Statistical methods*. 7th ed. Ames: Iowa State University Press.

Spielman, S. 1973. A refutation of the Neyman-Pearson theory of testing. *British Journal for the Philosophy of Science* 24:201–22.

Stolley, P. D. 1978. A review of data from the United States concerning the relationship of thromboembolic disease to oral contraceptives. In *Risks, benefits, and controversies in fertility control*, edited by J. Sciarra, G. Zatuchni, and J. Speidel. Hagerstown, Md.: Harper and Row.

Suppe, F. 1989. *The semantic conception of theories and scientific realism*. Urbana, Ill.: University of Illinois Press.

———. ed. 1977. *The structure of scientific theories*, 2d ed. Chicago: University of Illinois Press.

Suppes, P. 1969. Models of data. In *Studies in the methodology and foundations of science*, 24–35. Dordrecht, The Netherlands: D. Reidel.

Thompson, P. 1988. *The structure of biological theories*. Albany: State University of New York Press.

Thorne, K. S., and C. M. Will. 1971. Theoretical frameworks for testing relativistic gravity: I: Foundations. *Astrophysical Journal* 163:595–610.

Uhlenbeck, G. E., and L. S. Ornstein. 1954. On the theory of the Brownian motion. In *Selected papers on noise and stochastic processes*, edited by N. Wax, 93–111 (New York: Dover). First published in *Physical Review* 36(1954):823–41.

van Fraassen, B. 1980. *The scientific image*. Oxford: Clarendon Press.

———. 1989. *Laws and symmetry*. New York: Oxford University Press.

von Klüber, H. 1960. The determination of Einstein's light-deflection in the gravitational field of the sun. In *Vistas on astronomy*, vol. 3, edited by A. Beer, 47–77.

von Mises, R. 1957. *Probability, statistics and truth*. New York: Dover.

Wax, N. ed. 1954. *Selected papers on noise and stochastic processes*. New York: Dover.

Weinberg, S. 1992. *Dreams of a final theory*. New York: Pantheon.

Whewell, W. [1847] 1967. *The philosophy of the inductive sciences: Founded upon their history*. 2d ed. Vols. 1 and 2. Reprint, London: Johnson Reprint.

Wisdom, J. O. 1974. The nature of "normal" science. In *The philosophy of Karl Popper*, edited by P. A. Schilpp, 820–42. La Salle, Ill.: Open Court.

Worrall, J. 1978a. Research programmes, empirical support, and the Duhem problem: Replies to criticism. In *Progress and rationality in science*, edited by G. Radnitzky and G. Andersson, 321–38. Boston Studies in the Philosophy of Science, vol. 58. Dordrecht, The Netherlands: D. Reidel.

———. 1978b. The ways in which the methodology of scientific research programmes improves on Popper's methodology. In *Progress and rationality in science*, edited by G. Radnitzky and G. Andersson, 45–70. Boston Studies in the Philosophy of Science, vol. 58. Dordrecht, The Netherlands: D. Reidel.

———. 1985. Scientific discovery and theory-confirmation. In *Change and progress in modern science: Papers related to and arising from the Fourth International Conference on History and Philosophy of Science*, edited by J. C. Pitt, 301–32. Dordrecht, The Netherlands: D. Reidel.

———. 1989. Fresnel, Poisson and the white spot: The role of successful predictions in the acceptance of scientific theories. In *The uses of experiment:
Studies in the natural sciences*, edited by D. Gooding, T. Pinch, and S. Schaffer,
135–57. Cambridge: Cambridge University Press.

———. 1993. Falsification, rationality, and the Duhem problem. In *Philosophical problems of the internal and external worlds: Essays on the philosophy of Adolf
Grünbaum*, edited by J. Earman, A. Janis, G. Massey, and N. Rescher, 329–
70. Pittsburgh: University of Pittsburgh Press.

Zahar, E. 1973. Why did Einstein's programme supersede Lorentz's? Parts 1
and 2. *British Journal for the Philosophy of Science* 24: 95–125, 223–62.

Index

Abrams, S., xvi
Achinstein, P., 60n. 1, 249
Ackermann, R. 58, 62n. 2
actual significance level, 303–5, 310, 315–17, 343, 344, 350–52
after-trial analysis, 4, 16, 18, 129, 137, 144, 149, 205, 236, 448, 452
alternative hypothesis objection (*see also* severe tests; severity), 174–77, 187–89, 196–200, 207, 209, 212
Alzheimer's disease, 55, 190, 266–68; and ApoE, 267, 268n. 11; and beta amyloid, 55, 267n 11; and Dr. Allen D. Roses, 266–68
American Philosophical Association (symposium on Hempel), 112
ampliative inference, 9, 337, 435, 440, 443
amplifying error patterns, 6
analgesics, 344, 375
Andersen, B., 344
Anderson, A., 289–90; shadow effect, 287, 289–90
anomaly, 4, 25, 31, 43, 50–51, 55; assignment of blame for (*see also* Duhem's problem), 25, 107, 114, 147–48; unevadable, 51
anomalies, quantitative, 42–44, 48, 51
Anscombe, F., 343n. 16
argument from coincidence, 66–67, 216–17, 237, 454
argument from error (*see also* argument from coincidence; canonical; learning from error; severity): basic pattern of: 64, 404, 445; and an argument from coincidence, 66; and NP tests, 377, 411; and passing a severe test, 7, 157, 184–86, 197, 231, 234–35, 445, 460; and the significance question, 95, 197; as the aim of methodological rules, 19, 452n. 3; informal side of, 271, 448
argument from intentions, 346–48, 356n. 30, 363

Armitage, P., 341–45, 346n. 18, 349–57, 430
Armitage's example, 351–57
artifacts: distinguishing from real effects, xiii, 63, 66, 93, 94, 98, 122, 153, 162, 190–91, 216–17, 315, 457, 458, 461; ruling out 68, 110, 458
astrology, 32–35, 40, 49
Avirov, M., xvi
Avogadro's number, 27, 216–17, 222–23, 231–32, 234, 235, 248

Baccus, F., 76n. 14
Bailey, D., 154n. 10
Barnard, G. A., xv, 200, 282n. 18, 339, 341, 346, 381n. 8, 385, 393n. 18
Barnett, V., 359
Bartlett, M. S., 88–89, 341
Bayes, T., 112
Bayes factor, 116n. 7
Bayesian (*see also* Bayesian Way): account of acceptance of hypotheses, 87, 331n. 8; catchall factor, 109, 116–19, 148, 187–88, 190, 242, 324–25, 333, 340, 458; coherence, 76, 83, 86, 339, 340, 346, 357, 359, 363; conception of inductive inference, x, 99, 108, 321, 359; conditionalist, 75, 355–57, 391; convergence results, 84, 114; decision theory, 69n. 6, 87, 115, 119, 374; degrees of belief (*see also* Bayesian, subjective), 10, 79, 83, 99, 105, 107–9, 161, 242, 332, 409; Dutch Book arguments, 76; magic, 350–51; "objectivism," 406; objectivists, 72n. 8, 335, 358; personalist, 105, 119, 126, 187; principles (*see also* likelihood principle; Bayesian v. error probability principles), 125, 260, 320, 363; ratio of support (BR), 324, 327–30, 333; reconstructions of scientific episodes, 68, 90, 100, 106–9, 114, 147, 328, 331,

periments, 214–15, 226, 230, 233, 242–43, 245; of data, 135–36, 139–40, 157; provided by NP tests, 377, 406, 428; specifying, 172

Carnap, R., ix, 40n. 13, 73–74, 126; and logical probability, 73, 82, 126

Carnapian Bayesians, 73–74

Carnot's principle (see also second law of thermodynamics), 246–48

Cartwright, N., 58, 216, 244

catchall hypothesis (see also Bayesian catchall factor), 116

Cato, M., xvi

central limit theorem (CLT), 168, 170–72, 193n. 11, 206, 397, 439, 464

ceteris paribus, clause, 2, 15, 138, 239; conditions, 132, 139, 140, 144, 147, 206, 239, 243

Chandrasekhar, S., 229n. 14, 245n. 22

Chatfield, George, xv, xvi

Chatfield, Isaac, xvi

Chebychev's inequality, 169n. 17

checking assumptions: of data models, 13, 136–38, 160–61, 283n. 21, 364, 444; of experiment, 156–57, 205, 235–36, 238–39, 295, 435, 447, 461–62

chi-squared: test, 99, 228n. 13; distribution, 228

Clark, P., 248n. 23

classical thermodynamics, 51, 215, 217, 246–47

clinical trials, 375–76

Cochran, W., 167n. 16

cognitive science, 51n. 17

coin-tossing experiments, 12, 78, 150–51, 164–65, 167, 201, 365, 462

communications networks, 375–76

computed (or calculated) significance level, 303, 305, 311, 317, 328, 338–39, 341, 343–44

computer-aided heuristic searches, 306–8

conditional probability, 80, 81

conditionalist (see Bayesian conditionalist)

confidence interval estimation, 135, 272–74, 304, 354–57, 362, 395, 414, 426; confidence levels in, 379, 409n. 29; upper confidence bound, 409n. 29

controlled experiments, 453

conventions, 2, 19, 40, 51n. 18

conversion, theory change as (see also Kuhn), 22, 46, 50, 53

corroboration, 9–10, 14, 41

Cottingham, E. T., 278

counterfactual degrees of belief, 334–35

Cox, D. R., 341, 342

Cramér, H., 158n. 14, 171n. 21

critical discourse, 22–24, 31–32, 34–35, 41, 45, 113; unwarranted, 35, 47–49

Crommelin, A., 286, 287n. 25, 291

crucial experiments, 214, 217, 239

crud factor, 316n.9

curve-fitting problem, 200

da Vinci, Leonardo, 100–101, 274

Dancer, J. B., 218

Darden, L., 51n. 17

data analysis, 92, 136, 156, 235

data generation, 132, 138, 140, 144, 235, 238–39, 243, 444, 463

Davidson, C., 133, 136–37, 279, 285, 287n. 25

decision theory, 87n. 21

deductive logic, 85, 337, 415, 437

de Finetti, B., 71, 319, 362, 378n. 5

deflection of light, 87, 132–34, 136, 145, 188–90, 251, 264, 279, 280, 282–83, 286–89

DeGroot, M. H., 352n. 23

demarcating scientific inquiry, 36

dense bodies, 66, 121–22, 190, 237, 315, 454

Descartes, R., 252n. 2

difference in proportions, 142, 328, 384–85, 396

displacement distribution, 219, 222–24, 229–31

distance measure (see also statistical significance level), 134, 143, 158, 158n. 14, 160, 162–64, 172, 179n. 2, 228, 233, 300, 366, 390n. 15, 406

distribution-free methods, 157, 161

Dorling, J., 87, 103–6, 108, 110, 147, 335; and his homework problem 104–7, 111

double-blind techniques, 18, 149

double-counting of data, 257, 259, 285, 306, 327

"due to chance," 154, 173, 228, 299, 300, 344, 429–30, 446, 454